# Atlas of Trace Fossils in Well Core

Dirk Knaust

# Atlas of Trace Fossils in Well Core

## Appearance, Taxonomy and Interpretation

Springer

Dirk Knaust
Statoil ASA
Stavanger
Norway

ISBN 978-3-319-49836-2 ISBN 978-3-319-49837-9 (eBook)
DOI 10.1007/978-3-319-49837-9

Library of Congress Control Number: 2016958461



Cover image: Sectioned well-core sample of a cross-bedded sandstone with mud drapes and floating granulae, comprising the cross section of a large spreite burrow assigned to *Teichichnus zigzag*. Middle Jurassic (Bathonian) Tarbert Formation (tidally influenced delta front), Oseberg Sør Field, Norwegian North Sea (well 30/9-4S, ca. 3347.2 m). Image width ca. 9 cm. Photograph taken by the author.

Printed on acid-free paper

This Springer imprint is published by Springer Nature
The registered company is Springer International Publishing AG
The registered company address is: Gewerbestrasse 11, 6330 Cham, Switzerland

For my son and daughter,
*Per Ragnar and Solveig Alida*

# Foreword

It is not easy to tell which of the graduate students will become masters in their field. When I met Dirk Knaust in 1997 at the Fourth International Ichnofabric Workshop on San Salvador, he was just another graduate student, obviously very intelligent, but there were others with longer strings of publications. But as the years progressed, Dirk's work in ichnology accelerated, and it became clear that he is a perfectionist whose body of work is ripening at last. Building on his doctoral study of the Muschelkalk, he expanded his range to the North Sea and many other regions, and as we all know, the best ichnologist is the one who has seen the most trace fossils. In his maturity, he has produced a series of papers revising *Asteriacites, Rhizocorallium, Balanoglossites, Pholeus, Oichnus*, and other ichnogenera (Knaust 2002, 2008, 2013; Wisshak et al. 2015; Knaust and Neumann 2016), brought the world's attention to extraordinarily preserved tracemakers in the Muschelkalk (Knaust 2010a, b), and elucidated the connection between bioturbation and reservoir quality (Knaust 2009a, b; Knaust et al. 2014).

Dirk's most important effort in recent years has been, together with Richard G. Bromley, to organize and edit *Trace Fossils as Indicators of Sedimentary Environments*, a thick volume that brings ichnology to sedimentologists and other geologists—and sedimentology back to ichnologists (Knaust and Bromley 2012). He carried an advance copy of the book to Ichnia 2012, where it was a big hit among ichnologists on the long drives between field trip stops on the Avalon Peninsula of Newfoundland. The copy passed slowly from hand to hand to the rear of the bus and back to the front again. I think he must have sold to forty people on the bus within a few hours.

Currently, Dirk is at work with me on leading a revision of the trace-fossil volume of the *Treatise on Invertebrate Paleontology*. As he pointed out in a keynote address at Ichnia 2016, the number of named invertebrate ichnogenera has more than doubled since Walter Häntzschel's last revision in 1975. After Bertling et al. (2006) reached a consensus on which criteria were best for differentiating ichnotaxa, Dirk applied these ichnotaxobases systematically to the entire corpus of invertebrate trace fossils, using them to define categories within a key (Knaust 2012)—an effort that required consulting the diagnoses and major revisions of every ichnogenus. He reasoned that ichnogenera falling within the same category are potential synonyms, and indeed some have been found by this method. The key is effectively a massive test of the consensus criteria, generating predictions that might otherwise not be investigated. It is also a way to guide ichnologists toward a viable classification of trace fossils that utilizes ichnofamilies within the context of tracemaker behavior and paleoenvironments. In a recent paper, Dirk erected a new ichnofamily, the Siphonichnidae, to accommodate bivalvian activity as diverse as those recorded in *Parahaentzschelinia, Scalichnus*, and *Hillichnus* (Knaust 2015). He has also drawn attention to the meiobenthos as significant agents of bioturbation, even identifying fossil tracemakers in some cases (Knaust 2010a, b).

The book at hand, *Atlas of Trace Fossils in Well Core—Appearance, Taxonomy and Interpretation*, was developed over a period of years from the series of in-house workshops that Dirk has led from 2011 onwards. It is clear that this work comes from long gestation of ideas and presentation. It includes two sections: First, a brief introduction acquaints the reader with the fundamentals of ichnology, with special regard to their use in petroleum geology.

Second, in the main part of the book, 39 genera of trace fossils and associated features are discussed individually. For each ichnogenus, sections are given on their morphology, fill and size, ichnotaxonomy, substrate, appearance in core, similar trace fossils, producers, ethology (behavior), depositional environment, ichnofacies, age, and finally reservoir quality, accompanied by a generous number of illustrations. The treatment is condensed but draws on a very broad knowledge of the literature as well as extensive personal experience. Many of the illustrations are new; others are borrowed from the best in the literature. For many ichnogenera, the criteria for identification of the major ichnospecies are given, and close attention is paid to possible tracemakers and their behavior. Is such detail necessary? Not for every project, but the studies that pay attention to ichnospecies do yield finer details about depositional environments than those which identify trace fossils only to ichnogenus, or ignore ichnotaxonomy altogether.

The trace fossils that have been chosen for treatment in this volume include all the ichnogenera that are ordinarily found in cores, whether taken for academic or economic purposes, plus ichnogenera that are common in geological contexts with which Dirk is familiar. The result is that here is not only the most up-to-date compendium of information on trace fossils in cores, but also an advancement of science on trace fossils such as *Phoebichnus* that many ichnologists have not recognized in their own material.

Anyone who wishes to use trace fossils in core or outcrop can benefit from reading this book, but those who want information on how trace fossils determine the porosity and permeability of rocks will find it particularly useful (Knaust 2014). Dirk gives specific information about the properties that each ichnogenus may lend reservoir rocks, whether aiding or hindering the flow of pore fluids.

Need I remind readers that what controls petroleum flow also applies to groundwater? Hydrogeologists and environmental geologists may be pleasantly surprised to find that they can better understand the properties of aquifers and aquitards by reference to ichnology. This is a barely touched area of science, but one that will becoming increasingly important as water tables fall and engineering projects respond to environmental changes. The permeability of the limestone underlying Miami, for instance, is determined partly by its trace fossils. It will not be surprising if the stratigraphic distribution of trace fossils turns out to determine the fate of different areas of Florida as sea level rises.

You have in your hands an authoritative book, one that will have an influence for years to come throughout the field of invertebrate ichnology as well as aspects of sedimentology, petroleum geology, and environmental geology. It has built on decades of work by the ichnological community and Knaust's own, and is a fitting companion for the scientist who works with trace fossils.

July 2016

Andrew K. Rindsberg
University of West Alabama
Livingston

## References

Bertling M, Braddy SJ, Bromley RG et al (2006) Names for trace fossils: a uniform approach. Lethaia 39: 265–286

Knaust D (2002) Ichnogenus *Pholeus* Fiege, 1944, revisited. J Paleontol 76:882–891

Knaust D (2008) *Balanoglossites* Mägdefrau, 1932 from the Middle Triassic of Germany: part of a complex trace fossil probably produced by boring and burrowing polychaetes. Paläontologische Zeitschrift 82:347–372

Knaust D (2009a) Ichnology as a tool in carbonate reservoir characterization: a case study from the Permian–Triassic Khuff Formation in the Middle East. GeoArabia 14:17–38

Knaust D (2009b) Characterisation of a Campanian deep-sea fan system in the Norwegian Sea by means of ichnofabrics. Mar Pet Geol 26:1199–1211

Knaust D (2010a) Remarkably preserved benthic organisms and their traces from a Middle Triassic (Muschelkalk) mud flat. Lethaia 43:344–356

Knaust D (2010b) Meiobenthic trace fossils comprising a miniature ichnofabric from Late Permian carbonates of the Oman Mountains. Palaeogeogr Palaeoclimatol Palaeoecol 286:81–87

Knaust D (2012) Trace-fossil systematics. In: Knaust D, Bromley RG (eds) Trace fossils as indicators of sedimentary environments. Developments in Sedimentology, vol 64, pp 79–101

Knaust D (2013) The ichnogenus *Rhizocorallium*: classification, trace makers, palaeoenvironments and evolution. Earth Sci Rev 126:1–47

Knaust D (2014) Classification of bioturbation-related reservoir quality in the Khuff Formation (Middle East): towards a genetic approach. In: Pöppelreiter MC (ed) Permo-Triassic Sequence of the Arabian Plate. EAGE, pp 247–267

Knaust D (2015) Siphonichnidae (new ichnofamily) attributed to the burrowing activity of bivalves: ichnotaxonomy, behaviour and palaeoenvironmental implications. Earth Sci Rev 150:497–519

Knaust D, Bromley RG (2012) Trace fossils as indicators of sedimentary environments. Developments in Sedimentology, vol 64. Elsevier, Oxford, xxx+960 pp

Knaust D, Neumann C (2016) *Asteriacites* von Schlotheim, 1820—the oldest valid ichnogenus name—and other asterozoan-produced trace fossils. Earth Sci Rev 157:111–120

Knaust D, Warchoł M, Kane IA (2014) Ichnodiversity and ichnoabundance: revealing depositional trends in a confined turbidite system. Sedimentology 62:2218–2267

Wisshak M, Kroh A, Bertling M et al (2015) In defence of an iconic ichnogenus—*Oichnus* Bromley, 1981. Ann Soc Geol Pol 85:445–451

# Preface

This book provides the reader with a blend of high-quality photographs, figures, and accompanied text for the identification of trace fossils in well core and outcrop. Ichnological data are becoming more and more crucial in sedimentological and paleoenvironmental interpretation, not only in the exploration and exploitation of hydrocarbon but also in the characterization of aquifers and in scientific drilling. Key features include the identification and interpretation of trace fossils in core and outcrop, integrated sedimentological–ichnological core logging, and hydrocarbon reservoir characterization. It has been prepared for an audience in the fields of sedimentology, paleontology, and petroleum geoscience from academia (graduate students and professionals) and industry (reservoir geologists).

After an introduction to the study of trace fossils in well core and an outlining of ichnological basics, principles and concepts, this book offers a detailed description and interpretation of 39 commonly occurring ichnogenera together with recurrently associated features such as diffuse bioturbate texture, plant roots and their traces, borings and pseudo-trace fossils. The trace fossils are highlighted by their expression in well core, illustrated with numerous original photographs and supplemented with carefully selected schematic drawings from the literature. This unique information is complemented by examples of trace fossils in outcrop, as well as relevant key figures from existing work.

Each chapter is treated in a consistent manner, stating the ichnogenus name and author in the title, followed by sections on morphology, fill and size; ichnotaxonomy; substrate; appearance in core; similar trace fossils; producers; ethology; depositional environment; ichnofacies; age as well as reservoir quality. This book is rounded off with an extensive list of references for further reading. The material for the book originated from the author's continuous work with trace fossils, chiefly in core, over the last two decades.

The well-core examples selected for this book mainly originate from the Norwegian Continental Shelf, which has been subject to extensive exploration and exploitation for oil and gas over the last half-century, although data from other regions of the world have been added. Based on this, siliciclastic rocks are overrepresented as compared to carbonates, and the majority of material comes from Mesozoic strata; however, all major paleoenvironments are covered. The presented trace fossils and associated features are thus just examples of possible occurrences in core, and other regions or stratigraphical units may return other interesting ichnological data. It is my hope that this book will promote further studies in this field.

Many colleagues and friends have shared their ideas over the last years, as well as specimens and literature; these include, in alphabetical order:

Andrea Baucon (Milano), Zain Belaústegui (Barcelona), Markus Bertling (Münster), Richard G. Bromley (Copenhagen), Luis A. Buatois (Saskatoon), Richard H.T. Callow (Stavanger), Kevin J. Cunningham (Miami), H. Allen Curran (Northampton), Andrei V. Dronov (Moscow), Allan A. Ekdale (Salt Lake City), Christian C. Emig (Marseille), Christian Gaillard (Lyon), Jorge F. Genise (Buenos Aires), Jean Gérard (Madrid), Jordi de Gibert (Barcelona, deceased), Murray K. Gingras (Edmonton), Roland Goldring (Reading, deceased), Murray R. Gregory (Auckland), Hans Hagdorn (Ingelfingen), Geir Helgesen (Stavanger), William Helland-Hansen (Bergen), Günther Hertweck (Wilhelmshaven), Sören Jensen

(Badajoz), Jostein Myking Kjærefjord (Bergen), Christian Klug (Zurich), Kantimati Kulkarni (Pune), James A. MacEachern (Burnaby), M. Gabriela Mángano (Saskatoon), Anthony J. Martin (Atlanta), Allard Martinius (Trondheim), Duncan McIlroy (St. John's), Renata Meneguolo (Stavanger), Radek Mikuláš (Praha), Masakazu Nara (Kochi), Carlos Neto de Carvalho (Idanha-a-Nova), Renata G. Netto (São Leopoldo), Christian Neumann (Berlin), Jan Kresten Nielsen (Oslo), Eduardo B. Olivero (Ushuaia), Ørjan Berge Øygard (Bergen), S. George Pemberton (Edmonton), John E. Pollard (Manchester), Lars Rennan (Trondheim), Andrew K. Rindsberg (Livingston), Francisco J. Rodríguez-Tovar (Granada), Jennifer J. Scott (Calgary), Koji Seike (Tokyo), Adolf Seilacher (Tübingen, deceased), Andrew M. Taylor (Northwich), Roger T.K. Thomas (Lancaster), Alfred Uchman (Krakow), Lothar Vallon (Faxe), Michał Warchoł (Bergen), Andreas Wetzel (Basel), Max Wisshak (Wilhelmshaven), Beate Witzel (Berlin) and Lijun Zhang (Jiaozuo).

Several ideas originated during the International Congresses on Ichnology and International Ichnofabric Workshops and also during in-house core workshops, special projects, and teaching trace-fossil analysis. Statoil ASA, in particular Sture Leiknes (Bergen), Frode Hadler-Jacobsen (Trondheim), Kjell Sunde (Bergen), and Ole Jacob Martinsen (Bergen), is thanked for providing me with the opportunity to study trace fossils in so many cores and to enable the publication of selected parts of this knowledge. This book has greatly benefited from the thorough reviews provided by Andrew K. Rindsberg (Livingston) and Andreas Wetzel (Basel), whose timely suggestions are much appreciated.

Stavanger, Norway
August 2016

Dirk Knaust

# Contents

# About the Author

**Dirk Knaust** is Specialist in Sedimentology in Statoil's Research and Technology unit in Stavanger, Norway. After working as a geologist in an underground mine and graduating in Germany with a Ph.D. in Geology (Carbonate Sedimentology, Stratigraphy, and Paleontology in the Triassic), in 1997 he followed an offer from the Norwegian oil and gas industry working in Exploration and Field Development. Since then, he has been frequently exposed to a vast amount of well core for investigation. The author has studied various aspects of trace fossils and ichnology, which is documented in about 60 publications in scientific journals and edited volumes. He is co-editor of the book *Trace Fossils as Indicators of Sedimentary Environments* (Elsevier) and associate editor of the journal *Ichnos*, amongst others. Dirk is leading author of a revised version of the *Trace Fossils* volume to be published in the *Treatise of Invertebrate Paleontology*.

The author inspecting dinosaur footprints on a Jurassic carbonate mud flat in Portugal

# 1 Introduction

Driven by the need of a successful hydrocarbon exploration and exploitation, the value of ichnology has been recognized since an early phase of the oil and gas industry. It is therefore no wonder that trace-fossil analysis of well core samples and relevant outcrop analogs has played a key role in the geological interpretation of subsurface data. Trace fossils are increasingly regarded as important for different kinds of sedimentological interpretations, and a wealth of publications has been dedicated to this subject as summarized in a number of volumes (e.g. Crimes and Harper 1970, 1977; Frey 1975; Basan 1978; Ekdale et al. 1984; Curran 1985; Bromley 1990, 1996; Maples and West 1992; Pemberton 1992; Donovan 1994; Pemberton et al. 2001; McIlroy 2004, 2015; Seilacher 2007; Miller 2007; Bromley et al. 2007; MacEachern et al. 2007; Avanzini and Petti 2008; Hasiotis 2010; Buatois and Mángano 2011; Knaust and Bromley 2012).

From the beginning of industrial core sampling, for instance in the exploration for coal in Germany, trace fossils were commonly regarded together with other fossils from a biostratigraphical perspective (e.g. Gothan 1932, Jessen 1950; Fiebig 1956). About the same time, the boxcore sampling method was developed for shallow-marine deposits of the German North Sea by researchers from the Senckenberg Institute (e.g. Schäfer 1952, 1956; Reineck 1958, 1963), which enabled the observation of burrows along with their producers. This successful approach was later applied to the coastal systems (Sapelo Island, Georgia; e.g. Howard and Dörjes 1972; Howard and Frey 1973, 1985) and elsewhere (e.g. Howard and Reineck 1981; Howard and Scott 1983), which was closely tied to research in similar paleoenvironments in the U.S. Western Interior Basin by Weimer and Hoyt (1964), then by Howard and Frey (1973).

Several sources of data, including from core, facilitated the development of the seminal ichnofacies concept by Seilacher (1967). Ichnofacies interpretations are generally done on a broader scale with wide facies belts (e.g. shoreline, shelf, slope or deep sea; e.g. Knaust and Bromley 2012) and very well suit cases where data coverage is scant, such as new exploration areas. The ichnofacies approach is then complemented by ichnofabric analysis, a concept based on the detailed interweaves between sedimentary features and ichnological signals (Ekdale et al. 2012). It therefore becomes important in the meticulous description and interpretation of bioturbated sediments and sedimentary rocks, particularly in core.

Extensive scientific drilling began with the Deep Sea Drilling Project (DSDP, operated from 1968 to 1983) and its successors, the Ocean Drilling Program (ODP, operated from 1985 to 2004), and the Integrated Ocean Drilling Program (IODP, operated from 2003). These programs have provided another important impulse and comprehensive dataset for ichnological analysis based on core. Several hundred kilometers of core were obtained through these campaigns and resulted in dozens of reports with ichnological information (e.g. Warme et al. 1973; Chamberlain 1975; Ekdale 1977, 1980; Fütterer 1984; Wetzel 1987). Related to this exceptional dataset, Chamberlain (1978) delivered an overview of trace fossils in core, still building the fundamentals for modern trace-fossil analysis in core.

Building upon the eminent work done by Chamberlain (1978), the study of trace fossils in core soon emerged as a special field in the oil and gas industry. As an example, the Norwegian oil and gas company Statoil has a long history of trace-fossil analysis based on core. Comprehensive well core material from the giant gas and oil field Troll, offshore Norway, provided enough material for completing probably the first extensive trace-fossil atlas by Bockelie and Howard (1984), which at that time was prepared for the Research Centre in Norsk Hydro. This volume gives a good introduction to the study of trace fossils in core and treats 35 ichnotaxa with respect to morphology and paleoenvironment, nicely illustrated with numerous line drawings and core photographs. For more than three decades, this volume


D. Knaust, *Atlas of Trace Fossils in Well Core*,
DOI 10.1007/978-3-319-49837-9_1

became an in-house standard and has been the basis for many core descriptions, although only a small fraction of this study has reached the public domain (Bockelie 1991, 1994). In the 1990s, Frederic Bockelie and later also Jean Gérard from Elf (now Total) used to work together with Richard Bromley on Jurassic core material from the North Sea, which influenced the contents of Bromley's (1990) *Trace Fossils*, and was subsequently published in an ichnofabric atlas (Gerard and Bromley 2008). Similar work was done in the UK sector of the North Sea by Roland Goldring, John Pollard and their students (Duncan McIlroy, Andrew Taylor, Stuart Gowland, and Stuart Buck). In 1991, the First International Ichnofabric Workshop, based on core, was launched in Bergen and Oslo, an ongoing series of workshops organized every two years around the world (Ekdale et al. 2012).

Since the 1980s, modern ichnology has entered the oil and gas business as an integrated part of core descriptions and facies interpretations in exploration and production. In North America, the school around Robert Frey and James D. Howard in Georgia and their students and collaborators (S. George Pemberton, Andrew K. Rindsberg and Anthony J. Martin) developed the science of ichnology further. In particular, George Pemberton (Edmonton) and his students (e.g. Murray Gingras and James MacEachern) intensively worked on the Cretaceous of Alberta in an applied manner for the oil and gas industry (e.g. Pemberton 1992; Pemberton et al. 2001; MacEachern et al. 2007).

In recent years, the characterization of aquifers has become more important and advanced, and provides an additional source for well-core samples for ichnological analysis (e.g. Cunningham et al. 2009, 2012; Cunningham and Sukop 2011, 2012). Likewise, research in the oil and gas industry has been extended from the application of ichnology in facies reconstruction towards explaining varying reservoir quality in relation to bioturbation (e.g. Pemberton and Gingras 2005; Gingras et al. 2012; Knaust 2014).

Applications of trace-fossil analysis in core can be manifold but three main uses may be highlighted: (1) Historically, probably the most relevant application is the utilization of ichnological core work as part of integrated studies (stratigraphical, sedimentological, paleontological and petrological) with the aim of facies interpretation and reconstruction of paleoenvironments. (2) Trace fossils are essential for the identification and characterization of key bounding surfaces in the stratigraphical column and their use in sequence-stratigraphical subdivisions and paleoenvironmental reconstructions. (3) Currently evolving is the understanding of the impact of trace fossils and bioturbate texture on rock properties (such as porosity and permeability) and fluid behavior (connectivity), similar to the related and already established knowledge of bioturbation and sediment properties in biological and ecological science.

For all these reasons, identification of trace fossils with confidence has become more crucial than ever but also faces old challenges. Previous compilations of trace fossils in core have addressed these issues and presented a solid overview of the commonest forms (e.g. Chamberlain 1975, 1978; Bockelie and Howard 1984; Taylor 1995; Pemberton et al. 2001; Gerard and Bromley 2008). From a current perspective, however, beside easily recognized burrows are many other forms out of several hundred burrow ichnogenera (Knaust 2012), which of course also occur in cored rock samples. In addition to the more common ichnogenera (many of which are not covered by the outdated *Treatise on Invertebrate Paleontology* volume on *Trace Fossils*; Häntzschel 1975), the rare forms may bear important information and aid in the reconstruction of paleoenvironmental conditions.

## References


Avanzini M, Petti FM (eds) (2008) Italian ichnology. Studi Trentini di Scienze Naturali. Acta Geol 83:347 pp

Basan PB (ed) (1978) Trace fossil concepts. SEPM Short Course Notes, vol 5, 201 pp

Bockelie JF (1991) Ichnofabric mapping and interpretation of Jurassic reservoir rocks of the Norwegian North Sea. Palaios 6:206–215

Bockelie JF (1994) Plant roots in core. In: Donovan SK (ed) The palaeobiology of trace fossils. Wiley, Chichester, pp 177–199

Bockelie JF, Howard A (1984) Systematic description of Jurassic trace fossils as they would appear in cored sections. Norsk Hydro Research Centre, 337 pp (internal report, unpublished)

Bromley RG (1990) Trace fossils, biology and taphonomy. Unwin Hyman, London, 280 pp

Bromley RG (1996) Trace fossils: biology, taphonomy and applications. Chapman and Hall, London, 361 pp

Bromley RG, Buatois LA, Mángano G et al (eds) (2007) Sediment—organism interactions: a multifaceted ichnology, vol 88, 393 pp. SEPM Special Publication

Buatois LA, Mángano MG (2011) Ichnology. Organism-substrate interactions in space and time. Cambridge University Press, Cambridge, 347 pp

Chamberlain CK (1975) Trace fossils in DSDP cores of the Pacific. J Paleontol 49:1074–1096

Chamberlain CK (1978) Recognition of trace fossils in cores. In: Basan PB (ed) Trace fossil concepts. SEPM Short Course Notes, vol 5, pp 133–183

Crimes TP, Harper JC (eds) (1970) Trace fossils. Geol J, Special Issue 3, 547 pp

Crimes TP, Harper JC (eds) (1977) Trace fossils 2. Geol J, Special Issue 9, 351 pp

Cunningham KJ, Sukop MC (2011) Multiple technologies applied to characterization of the porosity and permeability of the Biscayne Aquifer, Florida. USGS, Open-File Report 2011–1037, 8 pp

Cunningham KJ, Sukop MC (2012) Megaporosity and permeability of *Thalassinoides*-dominated ichnofabrics in the Cretaceous karst-carbonate Edwards-trinity aquifer system, Texas. U.S. Geological Survey, Open-file report 2012–1021, 4 pp

Cunningham KJ, Sukop MC, Huang H et al (2009) Prominence of ichnologically influenced macroporosity in the karst Biscayne aquifer: stratiform "super-K" zones. GSA Bull 121:164–186

Cunningham KJ, Sukop MC, Curran HA (2012) Carbonate aquifers. In: Knaust D, Bromley RG (eds) Trace fossils as indicators of sedimentary environments. Developments in Sedimentology, vol 64, pp 869–896

Curran HA (ed) (1985) Biogenic structures: Their use in interpreting depositional environments. SEPM Special Publication 35, 347 pp

Donovan SK (ed) (1994) The palaeobiology of trace fossils. Wiley, Chichester, 308 pp

Ekdale AA (1977) Trace fossils in worldwide Deep Sea Drilling Project cores. In: Crimes TP, Harper JC (eds), Trace Fossils 2. Geol J, Special Issue 9, pp 163–182

Ekdale AA (1980) Trace fossils in Deep Sea Drilling Project Leg 58 cores. In: de Vries Klein G, Kobyashi K et al (eds), Initial Reports of the Deep Sea Drilling Project 58, pp 601–605

Ekdale AA, Bromley RG, Pemberton SG (1984) Ichnology: the use of trace fossils in sedimentology and stratigraphy. SEPM Short Course Notes 15:1–317

Ekdale AA, Bromley RG, Knaust D (2012) The ichnofabric concept. In: Knaust D, Bromley RG (eds) Trace fossils as indicators of sedimentary environments. Developments in Sedimentology, vol 64, pp 139–155

Fiebig H (1956) Einige Bemerkungen zum Vorkommen von *Planolites ophthalmoides* Jessen im Ruhroberkarbon. Neues Jahrbuch für Geologie und Paläontologie, Monatshefte 1956:214–221

Frey RW (ed) (1975) The study of trace fossils: a synthesis of principles, problems and procedures in ichnology. Springer, New York. xiv+562 pp

Fütterer DK (1984) Bioturbation and trace fossils in deep sea sediments of the Walvis Ridge, southeastern Atlantic, Leg 74. In: Moore TC, Rabinowitz PD et al (eds), Initial Reports of the Deep Sea Drilling Project 74, pp 543–555

Gerard JRF, Bromley RG (2008) Ichnofabrics in clastic sediments—application to sedimentological core studies: a practical guide. Jean R.F. Gerard, Madrid, 100 pp

Gingras MK, MacEachern JA, Dashtgard SE et al (2012) Estuaries. In: Knaust D, Bromley RG (eds) Trace fossils as indicators of sedimentary environments. Developments in Sedimentology, vol 64, pp 463–505

Gothan W (1932) Paläobotanisch-stratigraphische Arbeiten im Westen des Ruhrreviers (mit Ausblicken auf die Nachbarreviere). Arbeiten aus dem Institut für Paläobotanik und Petrographie der Brennsteine 2:165–206

Häntzschel W (1975) Trace fossils and problematica. In: Teichert C (ed) Treatise on invertebrate Paleontology, Part W, Miscellanea Supplement 1. Geological Society of America/University of Kansas Press, Boulder/Lawrence, pp W1–W269

Hasiotis ST (2010) Continental trace fossils. SEPM Short Course Notes 51:1–132

Howard JD, Dörjes J (1972) Animal-sediment relationships in two beach-related tidal flats; Sapelo Island, Georgia. J Sediment Petrol 42:608–623

Howard JD, Frey RW (1973) Characteristic physical and biogenic sedimentary structures in Georgia Estuaries. AAPG Bull 57: 1169–1184

Howard JD, Frey RW (1985) Physical and biogenic aspects of back-barrier sedimentary sequences, Georgia coast, U.S.A. Mar Geol 63: 77–127

Howard JD, Reineck H-E (1981) Depositional facies of high-energy beach-to-offshore sequence: comparison with low-energy sequence. AAPG Bulletin 65:807–830

Howard JD, Scott RM (1983) Comparison of Pleistocene and Holocene barrier island beach-to-offshore sequences, Georgia and Northeast Florida coasts, U.S.A. Sediment Geol 34:167–183

Jessen W (1950) "Augenschiefer"-Grabgänge, ein Merkmal für Faunenschiefer-Nähe im westfälischen Oberkarbon. Zeitschrift der Deutschen Geologischen Gesellschaft 101:23–43

Knaust D (2012) Trace-fossil systematics. In: Knaust D, Bromley RG (eds) Trace fossils as indicators of sedimentary environments. Developments in Sedimentology, vol 64, pp 79–101

Knaust D (2014) Classification of bioturbation-related reservoir quality in the Khuff Formation (Middle East): towards a genetic approach. In: Pöppelreiter MC (ed) Permo-Triassic sequence of the Arabian Plate. EAGE, pp 247–267

Knaust D, Bromley RG (2012) Trace fossils as indicators of sedimentary environments. Developments in Sedimentology, vol 64. Elsevier, Oxford, xxx+960 pp

MacEachern JA, Bann KL, Gingras MK et al (eds) (2007) Applied ichnology. SEPM Short Course Notes, vol 52, 380 pp

Maples CG, West RR (eds) (1992) Trace fossils. Short Courses in Paleontology, vol 5, 238 pp

McIlroy D (2004) Ichnofabrics and sedimentary facies of a tide-dominated delta: Jurassic Ile Formation of Kristin Field, Haltenbanken, offshore Mid-Norway. In: McIlroy D (ed) The application of ichnology to palaeoenvironmental and stratigraphic analysis, vol 228. Geological Society of London, Special Publications, pp 237–272

McIlroy D (ed) (2015) Ichnology: Papers from ICHNIA III, vol 9. Geological Association of Canada, Miscellaneous Publication, iii+272 pp

Miller W (ed) (2007) Trace fossils: concepts, problems, prospects. Elsevier, Amsterdam, 632 pp

Pemberton SG (ed) (1992) Applications of ichnology to petroleum exploration. A core workshop. SEPM Core Workshop, vol 17, 429 pp

Pemberton SG, Gingras MK (2005) Classification and characterizations of biogenically enhanced permeability. AAPG Bull 89:1493–1517

Pemberton SG, Spila MV, Pulham AJ et al (2001) Ichnology and sedimentology of shallow to marginal marine systems. In: Ben Nevis and Avalon Reservoirs, Jeanne d'Arc Basin. Short Course Notes vol 15. Geological Association of Canada, 343 pp

Reineck H-E (1958) Wühlbau-Gefüge in Abhängigkeit von Sediment-Umlagerungen. Senckenb Lethaea 39(1–23):54–56

Reineck H-E (1963) Sedimentgefüge im Bereich der südlichen Nordsee. Abhandlungen der Senckenbergischen Naturforschenden Gesellschaft 505:1–138

Schäfer W (1952) Biogene Sedimentation im Gefolge von Bioturbation. Senckenb 33:1–12

Schäfer W (1956) Wirkungen der Benthos-Organismen auf den jungen Schichtverband. Senckenb Lethaia 37:183–263

Seilacher A (1967) Bathymetry of trace fossils. Mar Geol 5:413–428

Seilacher A (2007) Trace fossil analysis. Springer, Berlin, 226 pp

Taylor AM (1995) Atlas of trace fossils in core-expression. Ichron. Statoil report 95/OO4/S, 28 pp (internal report, unpublished)

Warme JE, Kennedy WJ, Scheidermann N (1973) Biogenic sedimentary structures (trace fossils) in Leg 15 cores. In: Edgar NT, Saunders JB et al (eds), Initial Reports of the Deep Sea Drilling Project 15, pp 813–831

Weimer RJ, Hoyt JH (1964) Burrows of *Callianassa major* Say, geologic indicators of littoral and shallow neritic environments. J Paleontol 38:761–767

Wetzel A (1987) Ichnofabrics in Eocene to Maestrichtian sediments from Deep Sea Drilling Project Site 605, off the New Jersey coast. In: Hinte JE, Wise Jr SW et al (eds), Initial Reports of the Deep Sea Drilling Project 93, pp 825–835

# 2 Ichnological Basics, Principles and Concepts

## 2.1 Terminology and Definitions

*Ichnology* is the study of traces produced by organisms (animals, plants and microbes) on or within a substrate. It deals with all aspects related to modern (neoichnology) and fossil traces (paleoichnology), bioturbation and bioerosion, and is interdisciplinary in combining sedimentological, paleontological, biological and ecological methods (Bromley 1996). It complements and constrains sedimentological interpretations and serves as a powerful tool in reservoir characterization.

The subjects of paleoichnology are *trace fossils* (also called ichnofossils), which are fossilized structures produced in substrates ranging from unlithified sediment to sedimentary rock or organic matter (including shell, bone, wood and peat) by the activity of organisms. Traces of organisms can be grouped into categories, depending on the type of substrate and manner of origin (Fig. 2.1):

- *Burrows*: Most common trace-fossil category, comprising galleries, tunnels, shafts, chambers, etc., excavated by animals within an unconsolidated substrate.
- *Bioerosion trace fossils*: If the excavation takes place in a consolidated and lithified substrate, the resulting trace is a bioerosional trace fossil such as a boring or a scratch.
- *Trails*: Trails are surface features, in which the producer leaves a continuous path behind it while moving.
- *Trackways*: In contrast to trails, trackways are discontinuous paths which originate from walking animals. Individual imprints of the trackway are called tracks.
- *Plant-root traces*: Most traces are related to the activity of animals, although plants can also leave their traces by means of their roots.

There are many other categories of traces with less importance for the purpose of this book, of which coprolites (i.e. fossil feces) are probably the most important. An overview of accepted groups of traces is given by Bertling et al. (2006).

*Bioturbation* is the process by which the primary structure and properties of a sediment are modified by the activity of organisms living within it, which may result in sediment mixing (Bromley 1996). The latter expression is often loosely applied to the product of this process, which is better defined by the term bioturbate texture (Frey 1973). *Bioerosion*, in contrast, comprises processes of mechanical or biochemical destruction of hard substrates by organisms. From a sedimentological point of view, bioturbation, bioerosion, biodeposition and biostratification structures can be grouped together as *biogenic sedimentary structures* (Frey 1973).

*Ichnofacies* as concept was established by Seilacher (1967) based on his and others, earlier work. Trace-fossil communities (ichnocoenoses) were linked to an overall ocean profile, mainly related to the behavioral response of the tracemakers to a bathymetric gradient in food supply. Ichnofacies represents a powerful tool when working on a larger scale (e.g. basin scale) and screening new areas, where a rough interpretation of the paleoenvironment in terms of broad facies belts can be given. The ichnofacies concept has been continuously updated, refined and extended into the continental realm. Current overviews and discussions are provided by Buatois and Mángano (2011), MacEachern et al. (2012) and Melchor et al. (2012, for continental ichnofacies).

The *ichnofabric* concept regards all aspects of the texture and internal structure of a sediment that result from bioturbation at all scales (Ekdale and Bromley 1983, 1991; Bromley


D. Knaust, *Atlas of Trace Fossils in Well Core*,
DOI 10.1007/978-3-319-49837-9_2

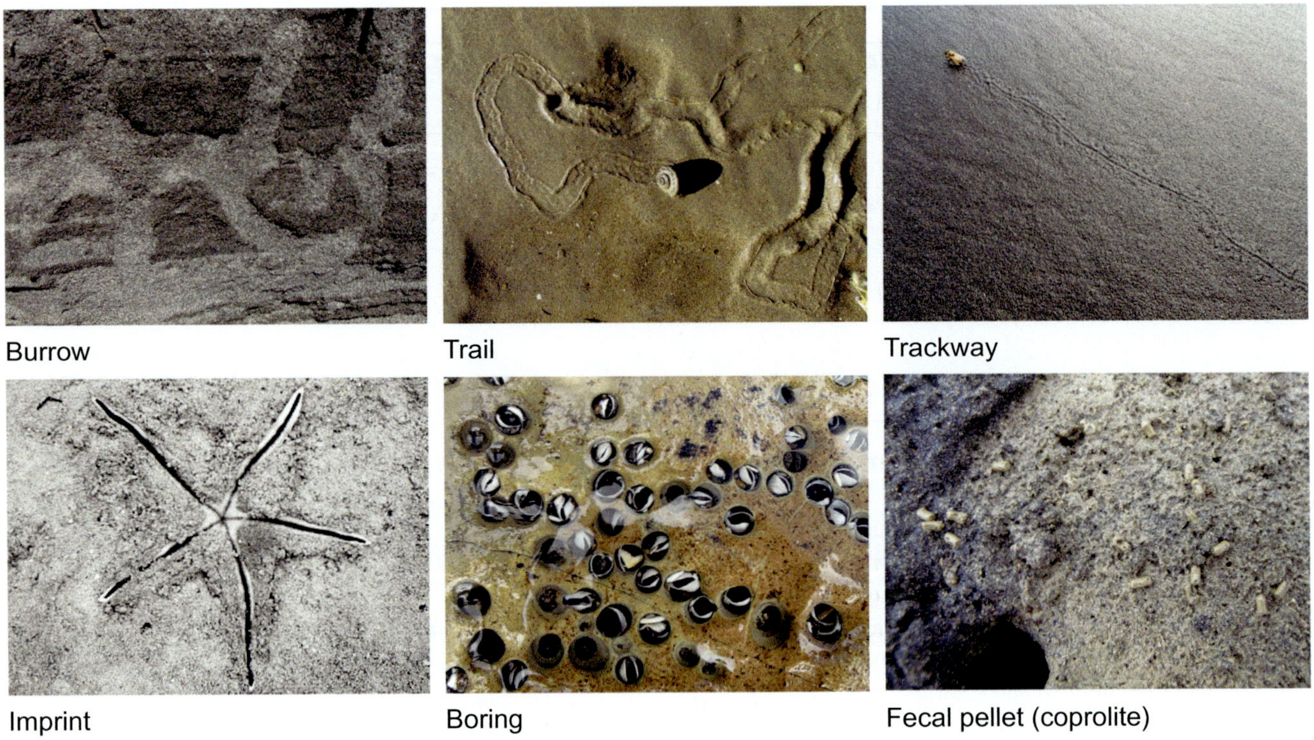


**Fig. 2.1** Major categories of traces that may become trace fossils if fossilized. Bedding-surface views except for burrow (*upper left*), which is vertical section. Imprint image courtesy of H. Allen Curran (Northampton)

and Ekdale 1986). It has been developed on sectioned rock faces, where different cross-cutting relations can be related to successive colonization or tiering, and changing degree of bioturbation can be analyzed. Compared with the ichnofacies concept, which puts emphasis on the recognition of recurrent ichnocoenoses and facies belts, the purpose of the ichnofabric concept is mainly the analysis of different stages contained in a particular piece of bioturbated rock. Therefore it is a valuable tool in the detailed interpretation of rock samples from core (Taylor et al. 2003; Ekdale et al. 2012).

## 2.2 Some Principles

The study of trace fossils is related to various challenges of which the following are highlighted.

- *One type of organism can produce many different traces*: For example, given a particular insect which is able digging a burrow into the substrate, leaving a trackway on the surface due to locomotion, or an imprint while resting, scraping hard substrate (such as wood) while feeding, building chambers while breeding, and leaving their excrement in form of fecal pellets. Other examples include many species of crustaceans and molluscs, able to produce different traces and burrows with contrasting characteristics (Fig. 2.2, see also Fig. 5.85).
- *Many different organisms can produce the same trace*: Simple vertical shafts without branching (e.g. *Skolithos*) would be a good example, because they can be produced by many different organisms such as priapulids, holothurians, polychaetes, phoronids, crustaceans, anthozoans, insects, spiders and even plant roots (Fig. 2.3).
- *The tracemaker is rarely known*: Particularly true for many trace fossils, the tracemaker is only preserved under rare circumstances (e.g. exceptionally preserved fossils or fossillagerstätten, Fig. 2.4; see also Figs. 5.103 and 5.133). In most cases, the producer can be inferred at a higher taxonomic rank with some uncertainty, for instance by analyzing particular features of the trace (e.g. architecture, scratches and fecal pellets) or reconstruction of its functional morphology.

**Fig. 2.2** Various traces produced by chiton (Polyplacophora)

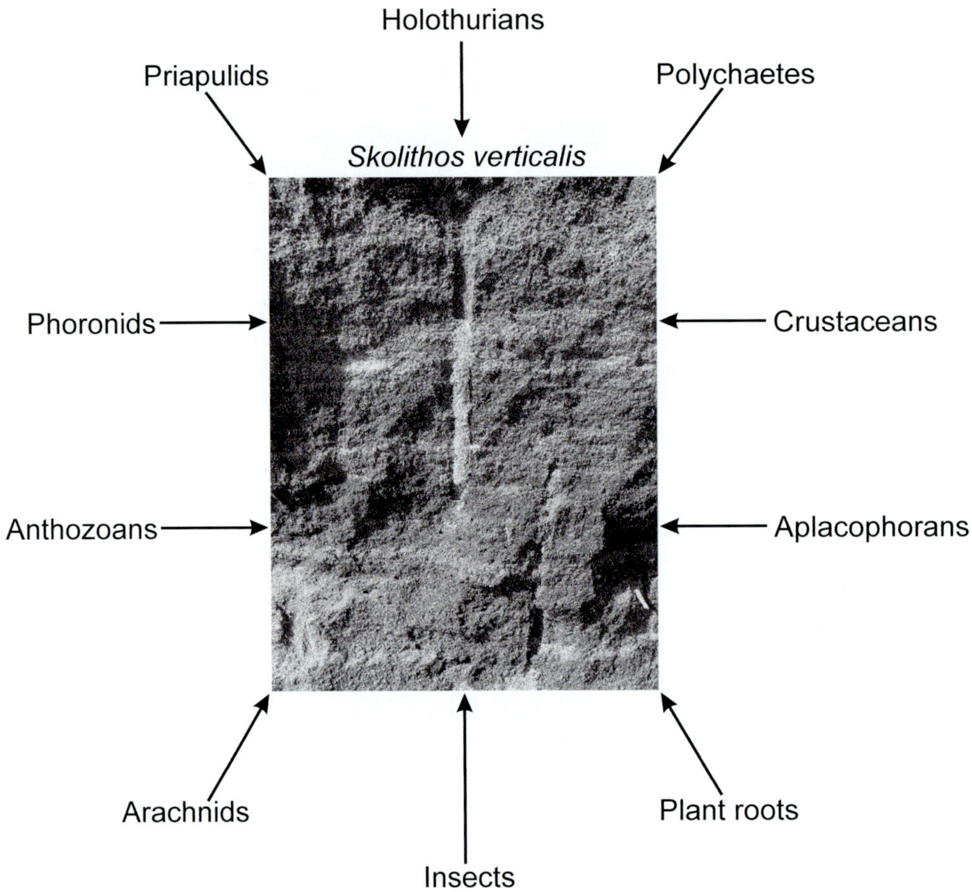


**Fig. 2.3** *Skolithos*, a very simple trace, can potentially be produced by a wide range of organisms of different phyla and environments

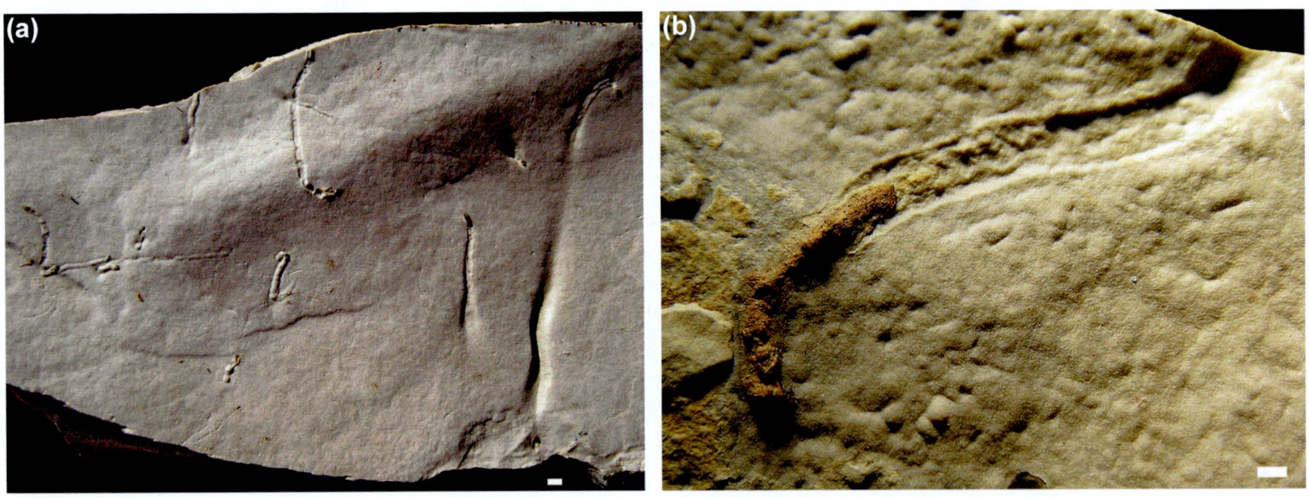


**Fig. 2.4** Examples of trace fossils (mainly trails and shallow burrows) that preserve their producers at the termination of the trace. The very fine-grained (micritic) sediment and favored taphonomic and diagenetic circumstances (e.g. microbial growth, lowered oxygenation) prevented a total loss of the organic material and promoted exceptional fossil preservation, which allows a determination of higher taxonomic categories. Middle Triassic (Anisian-Ladinian) Meissner Formation (Muschelkalk), Thuringia, Germany. *Scale bars* = 1 mm. For details see Knaust (2007, 2010, 2015). **a** Bedding surface with many trails and burrows, most of which preserve their producers at the termination in form of weathered sulfide aggregates (e.g. arthropods, nemerteans, nematodes) or calcite crystals (e.g. involutinidae foraminifers, turbellarian platyhelminthes). **b** Undulating bedding surface with pustules due to microbial modification with a trail occurring together with its supposed nemertean (ribbon worm) producer which is preserved as limonite aggregate. Note the slightly sinuous fecal string

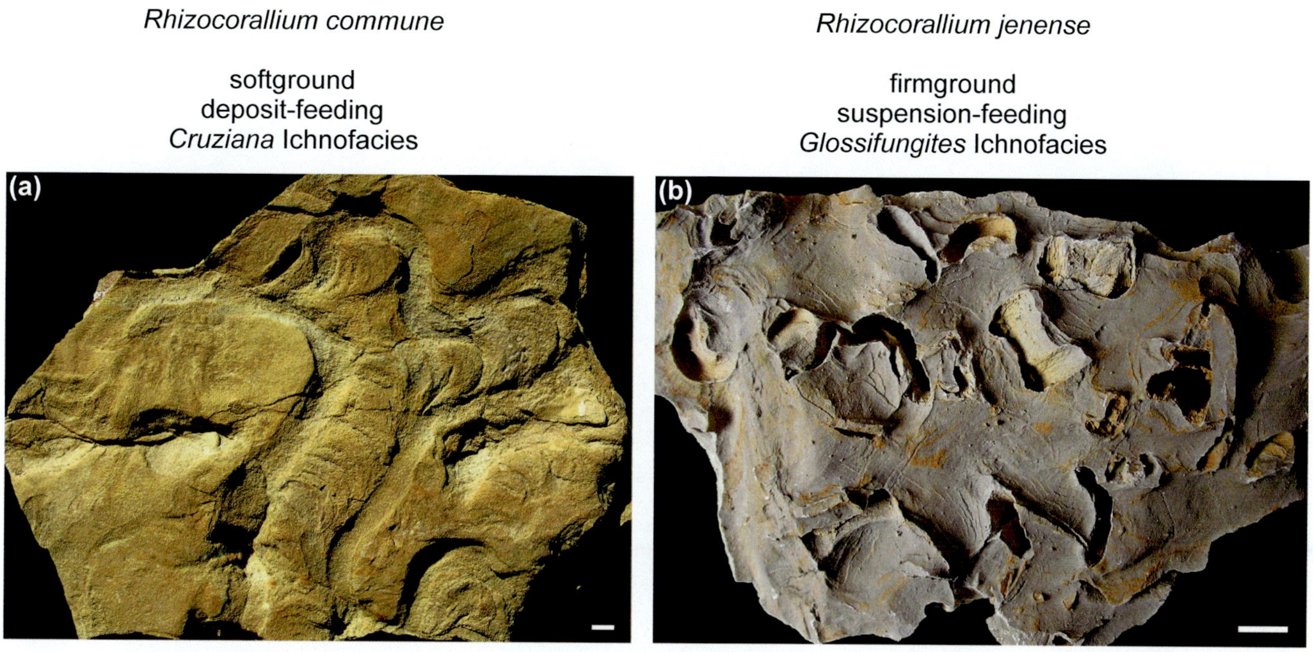


**Fig. 2.5** The only two valid ichnospecies of *Rhizocorallium*, with different morphological features mainly due to contrasting substrate conditions. *Scale bars* = 1 cm. **a** *R. commune* produced in softground, a wide horizontal burrow with actively created spreite, fecal pellets and occasional branching. **b** *R. jenense* produced in firmground, a narrow pouch-shaped and inclined burrow with passive fill and a dense pattern of scratches on the margin of burrow. After Knaust (2013), republished with permission of Elsevier; permission conveyed through Copyright Clearance Center, Inc. See also Fig. 5.117

- *The same trace becomes preserved differently in various substrates*: Different categories of substrate (e.g. softground, firmground and hardground) preserve the traces of the same producer in different ways. The ichnogenus *Rhizocorallium* is only one example (Fig. 2.5), where a probable polychaete makes extensive horizontal spreite burrows with occasional branching and active fill (*R. commune*, probably resulting from deposit feeding), while in firmground the burrows are shorter and inclined, unbranched and open or passively filled (*R. jenense*, probably resulting from suspension feeding).
- *Compound, composite and complex trace fossils*: Trace-fossil architecture can be complicated by the interaction of different tracemakers, or producers with contrasting behavior. Compound trace fossils are those comprising intergradational forms of ichnotaxonomically different parts, such as *Thalassinoides-Ophiomorpha-Spongeliomorpha* (Fig. 5.85). Composite trace fossils originate from the interpenetration of ichnotaxonomically different parts, which can be identified as such by their cross-cutting relationship. Complex trace fossils are morphologically complex structures, including compound trace fossils, which are characterized by their high degree of organization, for instance *Zoophycos* and *Hillichnus*.
- *Multiple colonization phases and surfaces, tiers and cross-cutting relationship lead to complex ichnofabrics*: Traces are rarely single, and interaction among different generations of traces with contrasting features is the norm (Fig. 2.6). This can result in partly or completely bioturbated substrate, which may preserve discrete traces on top of a diffuse (bioturbated) background. Interaction of benthic communities may also lead to reburrowing of existing burrows by subsequent producers.

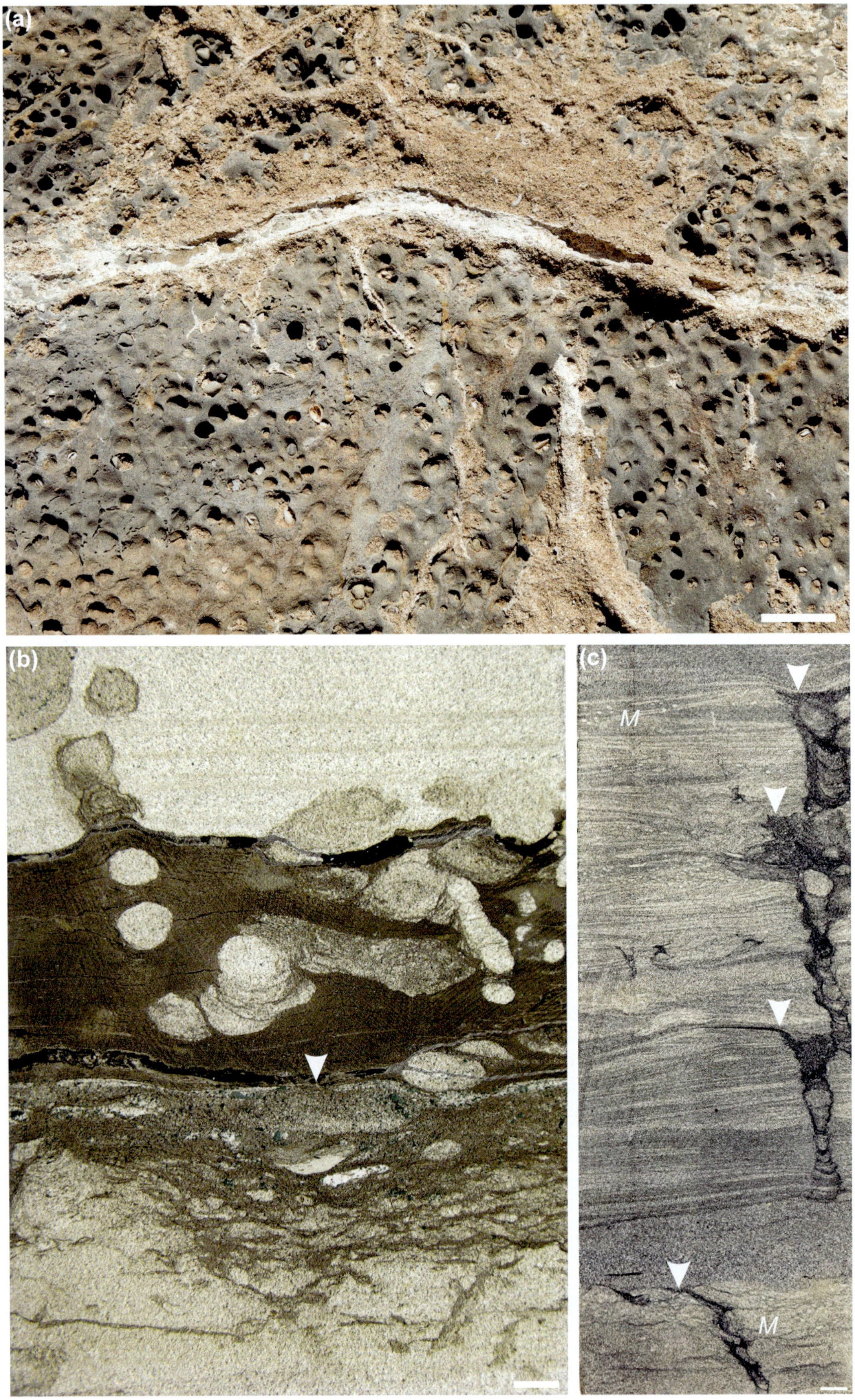
(a)
(b)
(c)
M
M

◀ **Fig. 2.6** Ichnofabrics with multiple colonization surfaces in outcrop and in core. *Scale bars* = 5 cm (**a**) and 1 cm (**b**, **c**). **a** Outcrop photograph of a limestone bedding plane showing hardground features with dense occurrence of *Gastrochaenolites* isp., produced by boring bivalves (in places preserved) in a shallow-marine environment. A second colonization of the same surface is documented by a network of calichified root traces, which belong to the overlying eolian sandstone. This surface is a regional angular unconformity between Cretaceous platform carbonates and Pliocene to Pleistocene eolian dune deposits (brownish patches of sand). Cliff section south of Taghazout, western Morocco. **b** Parts of two sandy turbidite layers interbedded with hemipelagic mudstone in vertical core section. The top surface of the lower turbidite served as a colonization surface (*arrow head*), which resulted in a mixed layer with incorporated green clay minerals beneath it due to repeated bioturbation. The mudstone layer above it contains large burrows actively filled with sand (*Thalassinoides* and *Ophiomorpha*) as part of deep-tier burrowing through the overlying sand (turbidite). Upper Cretaceous (Maastrichtian) Springar Formation (deepmarine), Norwegian Sea (well 6604/10-1, ca. 3647.5 m). **c** Ripple-laminated fine-grained sandstone with intervals containing *Macaronichnus segregatis* (*M*) in vertical core section. The displayed sandstone shows multiple colonization surfaces (*arrow heads*) from which the muddy spreite burrows *Teichichnus zigzag* penetrate the underlying sediment, indicating rapid deposition (sandy tidal-flat deposit). The succession is interrupted by a medium-grained sandstone layer in its lower part, probably a storm deposit (tempestite). Middle Jurassic (Bathonian) Tarbert Formation (sandy tidal flat), Oseberg Sør Field, Norwegian North Sea (well 30/9-F-26, ca. 4466.8 m)

## References


Bertling M, Braddy SJ, Bromley RG et al (2006) Names for trace fossils: a uniform approach. Lethaia 39:265–286

Bromley RG (1996) Trace fossils: biology, taphonomy and applications. Chapman and Hall, London, 361 pp

Bromley RG, Ekdale AA (1986) Composite ichnofabrics and tiering of burrows. Geol Mag 123:59–65

Buatois LA, Mángano MG (2011) Ichnology. Organism-substrate interactions in space and time. Cambridge University Press, Cambridge, 347 pp

Ekdale AA, Bromley RG (1983) Trace fossils and ichnofabric in the Kjølby Gaard Marl, uppermost Cretaceous, Denmark. Bull Geol Soc Den 31:107–119

Ekdale AA, Bromley RG (1991) Analysis of composite ichnofabrics: an example in the uppermost Cretaceous chalk of Denmark. Palaios 6:232–249

Ekdale AA, Bromley RG, Knaust D (2012) The ichnofabric concept. In: Knaust D, Bromley RG (eds) Trace fossils as indicators of sedimentary environments. Developments in Sedimentology, vol 64, pp 139–155

Frey RW (1973) Concepts in the study of biogenic sedimentary structures. J Sediment Petrol 43:6–19

Knaust D (2007) Meiobenthic trace fossils as keys to the taphonomic history of shallow-marine epicontinental carbonates. In: Miller W III (ed) Trace fossils: concepts, problems, prospects. Elsevier, Amsterdam, pp 502–517

Knaust D (2010) Remarkably preserved benthic organisms and their traces from a Middle Triassic (Muschelkalk) mud flat. Lethaia 43: 344–356

Knaust D (2013) The ichnogenus *Rhizocorallium*: classification, trace makers, palaeoenvironments and evolution. Earth Sci Rev 126:1–47

Knaust D (2015) Grazing traces (*Kimberichnus teruzzii*) on Middle Triassic microbial matground. In: McIlroy D (ed) Ichnology: Papers from ICHNIA III, vol 9. Geological Association of Canada, Miscellaneous Publication, pp 115–125

MacEachern JA, Bann KL, Gingras MK et al (2012) The ichnofacies paradigm. In: Knaust D, Bromley RG (eds) Trace fossils as indicators of sedimentary environments. Developments in Sedimentology, vol 64, pp 103–138

Melchor RN, Genise JF, Buatois LA et al (2012) Fluvial environments. In: Knaust D, Bromley RG (eds) Trace fossils as indicators of sedimentary environments. Developments in Sedimentology, vol 64, pp 329–378

Seilacher A (1967) Bathymetry of trace fossils. Mar Geol 5:413–428

Taylor A, Goldring R, Gowland S (2003) Analysis and application of ichnofabrics. Earth Sci Rev 60:227–259

# 3 Applications of Trace-Fossil Analysis

With respect to reservoir prediction and characterization, ichnology can contribute with three major themes, which are facies interpretation, stratigraphy, and reservoir quality.

## 3.1 Facies Interpretation

The interpretation of facies and depositional environments is probably the best recognized application for trace fossils and bioturbate textures. Depending on available data, the size of the study area and subject to be investigated, the ichnofacies and ichnofabric concepts can be utilized in combination with other methods, especially sedimentology (Fig. 3.1). For instance, in cases where a new basin is entered or a new petroleum play is explored, only a limited set of data is typically available, of which trace-fossil analysis can be used to delineate the overall setting (such as deep marine versus shelf). Smaller areas with better data coverage can be differentiated with respect to their paleoenvironments and summarized in paleogeographical maps. Great value can be gained by the identification of limiting factors such as oxygen depletion, salinity fluctuation and reduction, light availability, hydraulic energy, substrate consistency, availability of organic matter and so on.

**(a)** Ichnofacies (Seilacher, 1967)

**(b)** Ichnofabric (Bromley and Ekdale, 1986)

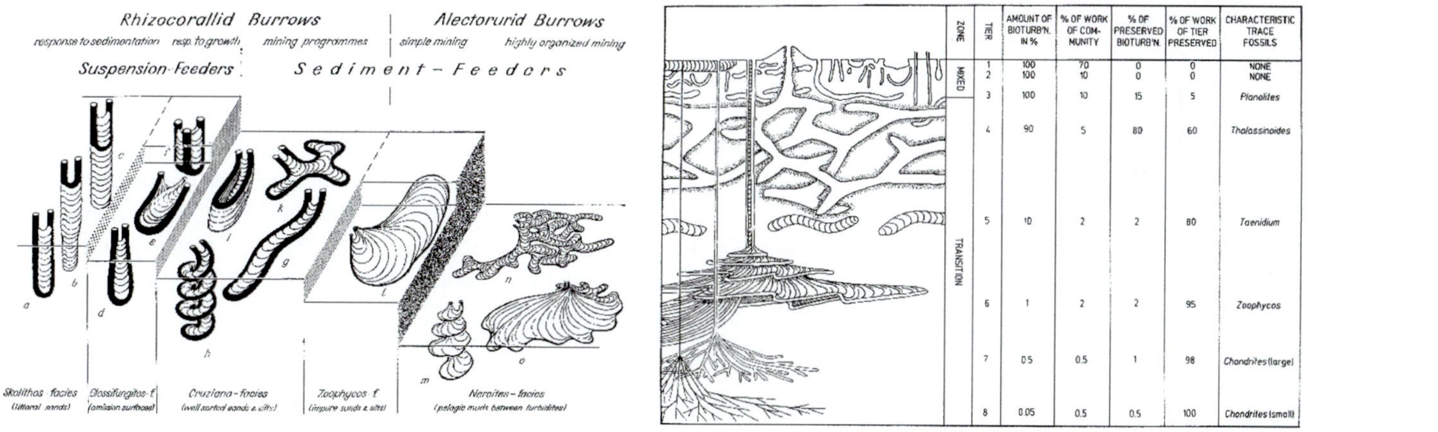


**Fig. 3.1** Ichnofacies (**a**) and ichnofabric concept (**b**), two brilliant, often used, and still debated approaches in the utilization of ichnological information. The ichnofacies concept was originally designed for trace fossils outlining broad facies belts and thus becomes applicable in areas where relative little data is available, whereas the ichnofabric concept integrates various ichnological information of a sectioned unit and thus provides a wide range of information. Reprinted from Seilacher (1967) (**a**), republished with permission of Elsevier, permission conveyed through Copyright Clearance Center, Inc.; and Bromley (1990) (**b**), reprinted with permission of Springer


D. Knaust, *Atlas of Trace Fossils in Well Core*,
DOI 10.1007/978-3-319-49837-9_3

## 3.2 Stratigraphy

In general, trace fossils are poor stratigraphical markers. However, some periods (e.g. in the lower Paleozoic) have been successfully dated by means of trace fossils (e.g. ichnospecies of *Cruziana*). This procedure might be helpful in exploration activities of large basins with relatively few data as demonstrated in North Africa. More common is the use of trace fossils and bioturbate textures in sequence stratigraphy, where particular intervals and key stratigraphical surfaces can be identified and correlated (Fig. 3.2). In addition to many other methods, ichnological analysis is important in constraining sequence-stratigraphical interpretations and to solve a wide range of problems in the following areas:

- Delineation of sequence-stratigraphic surfaces by means of their ichnological signal, which can be studied in outcrop and in subsurface datasets (e.g. cores and borehole images).
- Facies reconstruction and characterization of juxtaposed facies types to recognize changes in deposition and deviations from continuous facies transitions (e.g. Walther's Law).
- Vertical facies trends, reflecting shallowing- or deepening-upward in response to autocyclic or allocyclic processes, often recognizable by changing degree of bioturbation and trace-fossil content. (This is particularly crucial in relatively uniform lithologies such as in carbonate systems.)
- General aspects, such as degree of bioturbation, trace-fossil associations, ichnodiversity etc. in comparison to one another, which can contribute to the delineation of particular systems tracts.

## 3.3 Reservoir Quality

The flow behavior within a reservoir is dependent on its reservoir quality, which in turn is a function of porosity, permeability and connectivity. A number of factors play a role in the development of reservoir quality, including rock composition, sedimentary structures, cementation and dissolution, and fracturing (Fig. 3.3). In addition, bioturbation (organism/sediment interactions) and the resulting textures and trace fossils has proved to be an important agent in the postdepositional modification of sediments (Fig. 3.4).

The process of bioturbation results in the origin of highly variable textures and structures, which may increase or decrease the rock's porosity and thus reservoir quality (Fig. 3.5). In many cases, intense bioturbation leads to the homogenization of alternating sediments or even an increase of mud due to the introduction of clay minerals, which under certain circumstances may help protecting the existing porosity. More often, however, biogenically introduced mud reduces the porosity and diminishes the reservoir quality. The reorganized rock fabric and discrete trace fossils can act as preferred fluid paths for soluble mineral phases and promote early diagenetic cementation leading to reduced porosity. In contrast, open and sand-filled shafts and galleries of burrow systems can improve porosity significantly and to turn a fluid barrier into a reservoir. Finally, bioturbate textures and trace fossils are an important constituent of mudstone and shale and may lead to an increased heterogeneity due to the incorporation of silt and sand. This latter process plays an important role in the producibility and performance of shale gas reservoirs.

A wide range of infaunal organisms with contrasting burrowing styles is known and some create complex bioturbate textures and ichnofabrics. Some organisms (e.g. crustaceans) process the sediment and destroy the primary bedding, which leads to a uniform fabric with an isotropic character. Other animals (e.g. some kinds of polychaetes) are more selective and repeatedly rework the sediment whilst they selectively feed on the deposit. The result is a more anisotropic rock fabric with connectivity preferred in a particular direction (Fig. 3.6). More advanced architects (e.g. some worms, bivalves and shrimp) are responsible for rather complex burrow systems with three-dimensional branching, wall lining and active fill. Consequently, these bioturbators import a high degree of heterogeneity to the sediment.

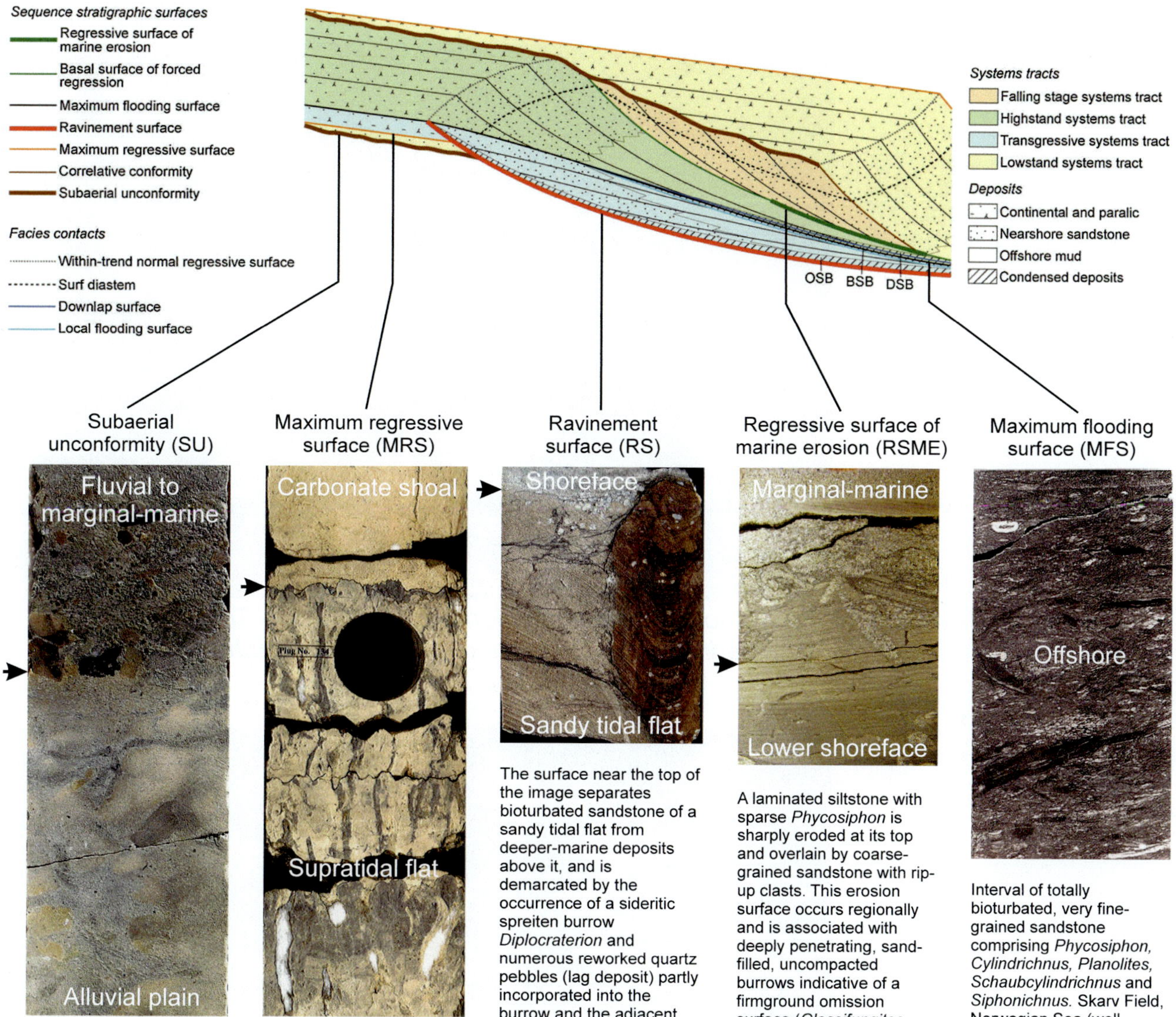


Boundary between Hegre Group (Upper Triassic) and Statfjord Group (Lower Jurassic). The Hegre Group in the lower part of the image consists of a mixed siliciclastic-carbonate rock with caliche features, fractures and complex burrow systems in the firmground, as well as *Taenidium* (not figured). It results from pedogenesis in an alluvial environment. The overlaying coarse-grained sandstone of the Statfjord Group is weakly bedded, contains a basal lag deposit with polymict clasts and granulae, and is without bioturbate textures. It results from fluvial processes. Johan Sverdrup Field, Norwegian North Sea (well 16/2-15, ca. 1972.7 m).

A dense root system (probably resulting from Cordaitales) is deeply penetrating from an erosive, blackened surface and is filled with organic-rich sediment, dolomite, and gypsum. It is sharply overlain by an oolitic limestone. Upper Permian carbonate platform of the Khuff Formation, South Pars Field, offshore Iran (well SP10, 2937.5 m).

The surface near the top of the image separates bioturbated sandstone of a sandy tidal flat from deeper-marine deposits above it, and is demarcated by the occurrence of a sideritic spreiten burrow *Diplocraterion* and numerous reworked quartz pebbles (lag deposit) partly incorporated into the burrow and the adjacent host rock due to bioturbation. Middle Jurassic Hugin Formation, Sleipner Field, Norwegian North Sea (well 15/9-1, ca. 3660.35 m).

A laminated siltstone with sparse *Phycosiphon* is sharply eroded at its top and overlain by coarse-grained sandstone with rip-up clasts. This erosion surface occurs regionally and is associated with deeply penetrating, sand-filled, uncompacted burrows indicative of a firmground omission surface (*Glossifungites* surface). Lower Jurassic Cook Formation, Norwegian North Sea (well 35/10-1, ca. 3655.8 m).

Interval of totally bioturbated, very fine-grained sandstone comprising *Phycosiphon, Cylindrichnus, Planolites, Schaubcylindrichnus* and *Siphonichnus.* Skarv Field, Norwegian Sea (well 6507/5-1, ca. 3538.6 m).

**Fig. 3.2** Interpreted core examples of key stratigraphical surfaces (indicated by *arrows, middle and lower parts*) and their relationship within the sequence-stratigraphic framework provided by Zesshin and Catuneanu (2013, *upper part*, republished with permission of Elsevier; permission conveyed through Copyright Clearance Center, Inc.). Width of core images is approximately 10 cm

**Fig. 3.3** Important processes and their products, which result in small-scale heterogeneities of reservoirs in sedimentary rocks. During the evolution of a reservoir, each process in isolation or combination with others may have an impact on reservoir quality. Modified after Knaust (2013)

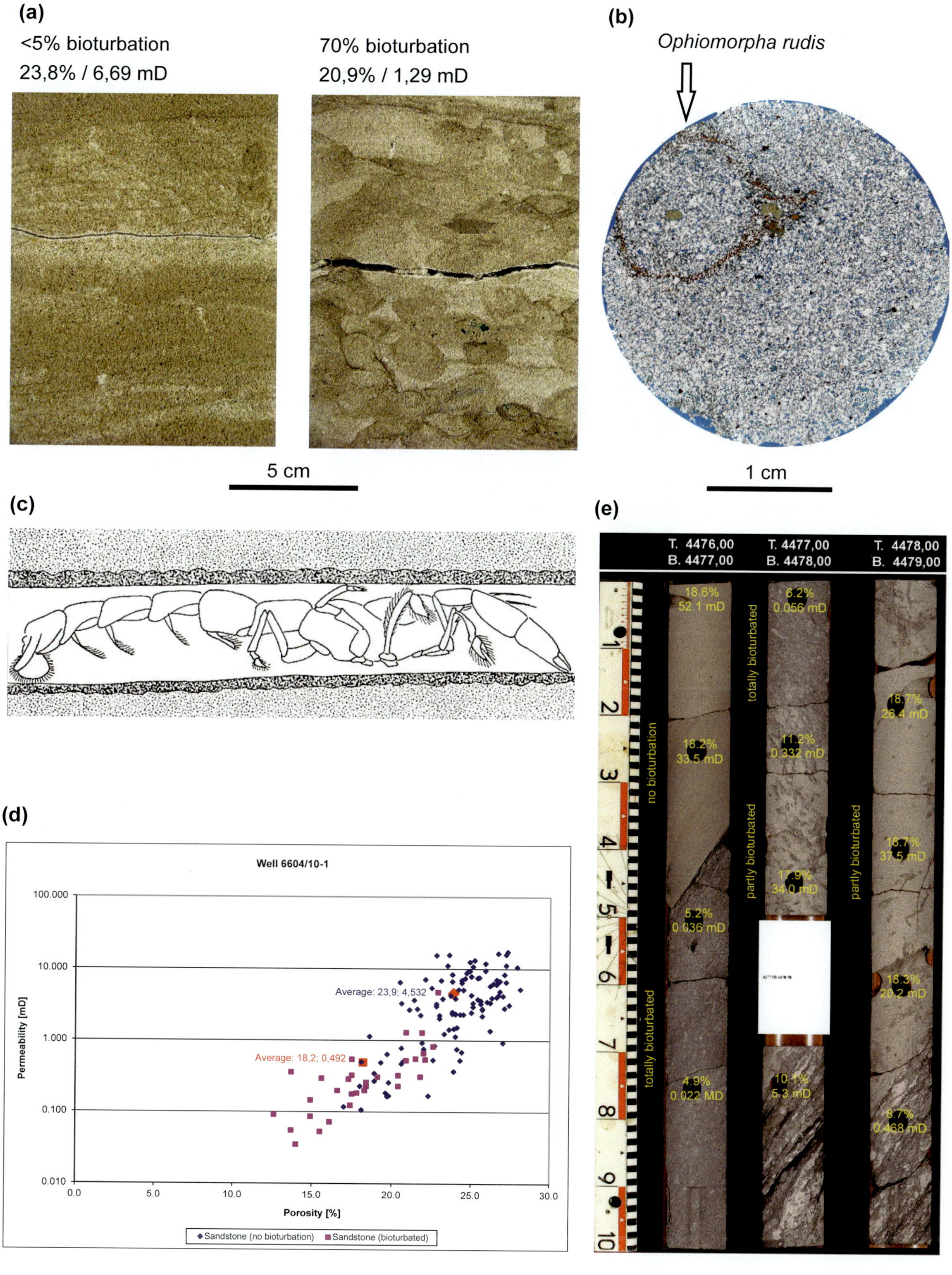
(a)
<5% bioturbation
23,8% / 6,69 mD
70% bioturbation
20,9% / 1,29 mD
5 cm
(b)
Ophiomorpha rudis
1 cm
(c)
(d)
Well 6604/10-1
100.000
10.000
1.000
0.100
0.010
Permeability [mD]
0.0
5.0
10.0
15.0
20.0
25.0
30.0
Porosity [%]
Average: 23,9; 4,532
Average: 18,2; 0,492
Sandstone (no bioturbation)
Sandstone (bioturbated)
(e)
T. 4476,00
B. 4477,00
T. 4477,00
B. 4478,00
T. 4478,00
B. 4479,00
18.6%
52.1 mD
18.2%
33.5 mD
5.2%
0.036 mD
4.9%
0.022 MD
6.2%
0.056 mD
11.2%
0.332 mD
17.9%
34.0 mD
10.1%
5.3 mD
18.7%
26.4 mD
18.7%
37.5 mD
18.3%
20.2 mD
8.7%
0.468 mD
no bioturbation
totally bioturbated
partly bioturbated

◀**Fig. 3.4** Impact of bioturbation on gas-reservoir quality based on examples from Upper Cretaceous deep-marine turbidite deposits of the Vøring Basin, Norwegian Sea. **a** Core examples of almost unbioturbated versus intensely bioturbated sandstone and their corresponding porosity and permeability measurements (based on horizontal core plug data), which indicate reduced properties for bioturbated sandstone. Springar Formation (Maastrichtian, well 6604/10-1). **b** Sandstone in thin section with homogeneous texture and incorporated clay minerals, some of which outline the wall of a discrete *Ophiomorpha rudis* burrow. Same well as in (**a**). **c** Callianassid shrimp in its lined burrow similar to *Ophiomorpha*. From Bromley and Asgaard (1972), republished with permission of GEUS. **d** Permeability and porosity cross plot showing the properties of bioturbated versus unbioturbated sandstone, as shown in (**a**) (based on horizontal core plug measurements). The bioturbated sandstone is clearly reduced in porosity and permeability, the latter of which is about one order of magnitude lower than measured for unbioturbated sandstone. **e** Thick unit of turbiditic sandstone with upward-increasing bioturbation (*Ophiomorpha rudis*) towards a sharp bedding surface at its top (ca. 4476.45 m), which goes hand in hand with a reduction of porosity and permeability due to incorporation of mud. The overlying sand unit is unbioturbated and has reasonably high properties. Kvitnos Formation (Santonian, well 6707/10-2A)

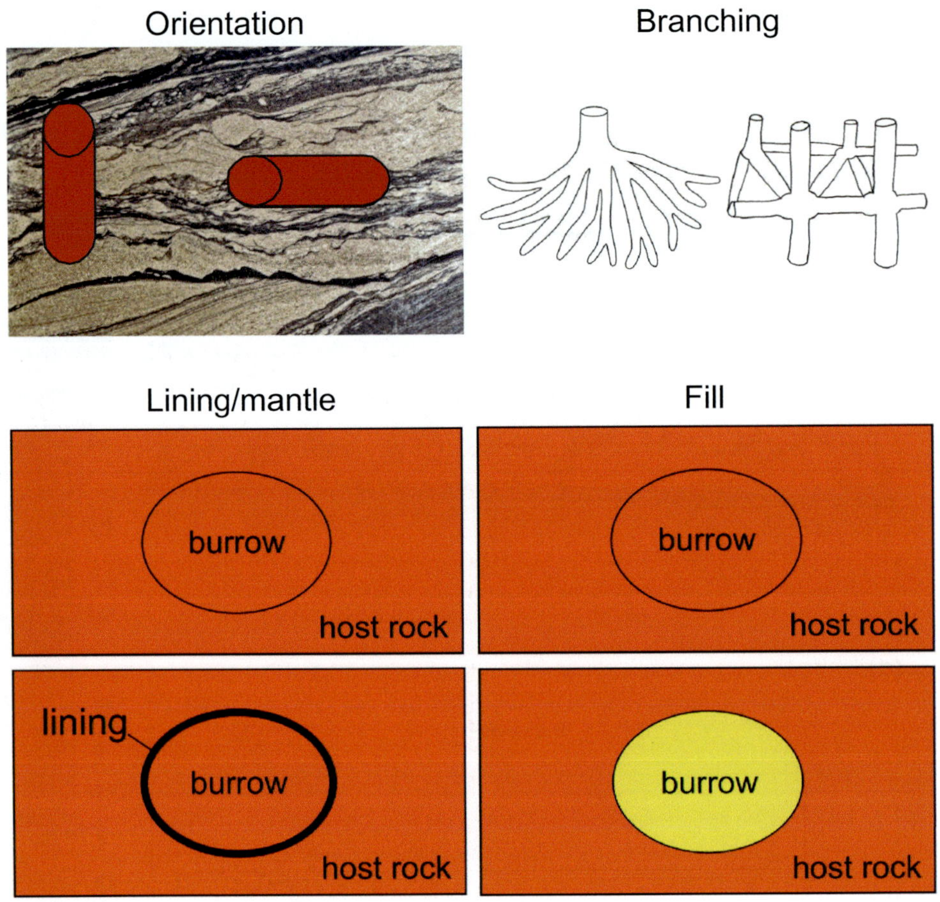


**Fig. 3.5** Architectural elements and characteristics of burrows, which may have an impact on reservoir quality, such as burrow orientation with respect to bedding (e.g. horizontal, vertical, oblique), branching (e.g. unbranched, branched, dichotomous, network, boxwork), presence or absence of lining or mantle, and burrow fill (e.g. active, passive) with respect to host rock (e.g. sand, mud). Modified after Knaust (2013)

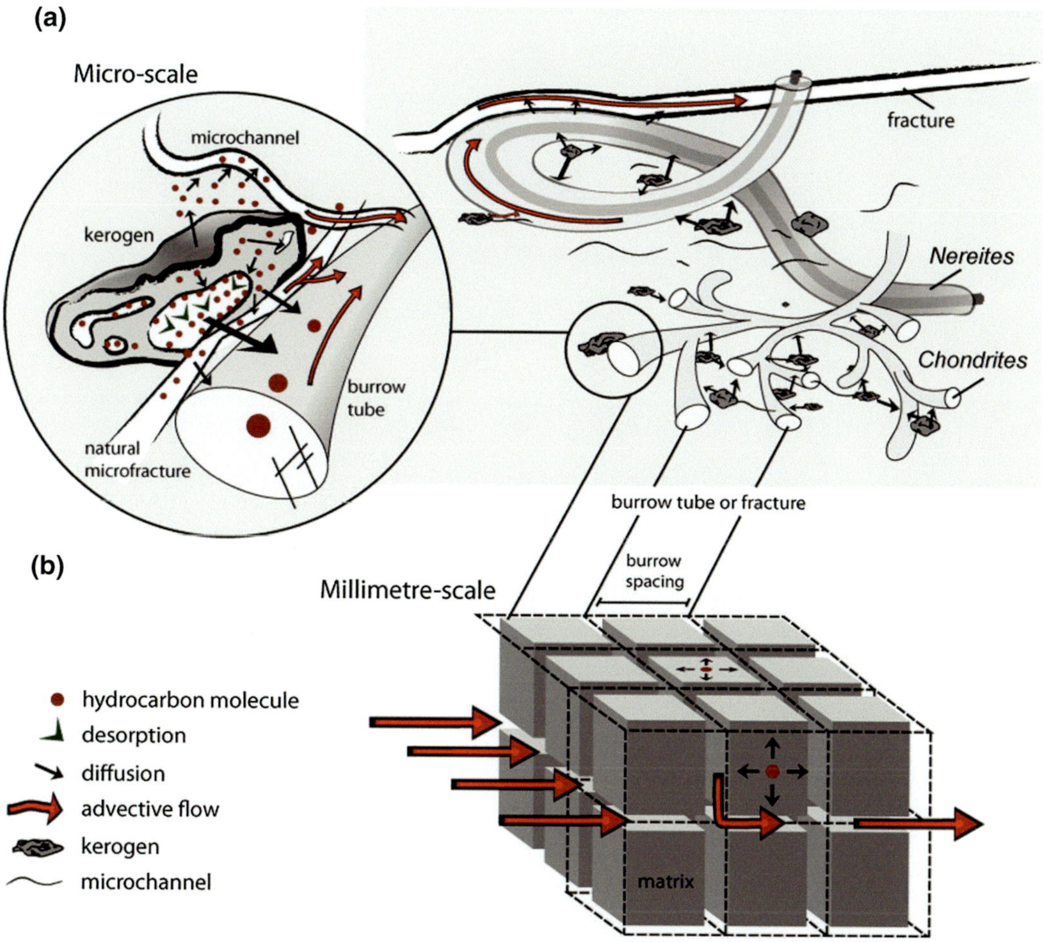


**Fig. 3.6** Schematic illustration presenting stages of the fluid flow within bioturbated, gas-charged tight mudstone. From Bednarz and McIlroy (2015), republished with permission of Elsevier; permission conveyed through Copyright Clearance Center, Inc. **a** Micro-scale. As pressure drops, hydrocarbon molecules enter the organo-porosity through desorption from pore walls and kerogen material. If the pore is connected to the (micro-) fracture or micro-channel, molecules travel through the conductive flow paths to the well bore. If there is no fracture or micro-channel connected to the organo-porosity, the molecules travel to the fracture network or permeable flow path through diffusion. **b** Millimetric scale. Efficiency of the flux of the diffusively migrating molecules into the fracture network is dependent on the distance from oil- or gas-charged pore to the closest fracture or permeable flow-path. Dense ichnofabric network minimises the distance of the diffusive flow through partitioning the hosting rock with permeable and brittle silty tubes and may also improve fracture spacing and/or complexity

## References


Bednarz M, McIlroy D (2015) Organism–sediment interactions in shale-hydrocarbon reservoir facies—Three-dimensional reconstruction of complex ichnofabric geometries and pore-networks. Int J Coal Geol 150–151:238–251

Bromley RG (1990) Trace fossils: biology and taphonomy. Unwin Hyman, London, 293 pp

Bromley RG, Asgaard U (1972) Notes on Greenland trace fossils, 1. Freshwater *Cruziana* from the Upper Triassic of Jameson Land, East Greenland. Grønlands Geologiske Undersøgelse, Rapport 49: 7–13

Bromley RG, Ekdale AA (1986) Composite ichnofabrics and tiering of burrows. Geol Mag 123:59–65

Knaust D (2013) Bioturbation and reservoir quality: towards a genetic approach. AAPG Search and Discovery Article #50900

Seilacher A (1967) Bathymetry of trace fossils. Mar Geol 5:413–428

Zesshin M, Catuneanu O (2013) High-resolution sequence stratigraphy of clastic shelves I: units and bounding surfaces. Mar Pet Geol 39:1–25

# 4 Methodology in Ichnological Core Logging

Well coring is an expensive but valuable method to obtain information about the geological situation in the subsurface, and a versatile analysis including ichnological study may lead to a robust interpretation. The amount of investigation and recommended workflow are dependent on the goal, the core (e.g. preservation, length, preparation, facies, etc.), the task at hand (e.g. what kind of information and in which detail should be extracted), and the skills of the investigator. In an optimal case, ichnological data is integrated with other relevant data and information with respect to regional geology, sedimentology, petrophysics and others. Based on my own experience, the following overall steps can be performed and account for different tasks but may build on each other.

1. Identification of bounding surfaces and quantification of bioturbation.
2. Identification and documentation of key trace fossils.
3. Analysis of burrow size and tiering patterns.
4. Quantification of ichnodiversity and ichnoabundance.
5. Advanced techniques and methods.
6. Neoichnological approach and analog studies.

Steps 1 and 2 are most relevant for (broad) ichnofacies analysis, whereas steps 3 and 4 have great potential in (narrow) ichnofabric studies. Steps 5 and 6 add some information for enhanced interpretation.

## 4.1 Identification of Bounding Surfaces and Quantification of Bioturbation

Deposition of sediment is a discontinuous process frequently interrupted by periods of nondeposition or erosion. These and other surfaces can be characterized with different methods, wherein the ichnological content can be investigated in conjunction with other features for their identification and characterization. Contrasting trace-fossil associations below and above a particular surface, and identifying firm- and hardgrounds on the basis of passively filled burrows (*Glossifungites* and *Trypanites* ichnofacies), are just two examples that help to subdivide the inspected core interval into genetically related units (see Sect. 3.2).

One of the very first steps performed by the logging sedimentologist is the quantification of bioturbation, which is an indicator of the conditions affecting the infaunal community. Different scales and methods are available, such as the semiquantitative field classification of ichnofabrics (Droser and Bottjer 1986), and bioturbation on bedding planes (Miller and Smail 1997). Widely applied and thus comparable from study to study is the bioturbation scale of Reineck (1963; see also Taylor and Goldring 1993), which ranges from grade 0 (no bioturbation) to 6 (bedding completely destroyed). These seven grades of bioturbation (0–6) contain very different values ranging from grade 0 (0%), grade 1 (1–4%), grade 2 (5–30%), grade 3 (31–60%), grade 4 (61–90%), grade 5 (91–99%) to grade 6 (100%) (Fig. 4.1). Because this scale was developed for the description of modern bioturbation encountered in boxcores and is explicitely limited to narrow intervals of the section (e.g. the decimeter range in boxcores), it is at best applicable to individual (small-scale) ichnofabrics which were not subject to repeated bioturbation, non-depositon, erosion and compaction. From a geological and paleoichnological perspective, however, the Reineck (1963) scale is limited in its application and a linear distribution of the different grades of bioturbation is recommended (Knaust 2012). The subdivision of such a linear scale into equally distributed categories depends on the logging scale and the length of the interval. Intervals with an increment of 10 or 20% of bioturbation structures would be common (e.g. in a scale of 1:10–1:50), whereas larger logging scales (e.g. 1:200) might well be served by a threefold subdivision. The chief advantage of a linear scale is its suitability for quantitative and statistical analysis of different units (e.g. Martin 1993; Knaust 2010).


D. Knaust, *Atlas of Trace Fossils in Well Core*,
DOI 10.1007/978-3-319-49837-9_4

## 4.2 Identification and Documentation of Key Trace Fossils

A second important step is the identification and documentation of trace fossils. This challenging task often sets a hurdle to the responsible worker and in the worst case this information is neglected. The outline of selected trace fossils in the following section will lower this hurdle and encourage the investigator in charge of to perform this task to a certain degree.

Of course, the number of described and named trace fossils is high: several hundred invertebrate ichnogenera, of which only a small fraction can be recognized with confidence in sectioned core, whereas the majority of traces escapes the conventional core logging process due to their occurrence along bedding planes, size issues and featureless appearance. Therefore it is important to be aware of the selectivity and bias of the collected trace-fossil data before utilizing them in advanced interpretations. It must be also stressed that the list of trace fossils may change from region to region and differ within stratigraphical units. Finally, the subjectivity and experience of the core logger and interpreter play major roles on the outcome of the trace fossils recorded.

The exposed surface area displaying burrows for investigation is important, such as full core or sectioned core. If sectioned, it may be worthwhile studying the counterpart as well, as it commonly reveals a slightly different section figure and thus facilitates three-dimensional reconstruction. Trace fossils having a complex morphology are only fractionally displayed and may be misidentified as other, simpler burrows. Even so, morphological details, for instance such related to the burrows fill, wall structure, cross section and pellets, may provide important clues for trace-fossil identification. The size limit imposed by cores is an important constraint when different size classes of trace fossils are addressed. Cores are commonly taken in a sub-vertical direction (except in inclined or horizontal drilling and inclined bedding) and thus overrepresent this orientation. In most cases, identification of trace fossils is restricted to the ichnogenus level if not more, specific diagnostic features are exposed. This fact has an impact on subsequent facies interpretation, which of course turns out to be less detailed than at the ichnospecies level. And, lest we forget, ichnological observations achieved from core are always limited in comparison with those obtained from outcrop.

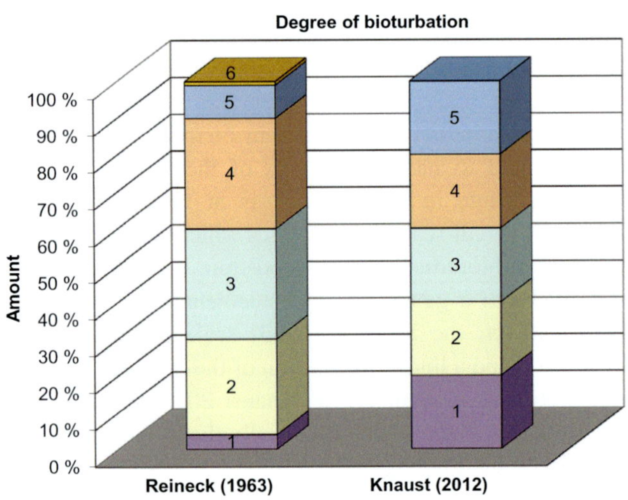


**Fig. 4.1** Quantification of bioturbation. The widely used bioturbation scale of Reineck (1963) with dissimilar proportions of categories compared with a linear distribution as proposed by Knaust (2012). Modified after Knaust (2012)

From a practical point of view, core logs (including trace fossils) can be drawn by hand, subsequently digitized using drawing software, or recorded directly in an application specifically designed for that task (Fig. 4.2). Displaying burrows goes hand in hand with the log of the bioturbation intensity (see above). While some workers prefer the common convention of displaying different symbols for each kind of trace fossil, others apply acronyms based on the ichnotaxon's name. Given that burrows in core are commonly fragmentary and often allow for different assignments, a good description and sketch should accompany the log entry. For more advanced studies, drawing entire ichnofabrics on a transparent sheet directly over the core section not only reveals the different burrows and their features, but also their cross-cutting relationships. Finally, high-resolution photography is a standard way of documentation, although it may not be as persuade as a detailed drawing in the trace-fossil analysis.

## 4.3 Analysis of Burrow Size and Tiering Patterns

After logging key trace fossils, additional information can be obtained from the core. Important information includes the analysis of trace fossils and bioturbate texture in relation to key stratigraphical surfaces, the succession of colonization, as well as cross-cutting relationships. The ichnofabric approach has become very useful in such investigations and is widely applied in core and in outcrop (e.g. Knaust 1998; Fig. 4.3).

It has turned out that burrow size may depend on environmental factors and thus can be crucial in interpretations, particularly in cases where other proxies are lacking. Any particular trace fossil (e.g. *Chondrites*) can occur in a

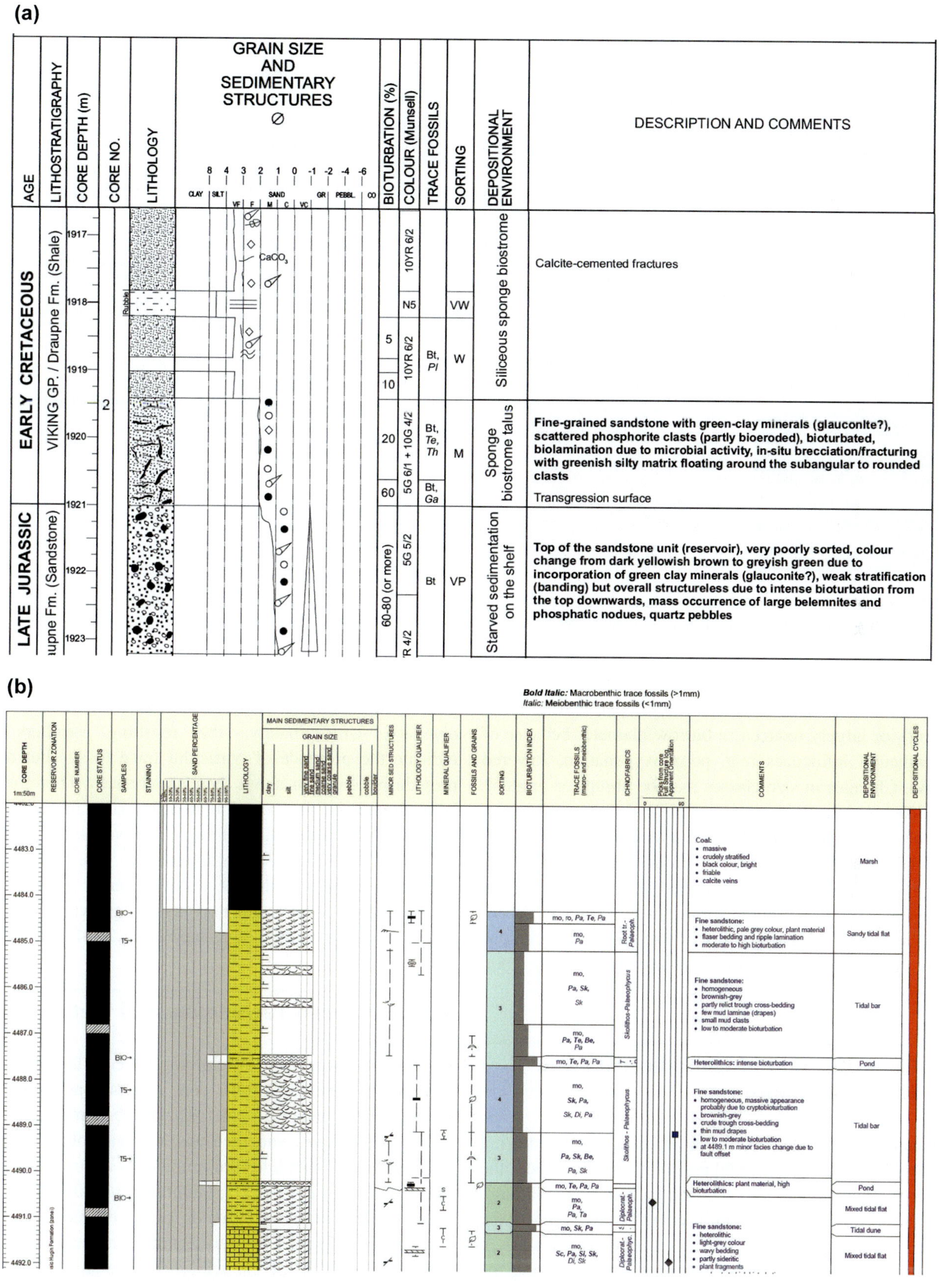


**Fig. 4.2** Examples of core logs including ichnological information. **a** Hand drawn and subsequently redrawn core section. **b** Core section logged directly in a core-logging application

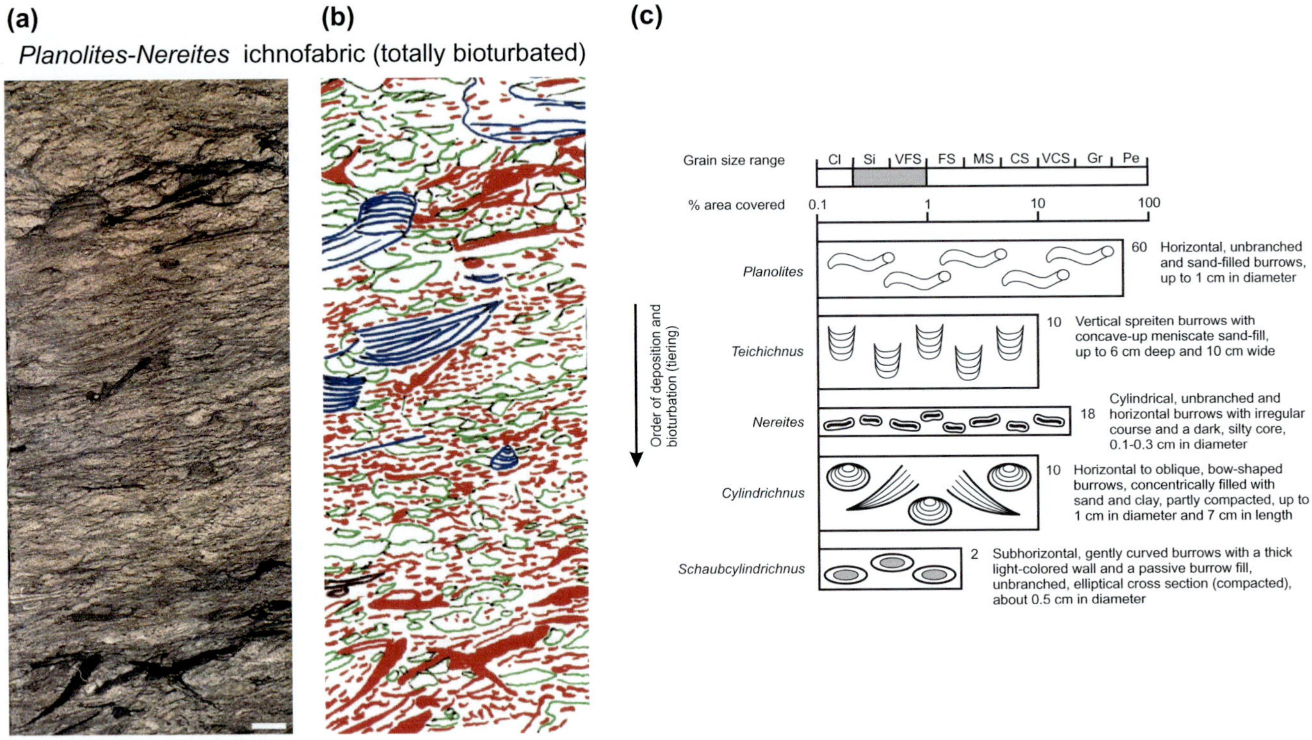


**Fig. 4.3** Example of an ichnofabric analysis based on a core section (**a**), *scale bar* = 1 cm, from which major burrows were drawn with different colors on an overlaid transparent sheet (**b**), and finally displayed in an ichnofabric constituent diagram (**c**)

"standard" size range, which needs to be qualified by calibration with "standard" conditions (e.g. open-marine environment), or largely reduced in burrow diameter because of environmental influences (e.g. poor oxygenation, lowered salinity). Changes in size classes may be abrupt or gradual, and in each case it is important to get it documented.

A relatively complete list of trace fossils in core is often beneficial for a particular project. In addition to a few key trace fossils which are easily identified, it can be worthwhile also to identify less common forms. In this way, the spectrum is increased and comparison of trace-fossil associations between different core intervals can be achieved with good confidence.

## 4.4 Quantification of Ichnodiversity and Ichnoabundance

After capturing the degree of bioturbation and identifying trace fossils, the next step could reveal information about the number and abundance of recognized trace fossils. Although ichnodiversity must not be confused with biodiversity, the number of contemporaneous burrows is a good indication of prevailing conditions at the sedimentary surface during the time of deposition. Ichnodiversity has been frequently evaluated in case studies and the conventional approach is to subdivide the total number of recognized trace fossils into grades, commonly resulting in three to six grades of ichnodiversity. Unfortunately, the resulting numbers only indicate the presence of particular trace fossils but do not provide information about their abundances. Therefore, the ichnoabundance can be calculated for particular core intervals, and is a combination of the ichnodiversity (e.g. number of trace fossils) and their abundance. According to the abundance of a logged trace fossil, different grades can be subdivided following an exponential growth factor (for details and case study, see Knaust et al. 2014; Fig. 4.4).

## 4.5 Advanced Techniques and Methods

The outlined procedure of ichnological core logging can be extended in cases where the core data is challenging or in studies where enhanced interpretation becomes necessary. Various techniques and methods are available for the different tasks and only a few common ones are mentioned here (see also Taylor et al. 2003; Gingras et al. 2011).

A semiquantitative estimation of the degree of bioturbation can be achieved by either putting a transparent sheet with a grid on top of the sectioned core surface or point-counting the number of hits where bioturbation is encountered (Marenco and Bottjer 2010; Knaust 2012). Grid size

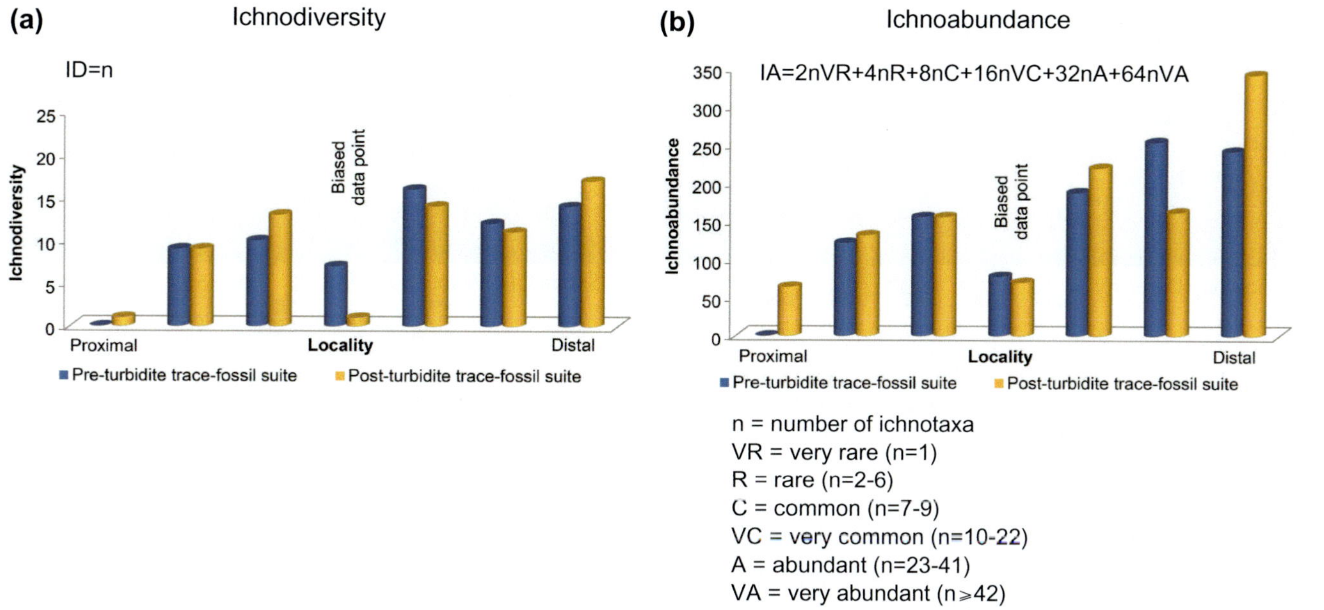


**Fig. 4.4** Logged ichnodiversity (**a**) versus calculated ichnoabundance (**b**) of a dataset from the confined deep-marine turbidite system of the Eocene Grès d'Annot Formation, southeastern France. The ichnodiversity is simply a count of ichnotaxa (richness) for each locality without regard for the abundance of each ichnotaxon (ichnogenera in this case). In contrast, the ichnoabundance is calculated by the sum of an exponential growth factor (reflecting the population dynamics of benthic organisms) multiplied with the abundance (frequency) of each grade (ranging from very rare to very abundant). Comparison of the results of the two methods reveals significant trends for ichnoabundance, such as an increase from a proximal to a distal position within the depositional system, which are hardly reflected in the ichnodiversity plot. Adapted from Knaust et al. (2014), republished with permission of Wiley; permission conveyed through Copyright Clearance Center, Inc.

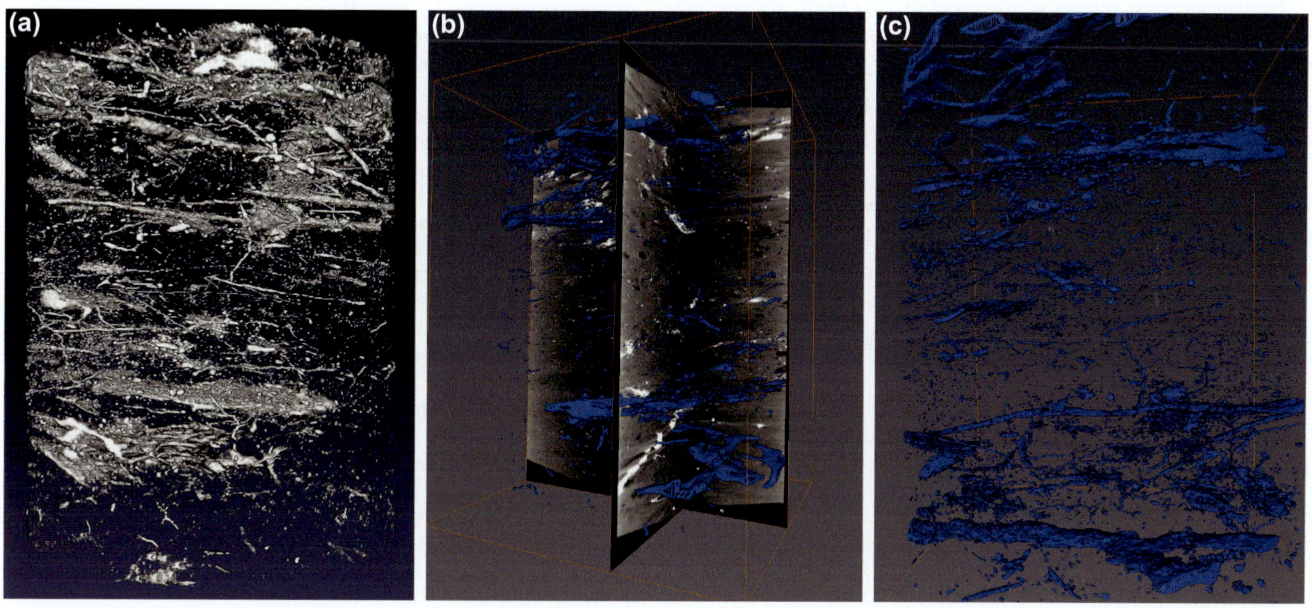


**Fig. 4.5** Core example of a completely bioturbated sandstone from the Paleocene Grumantbyen Formation (shelf) near Longyearbyen, Svalbard, treated with CT-scan and processed subsequently. Core diameter ca. 6 cm. Images courtesy of Lars Rennan (Trondheim) and Ørjan Berge Øygard (Bergen). **a** Overall CT-scan revealing numerous discrete burrows within a bioturbate texture due to density contrasts. **b** Orthogonally sliced CT-scan overlain by discrete burrow parts (*in blue color*). **c** Segmented burrows after removing small spots

(e.g. with centimeter scale) is somewhat dependent on the average size of the burrows present and the number of counts (e.g. 300) must be representative for the investigated interval. Another method of calculation is by means of digital treatment in image analysis software (Dorador and Rodriguez-Tovar 2015).

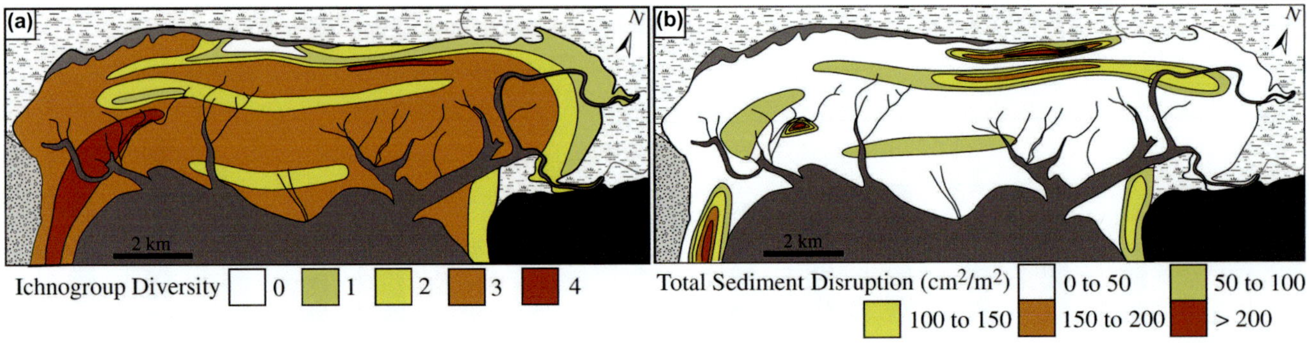


**Fig. 4.6** Example of a neoichnological dataset for comparison. Maps of a modern sandy tidal flat (delta front) of the Fraser River Delta (Boundary Bay), British Columbia, western Canada, showing the ichnological diversity (**a**) and sediment disruption resulting from infaunal activity (**b**). From Dashtgard (2011), republished with permission of Elsevier; permission conveyed through Copyright Clearance Center, Inc.

Displaying trace fossils and bioturbate texture might be difficult under certain circumstances, e.g. low contrast in color, grain size, lithology, and so on. Ultraviolet light instead of visible light might be a good way to do that, as well as CT-Scans (Fig. 4.5). At broader scale, borehole image analysis may reveal particular bioturbation patterns (Knaust 2012); while at a narrower scale, microfabric analysis based on thin sections could show disturbances in bedding and reorientation of grains related to bioturbation (Garton and McIlroy 2006). Finally, dedicated minipermeameter measurements have shown to be helpful in the evaluation of the reservoir properties of burrows versus matrix (Gingras et al. 2012).

## 4.6 Neoichnological Approach and Analog Studies

As a final step, links between the collected data and the subsequent interpretation can be made by comparing the ichnological findings in core with neoichnological evidence and similar cases described in literature, as well as applying analog studies with a focus on ichnology integrated with sedimentology (Fig. 4.6).

## References

Dashtgard SE (2011) Neoichnology of the lower delta plain: Fraser River Delta, British Columbia, Canada: implications for the ichnology of deltas. Palaeogeogr Palaeoclimatol Palaeoecol 307: 98–108

Dorador J, Rodriguez-Tovar F (2015) Ichnofabric characterization in cores: a method of digital image treatment. Ann Soc Geol Pol 85:465–471

Droser ML, Bottjer DJ (1986) A semiquantitative field classification of ichnofabric. J Sediment Petrol 56:558–559

Garton M, McIlroy D (2006) Large thin slicing: a new method for the study of fabrics in lithified sediments. J Sediment Res 76:1252–1256

Gingras MK, Baniak G, Gordon J et al (2012) Porosity and permeability in bioturbated sediments. In: Knaust D, Bromley RG (eds) Trace fossils as indicators of sedimentary environments. Developments in Sedimentology, vol 64, pp 837–868

Gingras MK, MacEachern JA, Dashtgard JE (2011) Process ichnology and the elucidation of physico-chemical stress. Sediment Geol 237:115–134

Knaust D (1998) Trace fossils and ichnofabrics on the Lower Muschelkalk carbonate ramp (Triassic) of Germany: tool for high-resolution sequence stratigraphy. Geol Rundsch 87:21–31

Knaust D (2010) The end-Permian mass extinction and its aftermath on an equatorial carbonate platform: insights from ichnology. Terra Nova 22:195–202

Knaust D (2012) Methodology and techniques. In: Knaust D, Bromley RG (eds) Trace fossils as indicators of sedimentary environments. Developments in Sedimentology, vol 64, pp 245–271

Knaust D, Warchoł M, Kane IA (2014) Ichnodiversity and ichnoabundance: revealing depositional trends in a confined turbidite system. Sedimentology 62:2218–2267

Marenco KN, Bottjer DJ (2010) The intersection grid technique for quantifying the extent of bioturbation on bedding planes. Palaios 25:457–462

Martin AJ (1993) Semiquantitative and statistical analysis of bioturbate textures, Sequatchie Formation (Upper Ordovician), Georgia and Tennessee, USA. Ichnos 2:117–136

Miller MF, Smail SE (1997) A semiquantitative field method for evaluating bioturbation on bedding planes. Palaios 12:391–396

Reineck H-E (1963) Sedimentgefüge im Bereich der südlichen Nordsee. Abhandlungen der Senckenbergischen Naturforschenden Gesellschaft 505:1–138

Taylor AM, Goldring R (1993) Description and analysis of bioturbation and ichnofabric. J Geol Soc, London 150:141–148

Taylor A, Goldring R, Gowland S (2003) Analysis and application of ichnofabrics. Earth Sci Rev 60:227–259

# 5 Selected Trace Fossils in Core and Outcrop

This section contains many well-known trace fossils but also several poorly understood ichnotaxa, some of which have just emerged during the last 10–15 years. It attempts to give a comprehensive overview of each treated trace fossil in a standardized way, starting with its morphology, fill and size; appearance in outcrop; an evaluation of ichnotaxonomical status; and preferred substrate. This first part is supplemented with key illustrations from outcrop, collections and literature. The main part of each section deals with the appearance of particular trace fossils in core, including a description and core image examples, as well as comparison with similar trace fossils. This part is mainly based on material from offshore Norway, but samples from elsewhere have been involved as well. The last part of each section outlines additional information, including the interpretation of the tracemaker, its behavior (ethology) and, quite importantly, notes on the depositional environment based on personal observations and on a review of published evidence. It rounds up with comments on ichnofacies, age range and impact of the trace fossil on reservoir quality. Figure 5.1 gives a summary of all ichnogenera dealt with herein, including their abundance, number of ichnospecies, similar trace fossils, inferred producers and paleoenvironment.

## 5.1 Classification of Burrows

Trace-fossil identification is not straightforward in two-dimensional core slabs, and often only limited information about the whole morphology, particularly of complex trace fossils, can be achieved. In contrast, different planes of sections through the burrows may very precisely reveal morphological details such as wall structure, fill, cross section, pellets, etc. These circumstances call for a nomenclature adapted for ichnotaxa identified in core, and in many cases, identification is limited to the ichnogenus level, which of course influences the precision of the paleoenvironmental interpretation. It therefore becomes obvious that ichnological observations achieved from core are not fully comparable with those obtained from outcrop. For instance, the trace fossils characteristic of subenvironments of deep-sea fan systems are well known from various outcrop studies, whereas vertical core sections from such systems display a very different suite of trace fossils. This is also related to the fact that in distal deep-sea deposits so-called graphoglyptids are preferably preserved at the base of turbidite beds, which are hard to study in core.

For hands-on identification of trace fossils in core, an experienced investigator may rely on his or her ability of pattern recognition and in this way may be able to identify the most common ichnotaxa. Another approach is to follow a determination key (Fig. 5.2), in which diagnostic criteria (e.g. ichnotaxobases) are arranged in a hierarchical order and thus allow trace-fossil identification based on major morphological features. This key follows the morphological classification of trace fossils as proposed by Knaust (2012a) and contains all burrows that are listed alphabetically and further explained in detail below. In addition to these ichnotaxa, the hierarchically higher trace-fossil categories bioturbate texture, plant roots and their traces, as well as pseudo-trace fossils are added subsequently. Trace-fossil associations and their relationship to particular ichnofacies on a basin profil are displayed in Fig. 5.3.


D. Knaust, *Atlas of Trace Fossils in Well Core*,
DOI 10.1007/978-3-319-49837-9_5

| Acronyms | Ichnogenera | Number of ichnospecies | Similar trace fossils | Inferred producers |
|---|---|---|---|---|
| *Art* | *Artichnus* | 1 | *Tei* | Holothurians |
| *Ast* | *Asterosoma* | ±5 | *Cyl, Ros, Art, Euflabella* | Polychaetes, crustaceans, holothurians, bivalves |
| *Cho* | *Chondrites* | 4+ | *Pilichnus, Skolichnus, Pragichnus, Phymatoderma, Vir* | Polychaetes, sipunculids, bivalves |
| *Cyl* | *Cylindrichnus* | 7 | *Ast, Ros, Laevicyclus, Pah,* root traces, *Glyphichnus, Catenarichnus* | Polychaetes, holothurians |
| *Mac* | *Macaronichnus* | 1 | *Pla, Gordia, Ner* | Polychaetes |
| *Ner* | *Nereites* | 6+ | *Phy, Mac, Cho* | Enteropneusts |
| *Oph* | *Ophiomorpha* | 8+ | *Tha, Spongeliomorpha, Psilonichnus, Pholeus, Sce* | Thalassinidean shrimp (e.g. callianassids) |
| *Pal* | *Palaeophycus* | ±20 | *Pla, Mac, Oph, Sip* | Annelids, arthropods |
| *Phy* | *Phycosiphon* | 1 | *Ner, Cho* | Worms |
| *Pla* | *Planolites* | ±10 | *Pal, Mac* | Worms, arthropods, bivalves |
| *Ros* | *Rosselia* | 1 | *Pah, Ast, Cyl, Lin, Sko* | Polychaetes, sea anemones |
| *Sch* | *Schaubcylindrichnus* | 1 | *Pal* | Enteropneusts |
| *Sip* | *Siphonichnus* | 1 | *Sko, Pal, Mac, Lin, Dip, Trypanites* | Bivalves |
| *Sko* | *Skolithos* | ±20 | *Cylindricum, Capayanichnus, Tri, Sip, Tha, Oph* | Worms, crustaceans, sea anemones, holothurians, insects, plant roots |
| *Tae* | *Taenidium* | ±7 | *Sce, Ancorichnus, Zoo, Rhi* | Insects, arthropods, worms |
| *Tei* | *Teichichnus* | ±12 | *Dip, Rhi, Phycodes, Art, Pad, Scal* | Annelids, arthropods, bivalves, holothurians |
| *Tha* | *Thalassinoides* | ±15 | *Spongeliomorpha, Psilonichnus, Parmaichnus, Cam* | Thalassinidean shrimp (e.g. callianassids), arthropods, sea anemones, worms |
| *Zoo* | *Zoophycos* | ±6 | *Rhi, Spirophyton, Echinospira, Tae* | Polychaetes, echiurans, sipunculans |
| *Are* | *Arenicolites* | ±15 | *Dip, Catenarichnus, Polykladichnus, Sko, Pal* | Polychaetes, crustaceans, holothurians, insect |
| *Ber* | *Bergaueria* | ±13 | *Con, Piscichnus* | Sea anemones |
| *Bor* | *Bornichnus* | 1 | Root traces, *Cho, Pilichnus, Vir* | Polychaetes, crustaceans |
| *Cam* | *Camborygma* | 4 | *Loloichnus, Oph, Tha, Spongeliomorpha, Psilonichnus, Pholeus* | Crayfish |
| *Con* | *Conichnus* | 3 | *Conostichus, Dip, Sca,* foot prints | Sea anemones, bivalves |
| *Dip* | *Diplocraterion* | 8 | *Rhi, Tei, Catenichnus, Tis, Sca, Oph* | Polychaetes, crustaceans, holothurians, echiurans |
| *Hil* | *Hillichnus* | 2 | *Ast, Lophoctenium, Pal, Sko, Pah, Tei* | Bivalves |
| *Lin* | *Lingulichnus* | 3 | *Sip, Con, Sko* | Lingulid brachiopods |
| *Pad* | *Paradictyodora* | 2 | *Dictyodora, Tei, Heimdallia, Stellavelum, Zavitokichnus, Euflabella* | Bivalves, polychaetes |
| *Pah* | *Parahaentzschelinia* | 2 | *Ros, Pad* | Bivalves |
| *Pho* | *Phoebichnus* | 4 | *Stelloglyphus, Oph, Mac, Sip,* root traces | Arthropods (e.g. crustaceans), worms |
| *Rhi* | *Rhizocorallium* | 2 | *Dip, Zoo, Tis, Tae, Sco* | Polychaetes, crustaceans, insects (e.g. mayflies) |
| *Sco* | *Scolicia* | ±5 | *Tae, Zoo* | Echinoids |
| *Sce* | *Scoyenia* | 2 | *Tae* | Millipeds, insects |
| *Tis* | *Tisoa* | 4 | *Dip, Rhi, Are, Paratisoa, Bathichnus* | Polychaetes |
| *Tri* | *Trichichnus* | 1 | *Polykladichnus, Sko* | Sipunculids |
| *Vir* | *Virgaichnus* | 1 | *Tha, Cho, Bor,* root traces | Nemerteans |

More common (green) / Less common (red)

Dominant (underlined) / Subordinate

**Fig. 5.1** Overview of the abundance of trace fossils observed in core and as described in this book, including some of their features

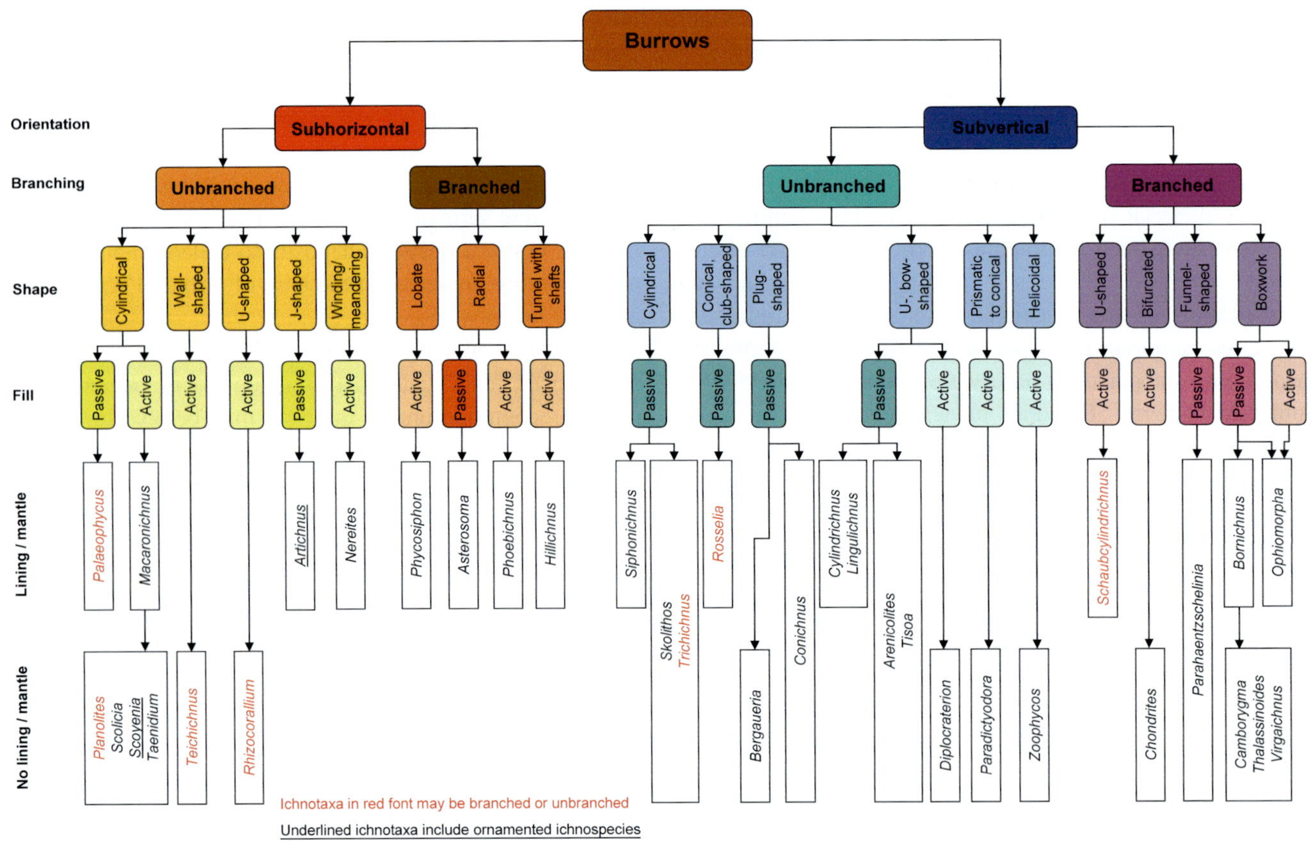


**Fig. 5.2** Morphological classification of burrows observed in core, following hierarchically arranged diagnostic criteria (modified after Knaust 2012a). The burrows being described and figured in the following sections are listed alphabetically

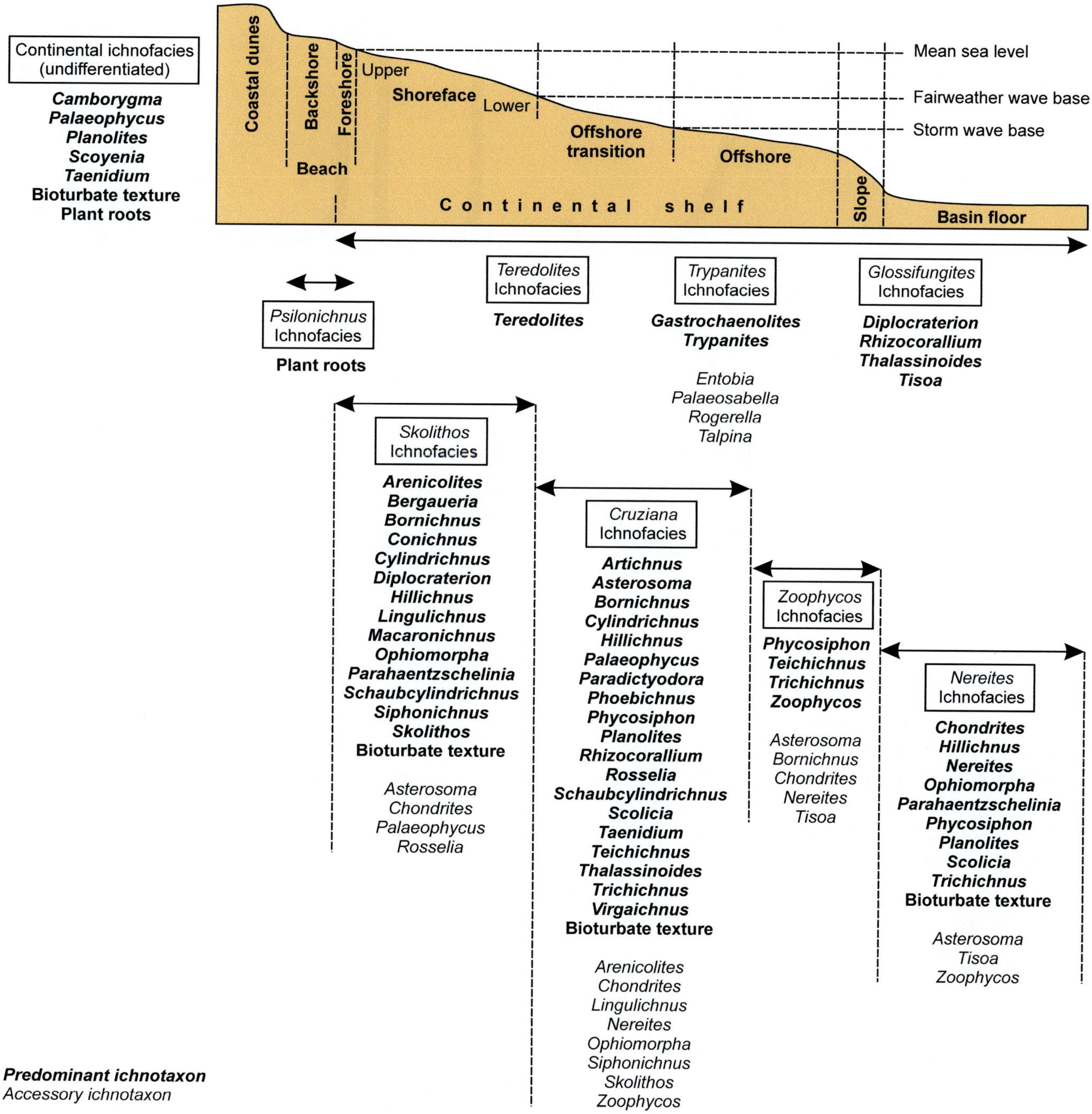


**Fig. 5.3** Trace-fossil associations as dealt with in this book and their occurrence in particular ichnofacies

## 5.2 *Arenicolites* Salter, 1857

*Morphology, Fill and Size: Arenicolites* refers to unbranched U-shaped burrows having a subvertical orientation, with or without lining and passive fill (Rindsberg and Kopaska-Merkel 2005; Bradshaw 2010; Figs. 5.4 and 5.5; see also Fig. 5.138c, d). Secondary successive branching may occur (e.g. *A. carbonarius*), while vertical adjustments of the tracemaker can lead to the development of a thin spreite-like structure. Collapsed *Arenicolites* can also lead to such spreite-like features and resembles *Diplocraterion*. Tube diameter typically is in the range of a few millimeters, and distance between the tubes as well as burrow depth ranges from a few to several centimeters.

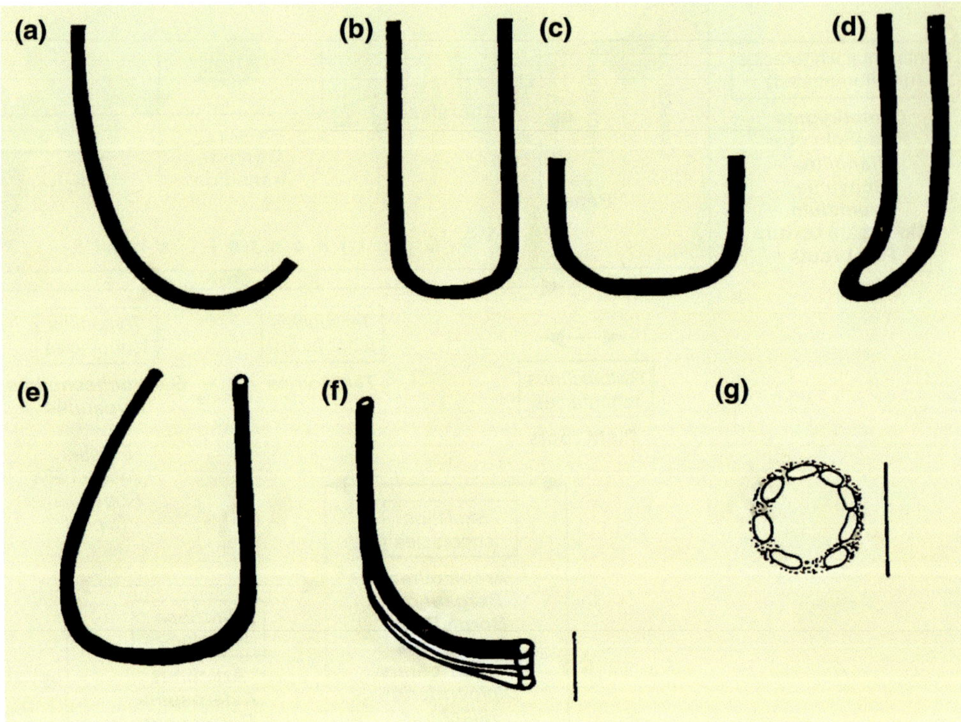


**Fig. 5.4** Morphological variability shown on *Arenicolites variabilis* (**a–f**, *scale bar* = 5 cm) and oolithic burrow wall (**g**, *scale bar* = 1 cm). From Fürsich (1974a)

*Ichnotaxonomy:* About 15 ichnospecies of *Arenicolites* have been described in the literature, many of them pending a critical review. Differentiation criteria include secondary branching (*A. carbonarius*) and orientation of the limbs (vertical in *A. sparsus*, the most characteristic ichnospecies; inclined in *A. curvatus*; variable in *A. variabilis*; subhorizontal in *A. longistriatus*), and presence or absence of lining (Rindsberg and Kopaska-Merkel 2005).

*Substrate: Arenicolites* is a characteristic burrow in sandy (siliciclastic and carbonate) substrate but also occurs in muddy sediment.

*Appearance in Core:* Its simple U-shaped morphology makes *Arenicolites* easy to recognize in core sections (Fig. 5.6, see also Fig. 5.146d). Although complete burrows are rarely exposed, individual burrow parts with vertical, J-shaped, elliptical and circular sections become visible in dependence on the section plane orientation with respect to the burrows. The burrows are passively filled and the occurrence of mud lining may enhance their appearance.

*Similar Trace Fossils:* The U-shaped morphology of *Arenicolites* is similar to that of *Diplocraterion*, which can be distinguished by its spreite; although some collapsed *Arenicolites* can resemble *Diplocraterion* very confusingly. Another U-shaped burrow, *Catenarichnus*, is a broad arc-like burrow (Bradshaw 2002). Partly preserved *Polykladichnus* (upper parts) could be mistaken for *Arenicolites*, while incomplete core sections of *Arenicolites* may resemble *Skolithos* (vertical) or *Palaeophycus* (horizontal). Compacted *Arenicolites* may resemble *Palaeophycus*.

*Producers:* Polychaete worms and amphipod crustaceans are the most common producers of modern *Arenicolites*-like burrows in marine settings (e.g. Gingras et al. 2008; Bradshaw 2010; Baucon et al. 2014; Fig. 5.7), although other organisms such as holothurians, sipunculans and echiurids may produce *Arenicolites* too (Smilek and Hembree 2012; Baucon and Felletti 2013; Baucon et al. 2014). Insects may produce similar traces in continental environments (Rindsberg and Kopaska-Merkel 2005).

*Ethology:* Most *Arenicolites* result from the dwelling (domichnial) activity of a suspension-feeding organism.

*Depositional Environment*: *Arenicolites* is known from a wide range of continental to deep-marine environments and often occurs in association with other trace fossils (Fig. 5.9b). *Arenicolites* is commonly associated with high-energy deposition, for instance stormdeposition (lower to middle shoreface) and migrating dunes and barforms. Mass occurrences of *Arenicolites* in low diversity is indicative of stressed environments, such as reduced and fluctuating salinity or increased organic productivity, and reflects opportunistic colonization (e.g. Price and McCann 1990; Bradshaw 2010).

*Ichnofacies: Arenicolites* belongs to the *Skolithos* Ichnofacies and, subordinately, is part of the *Cruziana* Ichnofacies. Opportunistic colonization of event beds (e.g., storm deposits)

**Fig. 5.5** *Arenicolites* in outcrop. *Scale bars* = 1 cm. **a** Vertical section of a sandstone-mudstone alternation containing numerous *Arenicolites*. Lower Triassic Buntsandstein Group (fluvial), Kahla, Thuringia, Germany. **b** Roof bed (lower bedding surface) of section in (**a**), showing numerous burrow apertures filled with sand. **c** Slab of bedding surface with *A.* cf. *variabilis* burrow apertures and weathered horizontal burrows, probably belonging to *Palaeophycus* isp. Upper Triassic Kågeröd Formation (fluvial), Bornholm, Denmark. From Knaust (2015b). **d** Same slab as in (**c**) (ca. 2–3 cm thick), lower surface with *A.* cf. *variabilis* (vertical burrows) and *Palaeophycus* cf. *alternatus* (horizontal burrows). From Knaust (2015b). **e** Vertical section of a sandstone displaying most of an *A. sparsus* burrow (post-turbidite). Eocene Grès d'Annot Formation (deep marine, confined turbidite system), Braux, southeastern France. After Knaust et al. (2014), republished with permission of Wiley; permission conveyed through Copyright Clearance Center, Inc. **f** Sandstone bedding plane preserving the paired plugs of the filled openings of *A.* cf. *sparsus* (pre-turbidite). Eocene Grès d'Annot Formation (deep marine, confined turbidite system), Chalufy, southeastern France. After Knaust et al. (2014), republished with permission of Wiley; permission conveyed through Copyright Clearance Center, Inc.

**Fig. 5.6** *Arenicolites* in sectioned core. *Scale bars* = 1 cm. **a**, **b** Cross-bedded sandstone with reactivation surfaces, from which mud-lined, passively filled, U-shaped *Arenicolites* (only partly seen) penetrate the substrate. They are accompanied by bioturbate texture and small equilibrium traces. Upper Triassic (Norian) Fruholmen Formation (fluvial to deltaic floodplain), Skavl Discovery, Johan Castberg area, Norwegian Barents Sea (well 7220/7-2S, ca. 1385.9 and 1395.9 m). **c** Heterolithic sandstone with ripple lamination and silty layers, displaying U-shaped burrows with funnel-shaped apertures. Upper Triassic (Rhaetian) to Lower Jurassic (Hettangian) Tubåen Formation (marginal marine, tidally influenced), Skavl Discovery, Johan Castberg area, Norwegian Barents Sea (well 7220/7-2S, ca. 1156.5 m). **d** Sandstone-mudstone alternation with mud-filled *Arenicolites* partly preserved and penetrating the sandstone layers from their top. Upper Jurassic (Kimmeridgian) Heather Formation (offshore), Fram Field, Norwegian North Sea (well 35/11-11, ca. 2577.5 m). **e** Dolomitic mudstone and wackestone having a horizon with mud-lined U-shaped burrows. Lower Triassic (Olenekian) Khuff Formation (carbonate platform with restricted lagoon), South Pars Field, Persian Gulf, Iran (well SP-9, ca. 2866.5 m). From Knaust (2009a), republished with permission of GulfPetroLink

was originally referred to the "*Arenicolites* Ichnofacies" by Bromley and Asgaard (1991) but such ichnocoenoses are now included in the *Skolithos* Ichnofacies.

*Age: Arenicolites* is known from the Early Cambrian (e.g. Crimes et al. 1977) to Holocene (e.g. Baucon and Felletti 2013).

*Reservoir Quality:* Given its passive fill, often consisting of sand, a greater abundance of *Arenicolites* may increase the reservoir quality. However, the occurrence of a mud lining along the burrow margin may affect negatively reservoir quality.

**Fig. 5.7** Examples of organisms producing *Arenicolites*-like burrows in modern sediments. **a** The polychaete lugworm *Abarenicola pacifica* in its U-shaped burrow. Note the fecal mound above the sediment surface (*arrow head*). From Dashtgard and Gingras (2012), republished with permission of Elsevier; permission conveyed through Copyright Clearance Center, Inc. **b** Monospecific assemblage of *Arenicolites*-like burrows together with their producer, the amphipod crustacean *Corophium volutator*. Bay of Fundy, Canada. From Gingras et al. (2012a), republished with permission of Elsevier; permission conveyed through Copyright Clearance Center, Inc. **c** Resin cast from a boxcore from the southern North Sea with *Arenicolites*-like burrows (upside down, up to 3 cm long) produced by *C. volutator*. Original from Hans-Erich Reineck, Senckenberg, Wilhemshaven, Germany. From Knaust (2012b), republished with permission of Elsevier; permission conveyed through Copyright Clearance Center, Inc.

## 5.3 *Artichnus* Zhang et al., 2008

*Morphology, Fill and Size: Artichnus* is a wide, J-shaped burrow with a narrow, upward tapering shaft and the distal end tapering to a blind termination (Figs. 5.8 and 5.9). The burrow lumen is thickly laminated and accompanied by a spreite, which is best developed in the lower part of the burrow. The size of *Artichnus* in the type material is quite variable and ranges from 5 to 15 cm in length and 2 to 5 cm in width (Zhang et al. 2008).

*Ichnotaxonomy: A. pholeoides* and *A. giberti* are the only described ichnospecies.

*Substrate: Artichnus* is common in mud-dominated substrates such as heterolithic deposits.

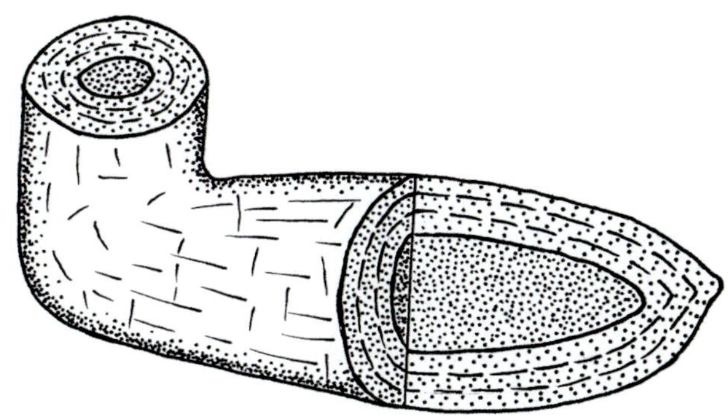

**Fig. 5.8** Schematic reconstruction of *Artichnus* based on observations of material from the Eocene Grès d'Annot Formation, southeastern France (see Fig. 5.9)

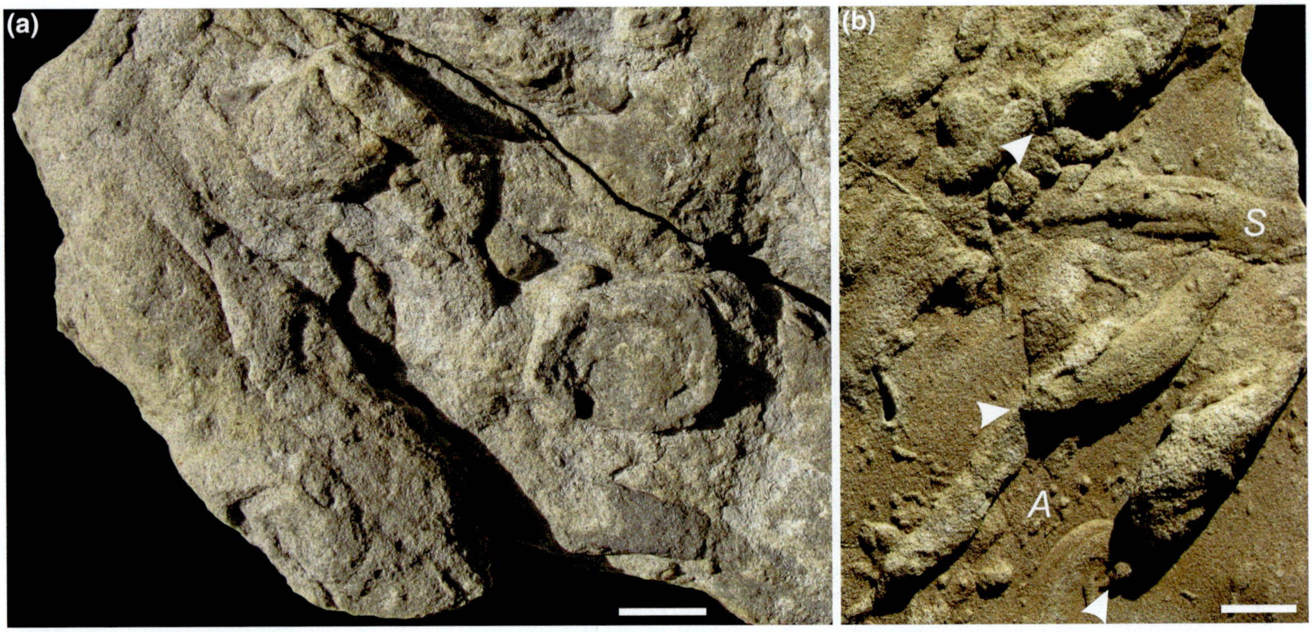


**Fig. 5.9** *Artichnus* in the Eocene Grès d'Annot Formation (deep marine, turbiditic), southeastern France. *Scale bars* = 1 cm. **a** Longitudinal burrow part (*lower left*) and two shafts (*centre*) displaying the thick lamination of the burrows. From Knaust et al. (2014), republished with permission of Wiley; permission conveyed through Copyright Clearance Center, Inc. **b** Three specimens with a spindle-shaped appearance (*arrow heads*) co-occurring with *Scolicia* (*S*) and *Arenicolites* (*A*)

*Appearance in Core:* Although the J-shaped morphology of *Artichnus* is hard to prove in two-dimensional core sections, these spreite burrows are typically expressed by their retrusive spreite element (often dominant) and the passively filled lumen (causative tube) included within it (Fig. 5.10; see also Ayranci and Dashtgard 2013; Ayranci et al. 2014). Longitudinal sections may hint at the J-shaped morphology, while in cross and oblique sections the lumen appears to be circular and elliptical, respectively. The lumen is only a few millimeters in diameter and reaches a length of a few centimeters. The entire burrow lies typically in the centimeter range.

*Similar Trace Fossils: Artichnus* strongly resembles the similar spreite burrow *Teichichnus* and has undoubtedly been mistaken as such in many cases. Distinction of both ichnotaxa in core is not straightforward and depends on the recognition of the overall morphology (J-shaped versus wall-shaped) as well as on the occurrence of a comparatively thick lumen surrounded by a few laminae. Moreover, *Artichnus* has generally a smaller vertical extent compared with *Teichichnus*.

*Producers*: *Artichnus* is interpreted to be produced by burrowing holothurians (sea cucumbers), probably belonging to the order Apodida (Zhang et al. 2008; Ayranci and Dashtgard 2013; Fig. 5.11).

*Ethology:* The J-shaped morphology of *Artichnus* with its thickly laminated rim suggests that it is a dwelling trace (domichnion) of a suspension- or detritus-feeding tracemaker.

*Depositional Environment:* Infaunal holothurians, such as Apodida, burrow in littoral to deep-sea environments. So far, related trace fossils have been described from deep-marine deposits (Zhang et al. 2008; Knaust et al. 2014). Ayranci and Dashtgard (2013) and Ayranci et al. (2014) document *Artichnus*-shaped traces from modern delta-front and prodelta deposits and hypothesize that in the rock record this trace could be used as evidence of stable euhaline conditions and, in particular, of deposition and colonization below storm-wave base.

*Ichnofacies: Artichnus* fits nicely within the *Cruziana* Ichnofacies.

*Age:* In addition to the above-mentioned Jurassic and Cretaceous occurrences, *Artichnus* has been recorded from Eocene and modern deposits. As this ichnogenus was only recently named, it seems likely that its range will be extended in the near future.

*Reservoir Quality:* A general reduction of reservoir quality can be assumed of *Artichnus*-bearing sedimentary rocks due to the predominance of the mud-filled laminae and spreite, especially where occurring in high densities.

## 5.4 *Asterosoma* von Otto, 1854

*Morphology, Fill and Size: Asterosoma* is a morphologically variable trace fossil. It consists of a number of arm-like, often bulbous burrows radiating outwards from a central axis

**Fig. 5.10** *Artichnus* in sectioned core of heterolithic sandstone. *Scale bars* = 1 cm. **a** Dense ichnofabric with numerous burrows displaying the lumen in longitudinal and cross section, with the dominance of retrusive spreite elements below it. Lower Jurassic (Pliensbachian-Toarcian) Cook Formation (shallow marine), Norwegian North Sea (well 34/5-1S, ca. 3658.5 m). **b** Thickly laminated burrows (*upper part*) accompanied with spreiten (*lower right*). Lower Jurassic (Toarcian-Aalenian) Stø Formation (offshore), Snøhvit Field, Norwegian Barents Sea (well 7120/8-3, ca. 2211.75 m). **c** Sandstone bed with numerous small, mud-dominated specimens. Lower Jurassic (Hettangian-Sinemurian) Nansen Formation (marginal marine), Oseberg Sør Field, Norwegian North Sea (well 30/9-16, ca. 3464.9 m). **d** Dense ichnofabric displaying individual burrows with thickly laminated lumina in longitudinal, oblique and cross sections. Upper Cretaceous San Antonio/San Juan Formation, La Vieja, Macal Tacata, Venezuela

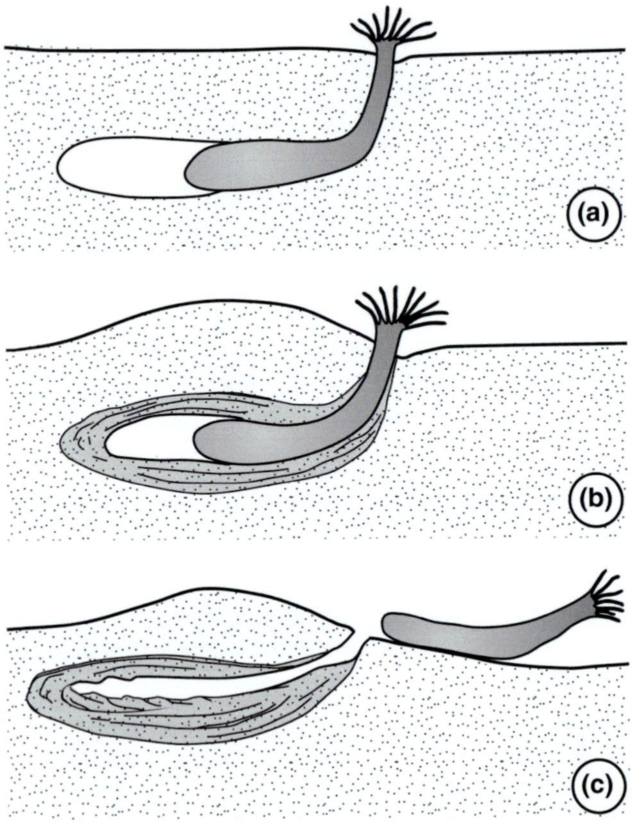


**Fig. 5.11** Schematic diagram of a holothurian of the order Apodida as potential producer of *Artichnus*. From Zhang et al. (2008). **a** Development of initial burrow. **b** Well-developed burrow showing the central lumen surrounded by a thick rim of laminae. **c** Abandonment of burrow with slight effect of compaction

and tapering towards blind extremities, which may be extended into very fine galleries (Schlirf 2000; Bromley and Uchman 2003; Bradshaw 2010; Fig. 5.12). The burrows are variable in dimension, number, and orientation and may show tension fractures as well as faint wall ornamentation (Fig. 5.13).

*Ichnotaxonomy:* About five ichnospecies of *Asterosoma* have been described so far, of which *A. radiciforme* and *A. ludwigae* are probably the most common (Figs. 5.14 and 5.15).

*Substrate:* *Asterosoma* occurs in siliciclastics and carbonates alike, although the tracemakers seem to prefer sandy substrates.

*Appearance in Core:* Only fractions of the entire burrow system are commonly exposed in core sections (Fig. 5.16). Those burrow parts typically appear in form of clusters of "bulbs", in which mud laminae surround a passively, often sand-filled lumen. Parts of the shaft and the ascending tubes may be associated with the bulbous burrow parts.

*Similar Trace Fossils:* Ichnospecies of *Asterosoma* in core may be confused with other burrows containing a concentrically laminated fill, such as *Cylindrichnus* (a bow-shaped burrow, commonly unbranched), *Rosselia* (a dominantly vertical burrow), *Artichnus* (J-shaped burrow), *Hillichnus* (sand-filled tubes, Fig. 5.64a) and *Euflabella* (palmate spreite burrow). Transitions between those forms may occur and may be related to the activity of the same or a similar producer.

*Producers:* As with *Cylindrichnus* and *Rosselia*, polychaete and other worms are potential producers of *Asterosoma* (Chamberlain 1971; Bromley 1996; Pemberton et al. 2001; Dashtgard and Gingras 2012; Monaco 2014). Based on the occurrence of superficial striae, other authors attribute *Asterosoma* to the activity of crustaceans (Müller 1971; Gregory 1985; Schlirf 2000; Neto de Carvalho and Rodrigues 2007). This interpretation is less conclusive because other organisms than crustaceans (including worms) are also

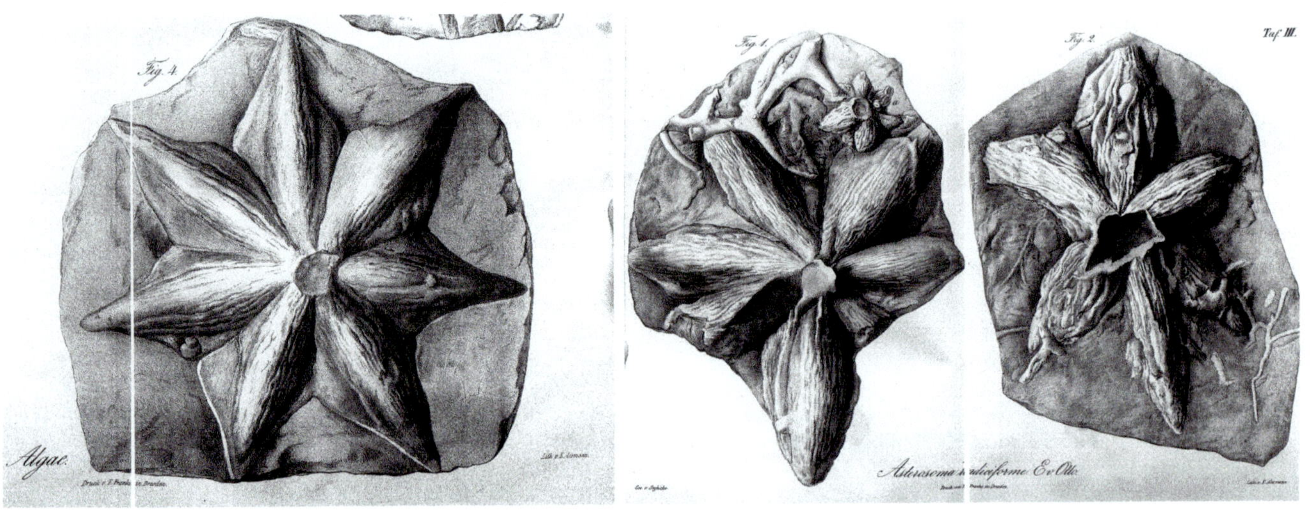

**Fig. 5.12** Historical figure of the type ichnospecies of *Asterosoma*, *A. radiciforme*, in its original description by von Otto (1854)

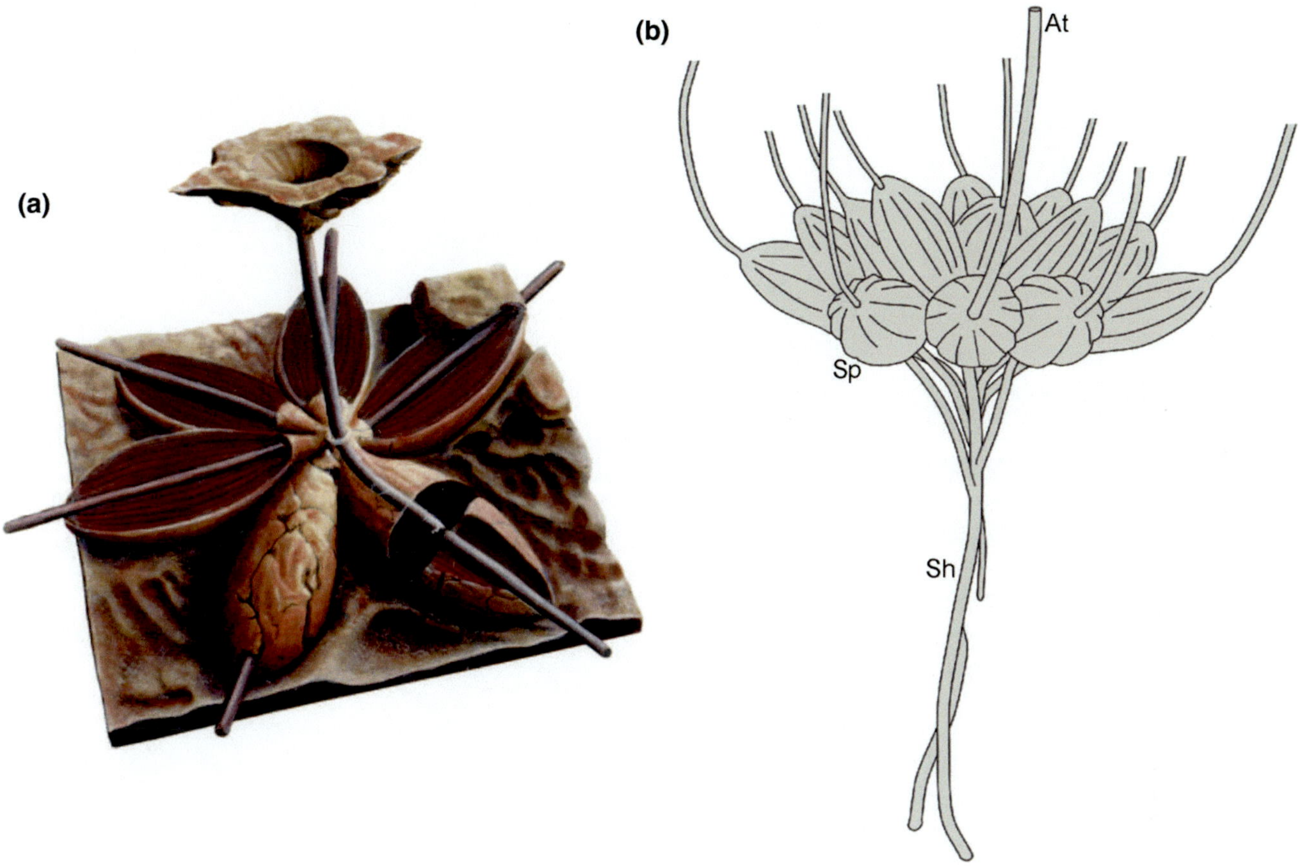


**Fig. 5.13** Reconstructions of *Asterosoma*. **a** Schematic diagram of the construction of *Asterosoma*. After Pemberton et al. (2001), republished with permission of the Geological Association of Canada. **b** *Asterosoma* from the Lower to Middle Jurassic of Bornholm, Denmark. A shaft (*Sh*) leads up to a fan-like group of intersecting, steeply inclined spindles (*Sp*), several of which show the central lumen emerging as an ascending tube (*At*). Total length is about 35 cm. Reconstruction after Bromley and Uchman (2003), republished with permission of the Geological Society of Denmark

capable producing scratches (see Knaust 2008), and such features are rather tension fractures than striae (Monaco 2014). In addition, other groups of organisms may be considered as tracemakers of *Asterosoma*. For instance, Ayranci and Dashtgard (2013) describe multiple diminutive burrows with a thick mud lining (*Artichnus*) produced by holothurians (sea cucumbers) from modern delta deposits; in sectioned core these burrows mimic *Asterosoma* (Dashtgard and Gingras 2012). In contrast, Percival (1981) interpreted star-shaped burrows similar to *Asterosoma* as the result of the feeding activity of tellinid bivalves.

*Ethology: Asterosoma* is commonly regarded as the feeding trace (fodinichnion) of a deposit-feeding (Schlirf 2003; Bradshaw 2010) or suspension-feeding animal (Neto de Carvalho and Rodrigues 2007).

*Depositional Environment: Asterosoma* has been frequently reported from a wide range of marine environments ranging from paralic to deep-marine settings. In many cases, the ichnotaxonomical determination does not go below the ichnogenus level and therefore, little is yet known about the environmental significance of individual ichnospecies. *Asterosoma* is a common constituent of the lower shoreface to offshore transition (or distal ramp setting in carbonate systems), but also occurs in other parts of the shelf (e.g. Farrow 1966; Howard 1972; Gowland 1996; MacEachern and Bann 2008; Joseph et al. 2012; Pemberton et al. 2012). Likewise, it is common in deltaic successions, where it occurs most frequently in delta front and prodelta deposits (e.g. McIlroy 2004; MacEachern et al. 2005; Gani et al. 2007; Carmona et al. 2008, 2009; Dafoe et al. 2010; Tonkin 2012). Some *Asterosoma* producers tolerate reduced and fluctuating salinities and occur in environments with brackish conditions, including estuaries, bayhead deltas and other paralic deposits (e.g. Greb and Chesnut 1994; Hubbard

**Fig. 5.14** *Asterosoma* in outcrop. *Scale bars* = 22 cm (**a**) and 1 cm (**b–f**). **a** Bedding plane with *A. radiciforme*. Upper Miocene (shallow marine), East Cape, North Island, New Zealand. **b** Vertical section showing a cluster of *Asterosoma* arms with a thick mud lining surrounding a thin, passively filled tube. Lower Jurassic (Hettangian) Höganäs Formation (nearshore), Helsingborg, southern Sweden. **c** Lower bedding plane with *A. ludwigae* (*between arrow heads*). Eocene, Grès d'Annot Formation (deep marine, turbiditic), southeastern France. From Knaust et al. (2014), republished with permission of Wiley; permission conveyed through Copyright Clearance Center, Inc. **d** Lower bedding plane with *A.radiciforme*. Eocene, Grès d'Annot Formation (deep marine, turbiditic), southeastern France. From Knaust et al. (2014), republished with permission of Wiley; permission conveyed through Copyright Clearance Center, Inc. **e**, **f** *A. ludwigae* on bedding plane of glauconitic sandstone. Upper Cretaceous (Coniacian-Santonian) Bavnodde Greensand (shelf), near Rønne, Bornholm, Denmark

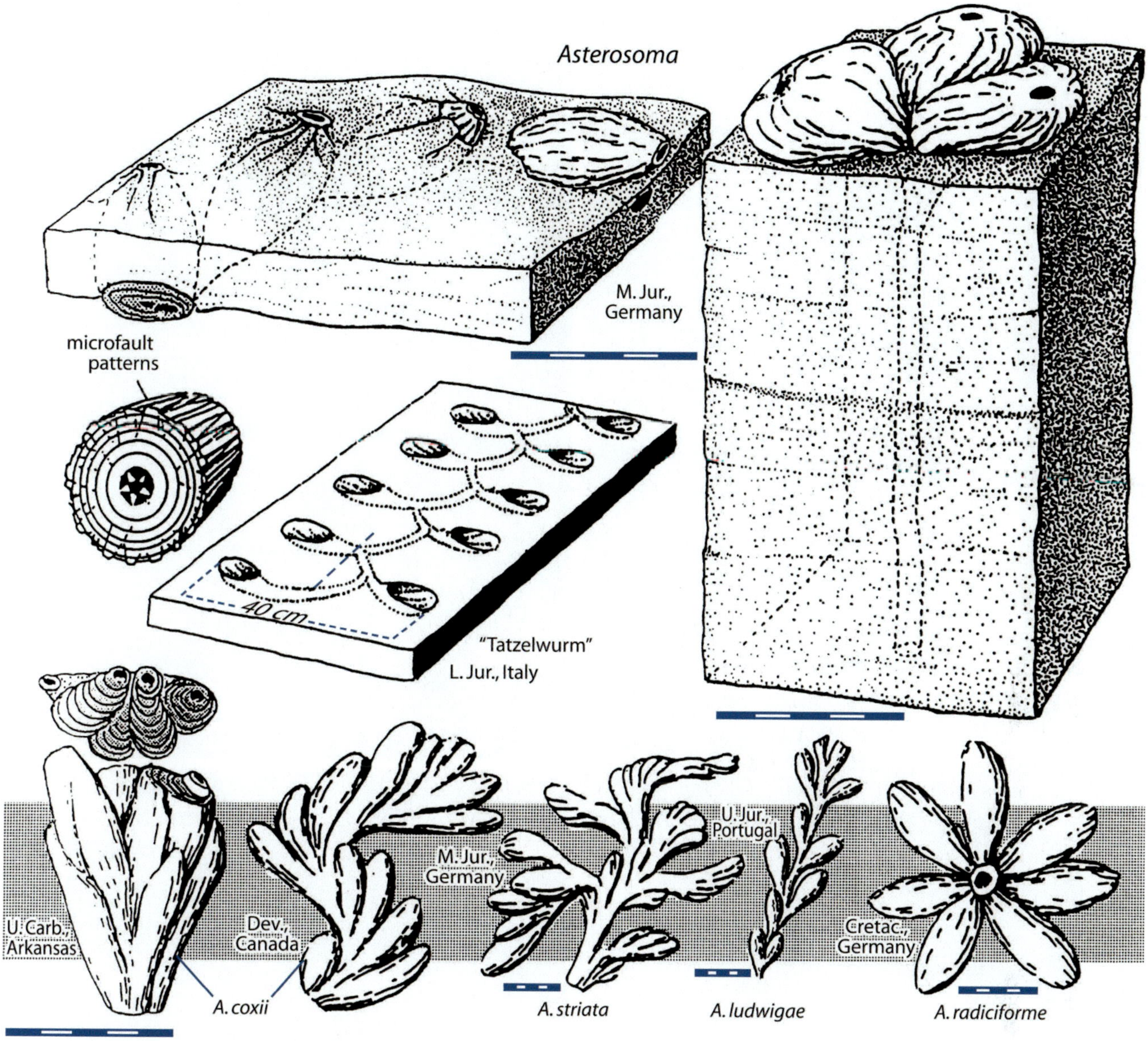


**Fig. 5.15** Different ichnospecies and preservational variants of *Asterosoma*. From Seilacher (2007), republished with permission of Springer

et al. 2004; MacEachern and Gingras 2007; Bradshaw 2010; Leszczyński 2010; Gingras et al. 2012a; Joeckel and Korus 2012; Pearson et al. 2013). *Asterosoma* even occurs on tidal flats (e.g. Miller and Knox 1985; Knaust 2009a; Knaust et al. 2012). Occurrences of *Asterosoma* in deep-marine environments are related to slope-, fan- and basin-floor deposits (e.g. Pickerill 1980; Powichrowski 1989; Heard and Pickering 2008; Uchman and Wetzel 2011; Hubbard et al. 2012; Knaust et al. 2014; Monaco 2014). In general, *Asterosoma* is related to well-oxygenated environments, although exceptions exist (e.g. Neto de Carvalho and Rodrigues 2007).

*Ichnofacies: Asterosoma* is mainly a component of the *Cruziana* Ichnofacies, although it occurs in the *Skolithos*, *Zoophycos* and *Nereites* ichnofacies as well.

*Age: Asterosoma* has been described from strata of the entire Phanerozoic from Early Cambrian (e.g. Desai et al. 2010) to Holocene (e.g. Dashtgard et al. 2008).

*Reservoir Quality: Asterosoma* is a burrow with an active (mud-dominated) fill, which is less favorable for reservoir quality and fluid migration (La Croix et al. 2013; Knaust 2014a).

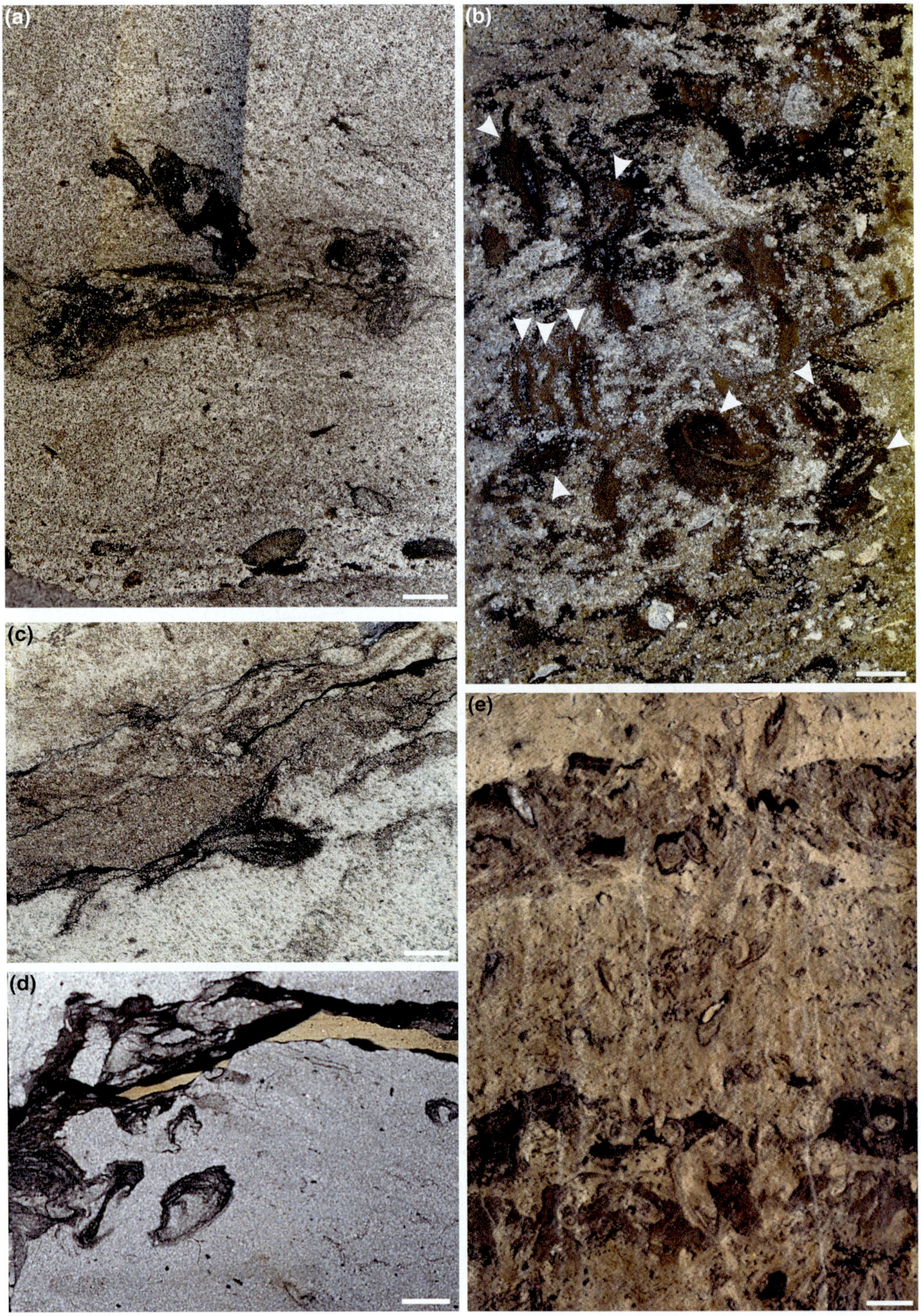
(a)
(b)
(c)
(d)
(e)

◀ **Fig. 5.16** *Asterosoma* in sectioned core. *Scale bars* = 1 cm. **a** Sandstone with basal pebble-lag deposit and a cluster of mud-lined, bulbous *Asterosoma*. Lower Jurassic (Pliensbachian) Nordmela Formation (tidal flat), Iskrystall Discovery, Norwegian Barents Sea (well 7219/8-2, ca. 3019.1 m). **b** Sandstone with admixed granulae and bioclasts (transgressive unit), partly calcite cemented and displaying an *Asterosoma* ichnofabric with horizontal and oblique burrow segments (some of which are indicated by *arrow heads*). Middle Jurassic (Bathonian-Oxfordian) Hugin Formation (shoreface), Ivar Aasen Field, Norwegian North Sea (well 16/1-16, ca. 2398.5 m). **c** Heterolithic and glauconitic sandstone showing parts of an *Asterosoma* system. Lower Jurassic (Pliensbachian-Toarcian) Tofte Formation (fan delta), Åsgard Field, Norwegian Sea (well 6506/12-I-2H, 4867.6 m). **d** Sandstone with clustered *Asterosoma* sectioned in oblique direction. Lower to Middle Jurassic (Toarcian-Aalenian) Ile Formation (tidal-influenced delta), Norwegian Sea (well 6406/8-1, ca. 4388.2 m). **e** Dolomitic limestone with a dense *Asterosoma* ichnofabric. Upper Permian Khuff Formation (muddy tidal flat to grain-shoal transition), South Pars Field, Persian Gulf, Iran (well SP9, ca. 3025.15 m). From Knaust (2009a), republished with permission of GulfPetroLink

## 5.5 *Bergaueria* Prantl, 1946

*Morphology, Fill and Size: Bergaueria* is defined as a hemispherical to shallow cylindrical, vertical trace fossil with a rounded base (Alpert 1973; Pemberton et al. 1988; Fig. 5.17). Its diameter is generally greater than or equal to its length. Burrow walls are smooth and unornamented; although a lining may be present, and the base may contain a shallow central depression and radial or biradial ridges (Alpert 1973; Pemberton et al. 1988). Burrow fill is generally structureless and most commonly attached to and genetically related to sediment from the overlying bed, forming convex hyporelief preservation (Pemberton et al. 1988; Mata et al. 2012). The size of *Bergaueria* ranges from less than a centimeter to more than a decimeter in burrow diameter and depth.

*Ichnotaxonomy:* About a dozen ichnospecies of *Bergaueria* have been erected based on morphological differences (Figs. 5.17 and 5.18).

*Substrate: Bergaueria* is common in siliciclastic deposits of softground origin and is most easily recognized at the base of sandstone beds. It also occurs in carbonates.

*Appearance in Core:* In core samples, *Bergaueria* typically appears as plug-shaped depressions along the base of sandstone layers (Fig. 5.19). If exposed in an axial position, a central peak at the base of the burrow may become visible (Fig. 5.19c). *Bergaueria* occurs solitarily or in colonies.

*Similar Trace Fossils:* Other plug-shaped burrows might be confused with *Bergaueria*. Especially *Conichnus* can be similar to *Bergaueria* but is conical with downward tapering laminae, is commonly larger than *Bergaueria*, and probably results from a similar maker and mode of construction. Another burrow similar in shape to *Bergaueria* is *Piscichnus*, which results from the nesting and yet-feeding activity of rays and thus is larger than *Bergaueria*. In addition to burrows, there are also sedimentary erosion features such as potholes (or pot casts), which may resemble *Bergaueria*. However, pot casts are not only larger than *Bergaueria* but

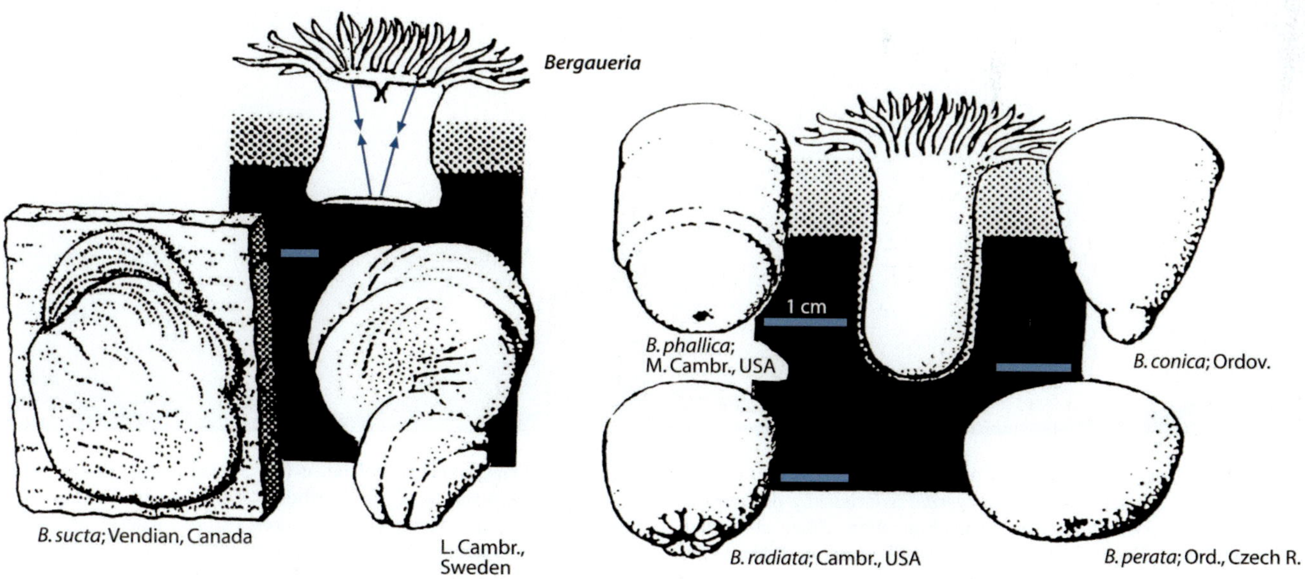


**Fig. 5.17** Examples of ichnospecies of *Bergaueria*. From Seilacher (2007), republished with permission of Springer

**Fig. 5.18** *Bergaueria* preserved in positive hyporelief from outcrops. *Scale bars* = 1 cm. **a** Holotype of *B. perata* as originally designed by Prantl (1946). Upper Ordovician Letná Formation, Zdice at Beroun, Czech Republic. Original in the National Museum in Praha. After Mikuláš (2006), republished with permission of the author. **b** *B. sucta* on a sandstone bed from the Uper Cambrian of Wiśniówka Duża, Holy Cross Mountains, southern Poland. **c–f** *B. perata* (**c**), *B. hemispherica* (**d**) and *B. elliptica* (**e**, **f**) from Middle Triassic (Anisian-Ladinian) carbonates of the Meissner Formation (Muschelkalk), Thuringia, Germany. Note slight affection by pressure solution, particularly in (**e**) and (**f**). From Knaust (2007b), republished with permission of SEPM

**Fig. 5.19** *Bergaueria* in sectioned core. *Scale bars* = 1 cm. **a** Heterolithic sandstone with ripple lamination and a shifted depression along a bedding plane, which results from the vertical adjustment of the *Bergaueria* tracemaker. Lower Jurassic (Sinemurian-Pliensbachian) Tilje Formation (sandy tidal flat), Heidrun Field, Norwegian Sea (well 6507/7-A-38, ca. 2789.5 m). **b** Cross-bedded, fine-grained sandstone overlain by medium-grained sandstone and *Bergaueria* along the interface. Lower Jurassic (Sinemurian-Pliensbachian) Tilje Formation (sandy tidal flat), Heidrun Field, Norwegian Sea (well 6507/7-A-27, ca. 3195.5 m). **c** Silty mudstone with *Phycosiphon*, which is truncated with *Bergaueria* and overlain by cross-bedded sandstone (also with *Phycosiphon*). Note the central depression at the base of the burrow. Lower Jurassic (Hettangian-Pliensbachian) Amundsen Formation (lower shoreface), Fram area, Norwegian North Sea (well 35/10-1, ca. 3655 m). **d** Ripple-laminated sandstone with mud drapes and a *Bergaueria*, that is accompanied by synaeresis cracks. Middle Jurassic (Bajocian) Ness Formation (delta plain), Valemon Field, Norwegian North Sea (well 34/10-23, ca. 4197.8 m)

**Fig. 5.20** Dwelling sea anemones (Actiniaria) as potential producers of *Bergaueria* on a sandy tidal flat on the Pacific coast of California. **a** Single specimen with only tentacles emerging on the sandy surface. **b** Two specimens accumulating debris in a ring-like structure due to the activity of the tentacles

also differ from it by their tendency to be enlarged at their base and having a more irregular shape.

*Producers:* Sea anemones (Actiniaria) produce *Bergaueria*-like traces in modern sediments (Schäfer 1962; Bromley 1996). *Cerianthus*, for instance, can deeply burrow below the sediment-water interface while extending its tentacles and mouth above the sediment surface to feed (Frey 1970a; Dashtgard and Gingras 2012; Fig. 5.20). In case of disturbance (e.g. due to storm), the animal is able to retract most of its body down into the sediment for shelter (Fig. 5.21). Varying burrowing strategies are utilized by the anemones, including vertical adjustment, which leads to collapsed, chevron-like traces, and attachment to a buried hard substrate (such as gravel) with only limited vertical shifting (Dashtgard and Gingras 2012).

*Ethology: Bergaueria* is interpreted as the dwelling or resting trace (domichnion or cubichnion) of sea anemones (e.g. Pacześna 2010).

*Depositional Environment*: *Bergaueria* is commonly related to high-energy nearshore environments such as beaches and sandy tidal flats, but does also occur in other marine environments down to the deep sea (e.g. Książkiewicz 1977).

*Ichnofacies: Bergaueria* typically occurs in the *Skolithos* Ichnofacies.

*Age: Bergaueria* is known from the Lower Cambrian (e.g. Pacześna 2010; Mata et al. 2012) to the Holocene (Frey 1970a). Some dubious forms are reported from Ediacaran deposits (e.g. Fedonkin 1981; Narbonne and Hofmann 1987; Seilacher 2007).

*Reservoir Quality:* Because of its passive sand fill, the occurrence of *Bergaueria* generally has a slight positive effect on reservoir quality.

## 5.6 *Bornichnus* Bromley and Uchman, 2003

*Morphology, Fill and Size:* Small burrows composed of a crowded tangle of millimetric lined tubes in that are closely and tortuously branched (Fig. 5.22). The whole trace fossil occupies an ovoid region of sediment a few centimeters in size (Bromley and Uchman 2003), while the tube diameter typically lies within the size range of 1–2 mm. In addition, more loosely arranged branched burrows with less tortuous but more linear tubes are also included in *Bornichnus*.

*Ichnotaxonomy: B. tortuosus* is the only described ichnospecies. Loosely organized burrows with similar character and size occur too and could probably be assigned to a new ichnospecies.

*Substrate*: *Bornichnus* is known from sandy substrate.

*Appearance in Core:* These small trace fossils with a tube diameter of only 1–3 mm are easily overlooked but appear in core as clusters (ovoid areas) or loosely concentrated areas containing dark-gray burrows (Figs. 5.23 and 5.118a). Thick mud lining may result in entirely mud-filled burrows, although in some instances the passive fill becomes clearly visible. Morphology and branching pattern is quite variable and includes loosely winding burrow elements with T-shaped branching points and bifurcation. Association with *Ophiomorpha* burrows is common.

*Similar Trace Fossils: Bornichnus* resembles other small trace fossils with which it could be confused, such as rootlets (plant-root traces), *Chondrites* (dichotomously branched burrows with active fill and no lining), *Pilichnus* (horizontal burrows with dichotomous branching), and *Virgaichnus* (boxwork with pinching and swelling burrow diameter).

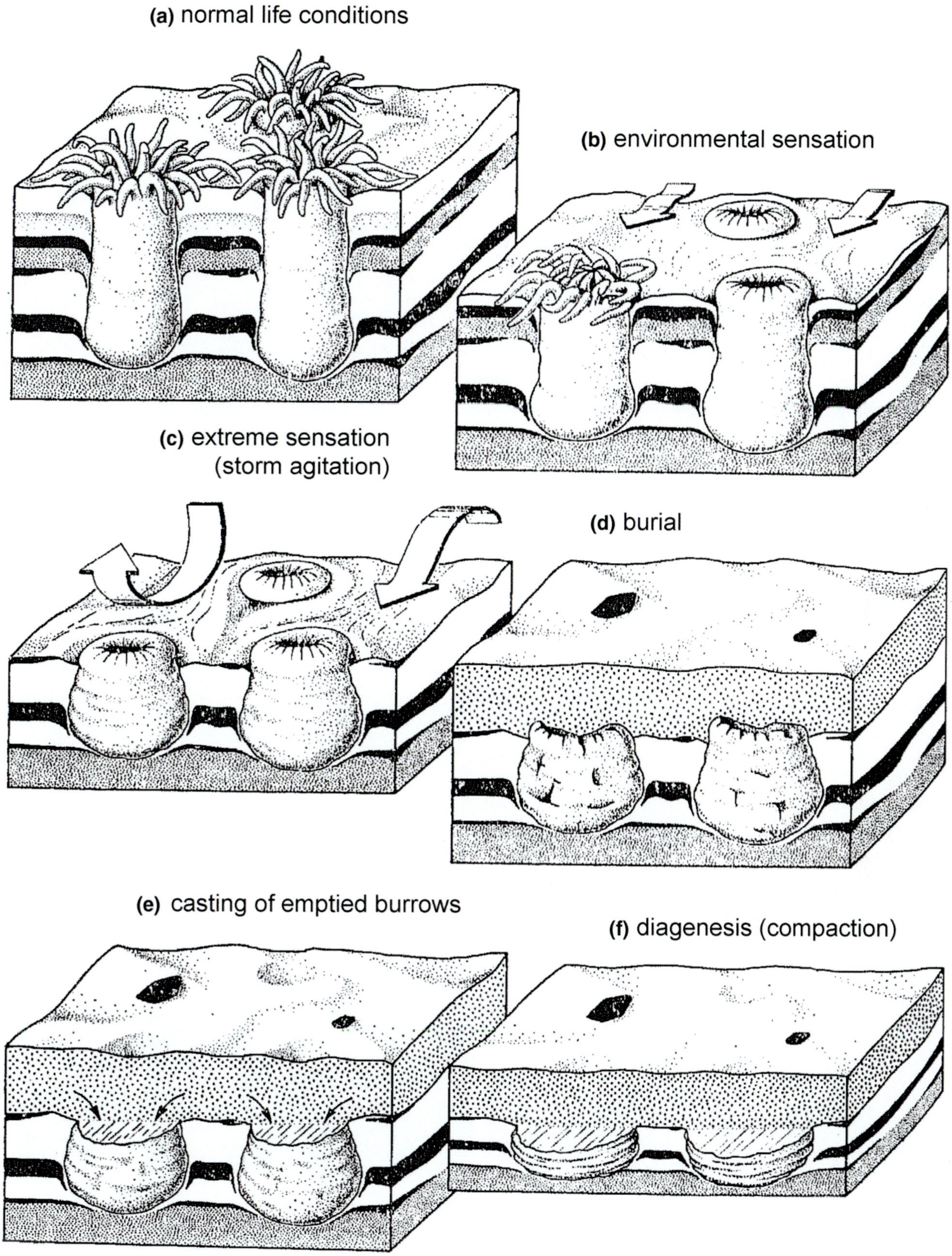


**Fig. 5.21** Environmental and taphonomic history of *Bergaueria* (="*Alpertia*") as a result of burrowing sea anemones. Based on observations on Middle Devonian strata of Poland. After Orłowski and Radwański (1986)

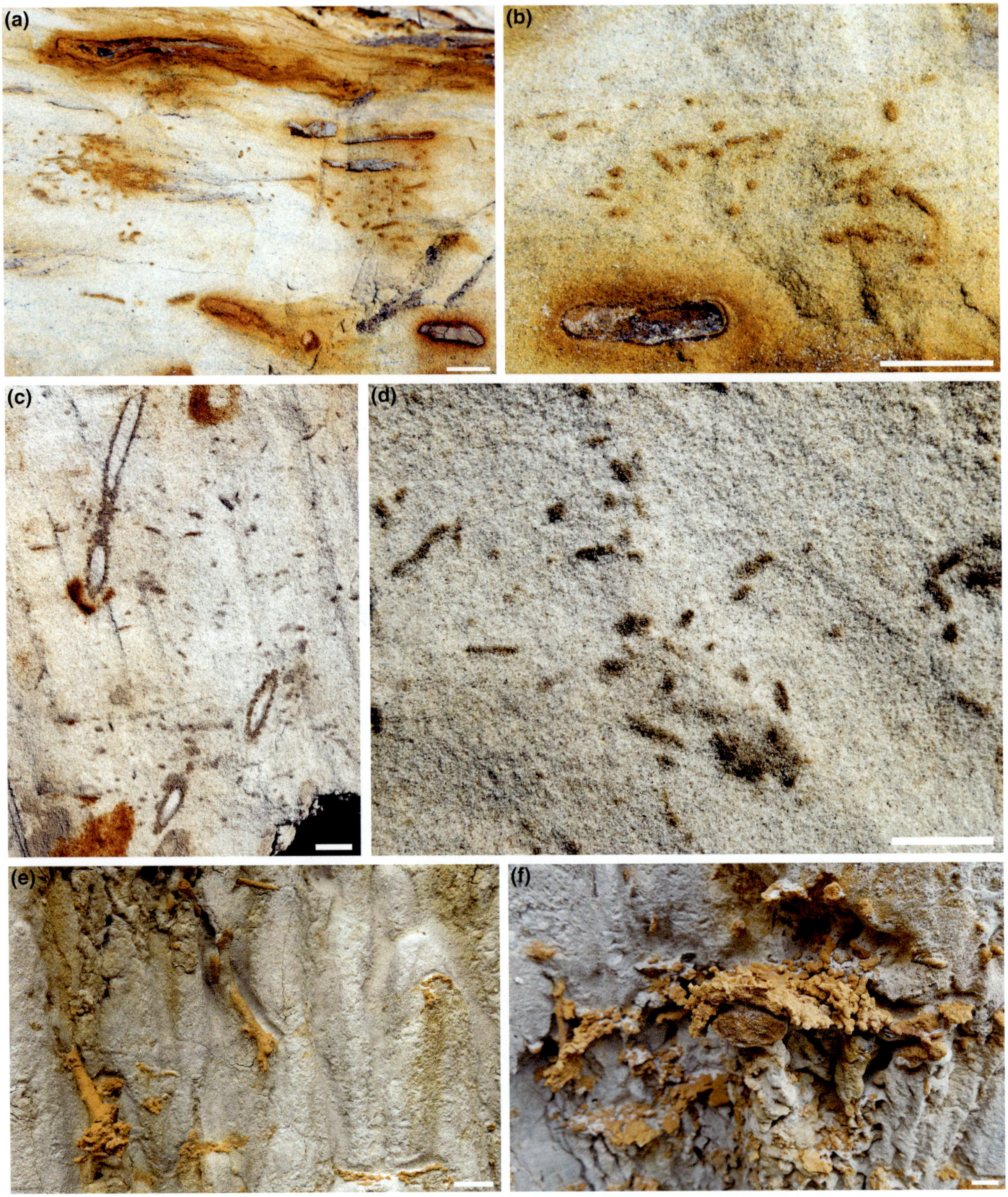


**Fig. 5.22** *Bornichnus* in outcrop. *Scale bars* = 1 cm. **a** *B. tortuosus* in vertical section from its type locality. Two ovoid clusters with loosely tangled burrows within glauconitic, cross-bedded sandstone with limonitic mud flasers (sandy tidal flat). Lower Jurassic (Pliensbachian-Toarcian) Sorthat Formation (marginal marine), near Rønne, Bornholm, Denmark. See Bromley and Uchman (2003) for details. **b** As in (**a**), close-up view of an ovoid cluster consisting of numerous tiny burrows enhanced by limonite. **c** Vertical section of sand with *Ophiomorpha nodosa* shafts and associated *B. tortuosus*. Lower Cretaceous (Berriasian) Robbedale Formation (shallow marine), Madsegrav near Rønne, Bornholm, Denmark. See Nielsen et al. (1996). **d** Tangled burrows with thick lining. Same locality as in (**c**). **e** *B. tortuosus* in vertical section occurring in isolation or in connection with crustacean burrows (e.g. *Thalassinoides*, *Gyrolithes* and *Spongeliomorpha*). The thickly lined burrows are impregnated with ferruginous material and thus readily visible. Upper Miocene near Lepe, Huelva, southwestern Spain. See Belaústegui et al. (2016b) for geological setting. **f** Crowded tangle of thickly lined tubes clustered around a crustacean tunnel (middle). Same locality as in (**e**)

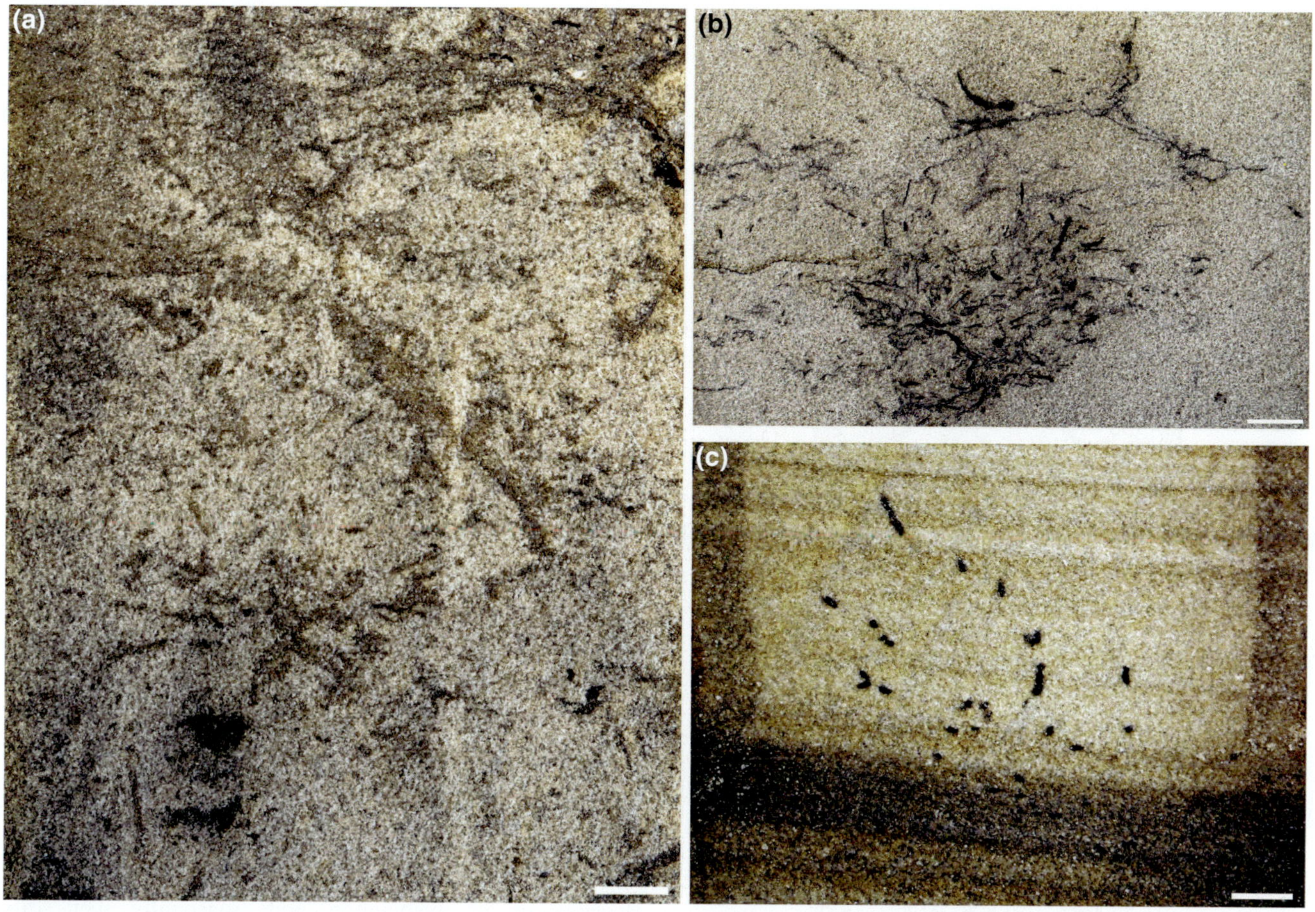


**Fig. 5.23** *Bornichnus* in sectioned core. *Scale bars* = 1 cm. **a** *Bornichnus* ichnofabric with clusters of thickly mud-lined, partially mud-filled tiny burrows. Middle Jurassic (Bajocian-Oxfordian) Hugin Formation (shallow marine), Norwegian North Sea (well 25/10-12ST2, ca. 2161 m). **b** *B. tortuosus* with an ovoid region with tangled and branched burrows, some of which show passive sand fill surrounded by a thick mud lining. Middle Jurassic (Bajocian-Oxfordian) Hugin Formation (shallow marine), Sleipner Vest Field, Norwegian North Sea (well 15/9-5, ca. 3592.5 m). **c** Cluster of *Bornichnus* with thick mud lining in turbiditic sandstone. Paleocene (deep marine, lobe complex), off Tanzania

*Producers:* Various species of modern polychaetes are known to produce *Bornichnus*-like traces, such as *Capitomastus* cf. *aciculatus*, *Scoloplos armiger* and *Heteromastus filiformis* (e.g. Schäfer 1962; Hertweck 1972; Hertweck et al. 2007; Fig. 5.24). However, co-occurrence of *Bornichnus* and *Ophiomorpha* may suggest a close relationship between these trace fossils. For instance, small burrows similar to *Bornichnus* were described originating from brooding chambers connected with large *Ophiomorpha* burrows and were produced by juvenile shrimp (Forbes 1973; Bromley and Frey 1974; Curran 1976; Verde and Martinez 2004; Fig. 5.25). Alternatively, such relationship between small and large burrows could also be the result of commensalism, e.g. between polychaetes and crustaceans as suggested by de Gibert et al. (2006).

*Ethology:* Deposit-feeding (fodinichnial) behavior of the *Bornichnus* tracemaker can be inferred.

*Depositional Environment: B. tortuosus* was originally described from tidal-flat deposits (Bromley and Uchman 2003). The examples presented in Fig. 5.23 extend this range into the shoreface, shelf and slope, where the producer has colonized sandy event deposits (e.g. tempestites and turbidites).

*Ichnofacies:* Too little data is available for an unequivocal ichnofacies assignment of *Bornichnus*, but existing findings indicate the preferred occurrence within the *Skolithos* and *Cruziana* ichnofacies, subordinately also in the *Zoophycos* Ichnofacies.

*Age:* The new records have extended the original Lower Jurassic (Pliensbachian to Toarcian) age to the Paleocene.

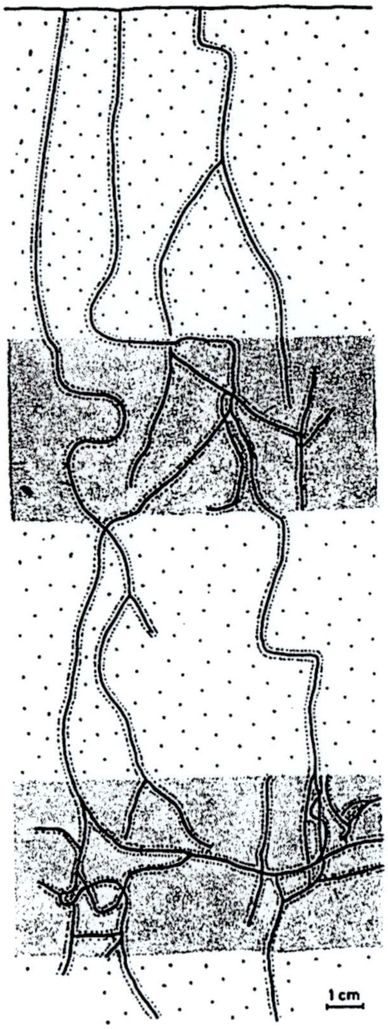


**Fig. 5.24** Vertical section with a fine sand-silt intercalation and burrows of the modern polychaete *Capitomastus* cf. *aciculatus* from the nearshore (2–10 m water depth) of the Georgia coastal region, Sapelo Island, USA. From Hertweck (1972), republished with permission of Schweizerbart (www.schweizerbart.de/home/senckenberg)

*Reservoir Quality:* Due to the high amount of mud within the burrow, *Bornichnus* has the potential of decreasing the reservoir quality of otherwise clean sandstone.

## 5.7 *Camborygma* Hasiotis and Mitchell, 1993

*Morphology, Fill and Size: Camborygma* includes simple to complex burrow systems with subvertical shafts, subhorizontal tunnels, and chambers (Hasiotis and Mitchell 1993; Fig. 5.26). They can be simple vertical shafts with a basal chamber, or laterally to downward-branched systems (Hasiotis 2010; Fig. 5.27). Depending on the complexity of the burrow architecture, their length can range from 10 to more than 400 cm, with a burrow diameter from 1 to 14 cm (Hasiotis 2010). The burrow wall can have bioglyphs (e.g. scratches, striae). In exceptional cases, a tower (chimney) consisting of pelletoidal sediment can be preserved above the paleosurface.

*Ichnotaxonomy:* Based on their burrow architecture, four ichnospecies of *Camborygma* are discriminated (Hasiotis and Mitchell 1993; Fig. 5.26).

*Substrate:* Crayfish burrows (*Camborygma* and similar ichnogenera) are typically related with indurated and carbonate-rich paleosols where they occur in a variety of clastic substrates (e.g. sandstone, mudstone, pyroclastic deposits, etc.).

*Appearance in Core:* Crayfish burrows such as *Camborygma* have been broadly overlooked in core and are only sporadically recognized as such (e.g. Hasiotis 2010) because only fractions of the complex burrow systems are typically exposed in slabbed core (Fig. 5.28). The burrows are characterized by a large diameter (1 cm or more), often a sharp margin (because of the firm paleosol substrate), and passive fill (debris) contrasting from the surrounding host rock. Circular to slightly elliptical tunnel cross sections are common, occasionally accompanied by vertical shafts and, in rare cases, chamber-like extensions. Because of its association with seasonally wet soils, *Camborygma* typically is associated with root traces and *Taenidium*, and some burrow parts may be affected by calichification.

*Similar Trace Fossils:* Because of its complex burrow architecture, *Camborygma* may be confused with morphologically similar crustacean burrows, making a clear distinction in core impossible. *Loloichnus* is another crayfish burrow from continental deposits which differs from *Camborygma* by the absence of chambers, common tunnels and multiple shafts (Bedatou et al. 2008). Other decapod crustacean burrows (e.g. *Ophiomorpha*, *Thalassinoides*, *Spongeliomorpha*, *Psilonichnus*, *Pholeus* and others) may partly resemble crayfish burrows (e.g. Martin et al. 2008) but are restricted to marine environments.

*Producers: Camborygma* is the burrow of crayfish (continental decapod crustaceans), which have modern counterparts (Fig. 5.29).

*Ethology: Camborygma* is a trace serving for dwelling and reproducing (domichnion; Hasiotis 2010).

*Depositional Environment: Camborygma* is a continental trace fossil typically associated with weakly- to well-

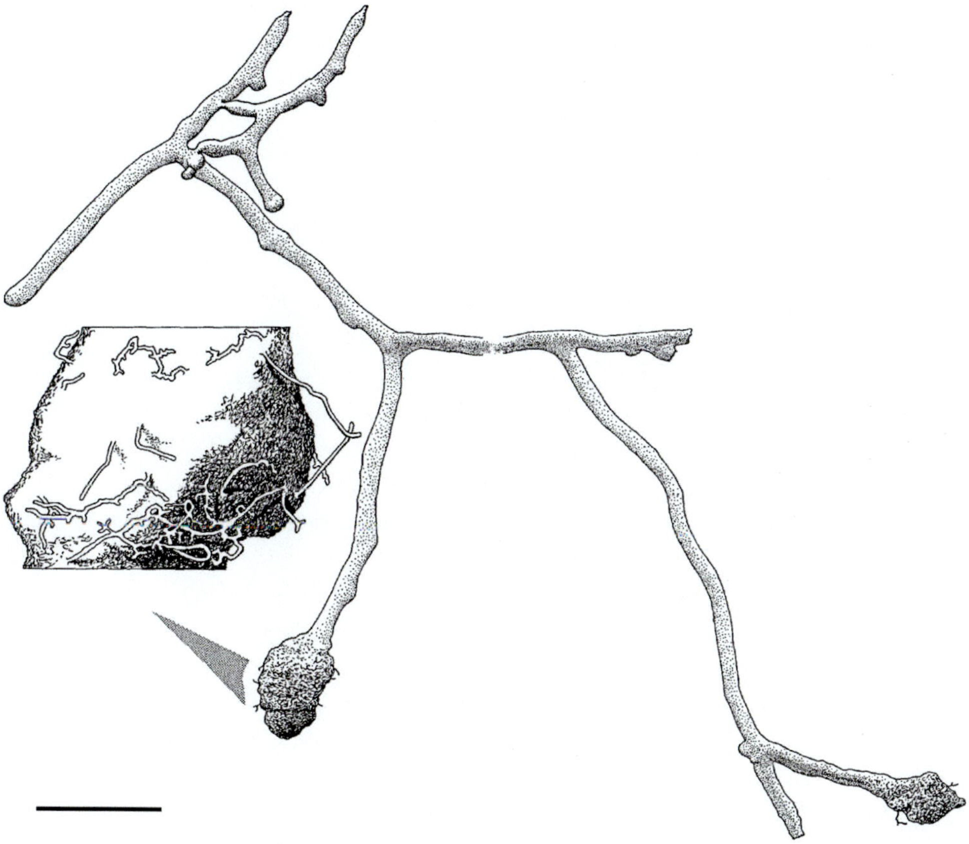

**Fig. 5.25** Resin cast of the burrow of the modern shrimp *Upogebia affinis* (similar to fossil *Ophiomorpha* and *Thalassinoides*) sampled in tidal creeks on Sapelo Island, Georgia. Some burrow terminations are swollen chambers with a rough surface, from which numerous minute burrows (ca. 1 mm in diameter) emerge (*inset*). *Scale bar* = 10 cm. After Bromley and Frey (1974), republished with permission of the Geological Society of Denmark

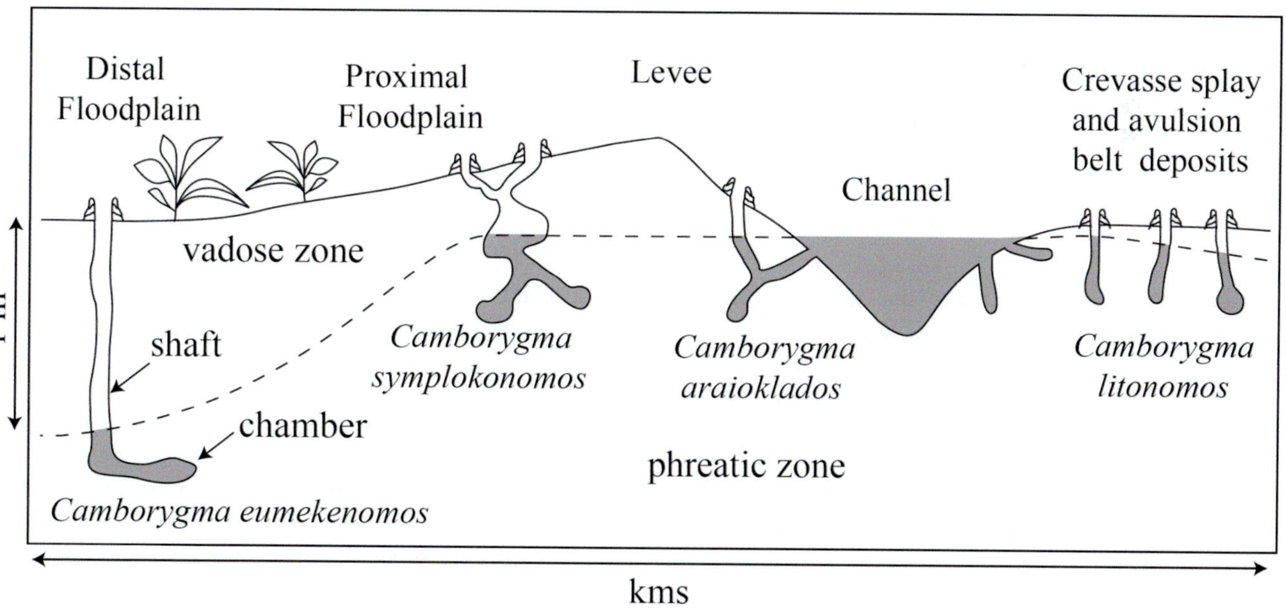


**Fig. 5.26** Architectural morphologies of the different ichnospecies of *Camborygma* with respect to position on the floodplain and depth to the water table. *C. litonomos* composed of simple shafts with little branching and few chambers imply high water table, periodic connectivity to open water sources, and shorter-term burrow occupation. Complex burrows with multiple shafts and chambers or long burrows that branch at depth indicate low water tables, decreasing access to surface waters, and long-term burrow occupation. Modified from Hasiotis and Honey (2000), Smith (2007)

**Fig. 5.27** *Camborygma* in caliche pebble-bearing sandstone to conglomerate (alluvial deposits on a floodplain) of the Upper Triassic (Carnian) Kågeröd Formation at Risebæk, Bornholm, Denmark. *Scale bars* = 5 cm. After Knaust (2015b). The burrows are passively filled with green mud that is cracked due to desiccation. **a** Oblique shaft associated with horizontal tunnels in longitudinal section (*left*) and cross section (*right*). **b** Vertical shaft. **c** Oblique shaft with Y-shaped branching (*upper part*) and chamber-like extension (*lower part*). **d** Burrow cross section

developed paleosols formed in proximal to distal alluvial and marginal-lacustrine environments (channel, levee and overbank, floodplain). A high abundance of burrowing crayfish occurs in humid to hot, seasonal wet climates (Hasiotis 2010), although continental and semiarid occurrences are reported too (Knaust 2015b; Fiorillo et al. 2016). Starting from a discontinuity surface at the top, the burrow shafts penetrate the vadose zone and typically branch below the water table with the chamber situated in the phreatic zone (Figs. 5.26 and 5.30). Architecture and depth of the hydrophilic burrows reflect the depth and fluctuation of the ancient water table (Hasiotis and Mitchell 1993). Burrow density and size variations are other aspects in the reconstruction of paleoenvironments. Lateral variation in burrow density may reflect spatial heterogeneity in water table and soil moisture levels, while crayfish size increases from the fluvial channel towards the overbank (Kowalewski et al. 1998). This size segregation along an environmental gradient may directly be used in reservoir prediction as long as the data density is sufficient for that.

**Fig. 5.28** *Camborygma* in sectioned core from the Upper Triassic (Norian-Rhaetian) Lunde Formation (fluvial) in the Snorre Field (**a**–**d**) and the Upper Triassic to Lower Jurassic Hegre Group (alluvial) in the Johan Sverdrup Field (**e**). *Scale bars* = 1 cm. **a** Fine-grained sandstone containing calichified tunnel elements (*t*) and a sand-filled shaft (*s*) (well 34/7A-4H, ca. 2872.5 m). **b** Mudstone paleosol with root traces and sand-filled shaft (*s*) and tunnels (*t*) (well 34/7A-9H, ca. 2736.5 m). **c** Sandstone paleosol with passively filled shaft (*s*) and chamber-like extended tunnel (*t*) (well 34/7A-9H, ca. 2595.5 m). **d** Mudstone paleosol with root traces and calichified tunnel (*t*) and shaft (*s*) with basal chamber (*c*) (well 34/7-1, ca. 2532.5 m). **e** Complex burrow system following a fractured limestone, displaying a long vertical shaft with horizontal branches filled with green mudstone within a caliche deposit (well 16/2-17S, ca. 2028.5 m)

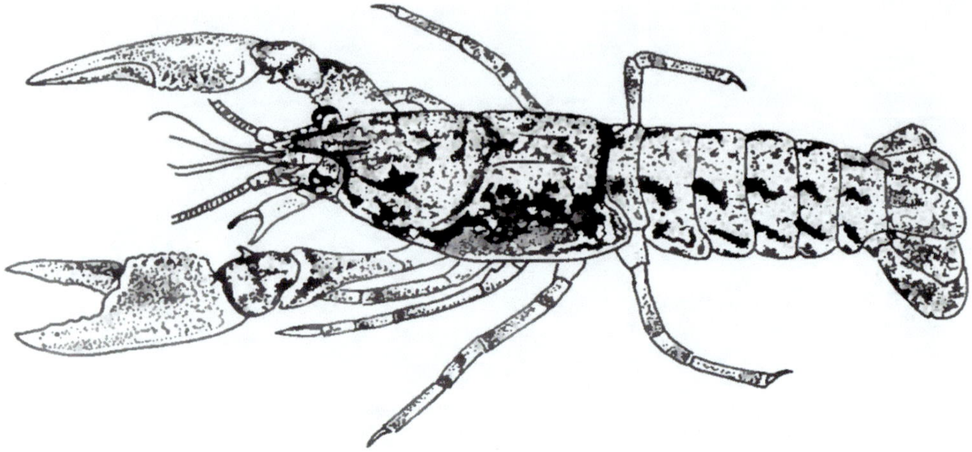

**Fig. 5.29** *Cambarus (Puncticambarus) georgiae* Hobbs, 1981, a modern burrowing crayfish from Georgia, USA (from Hobbs 1981). Entire length ca. 7–8 cm

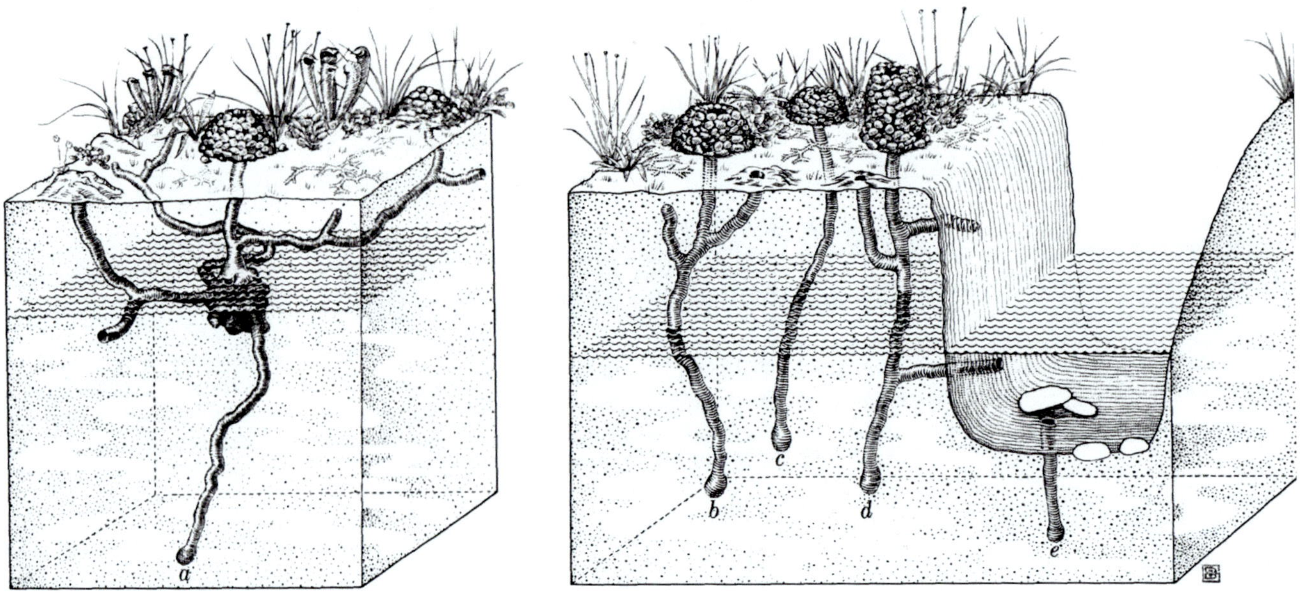


**Fig. 5.30** Generalized crayfish burrows with respect to the water table (from Hobbs 1981)

*Ichnofacies: Camborygma* is a common constituent of continental ichnofacies, including *Scoyenia*, *Coprinisphaera* and *Celliforma* ichnofacies (Buatois and Mángano 2011; Melchor et al. 2012).

*Age:* Crayfish evolved in the Permian and crayfish burrows such as *Camborygma* are recorded from continental deposits of Permian to Holocene age (Hasiotis 2010).

*Reservoir Quality:* Given their large dimensions and passive fill, *Camborygma* generally contributes to an improved reservoir quality and connectivity (e.g. Fig. 5.28b, c). The burrows may increase lateral connectivity of channel-margin deposits (e.g. levee and overbank deposits) with sand-filled channel deposits. However, due to the process of pedogenization, such burrows may also act as conduits for carbonate-rich solutions that become precipitated in form of pedogenic carbonate (e.g. Fig. 5.28a, d). In that case, calichified burrows would reduce reservoir quality and connectivity.

## 5.8 *Chondrites* von Sternberg, 1833

*Morphology, Fill and Size: Chondrites* is one of the most common and widely distributed trace fossils; due to its rootlike appearance it was originally interpreted as a plant fossil (Fig. 5.31). It consists of tunnel systems possessing a single or a small number of master shafts, presumably open to the surface, which ramifies with depth under acute angle to form a dendritic or root-like system (Osgood 1970; Fu 1991; Fig. 5.32). Most of the burrows show an active fill, sometimes with portions preserving a meniscate structure. The burrows are unlined. The tunnel diameter remains constant in different parts of the burrow and typically is in the range of less than 1 mm to a few millimeters. The lateral

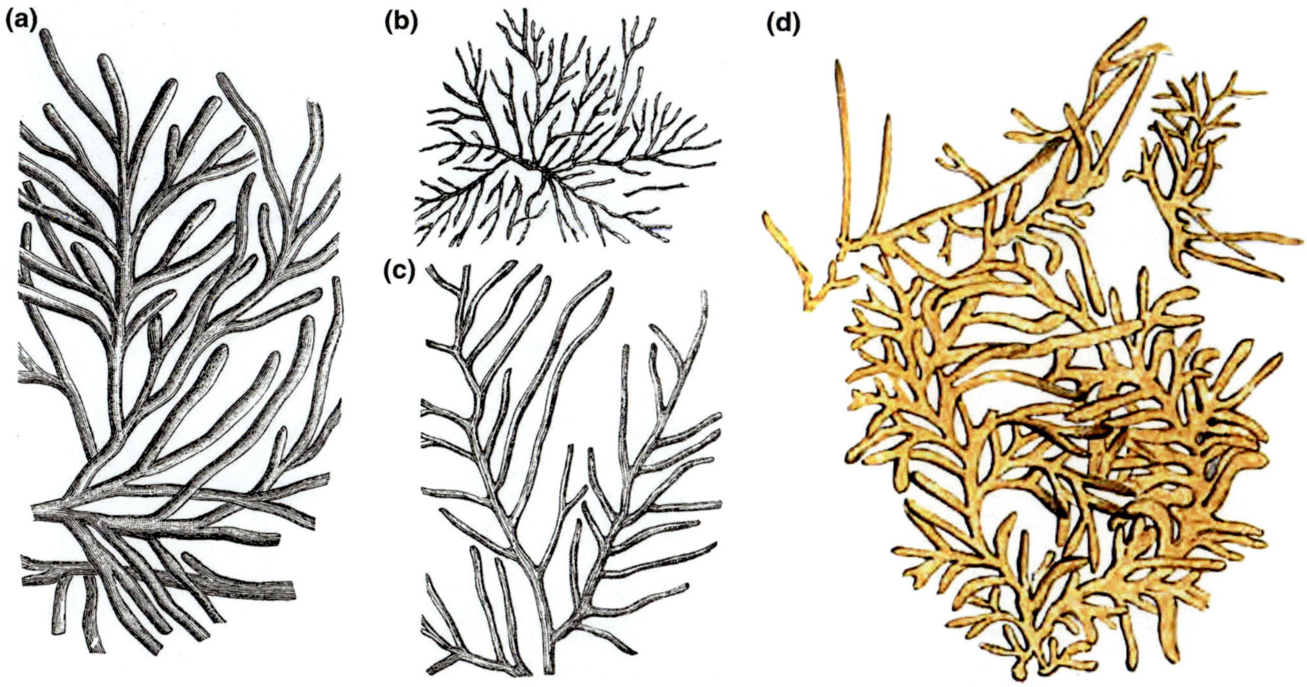


**Fig. 5.31** Historical figures of specimens of *Chondrites* originally interpreted as plant fossils (algae, fucoids). From Steinmann (1907, **a–c**) and Saporta (1873, **d**). Burrow diameter is typically in the size range of few millimeters. **a** *C. bollensis* (now *Phymatoderma granulata*), Lower Jurassic, Germany. **b** *C. intricatus*, Oligocene, Switzerland. **c** *C. targionii*, Oligocene, Switzerland. **d** *C. bollensis* (now *P. granulata*), Jurassic, France

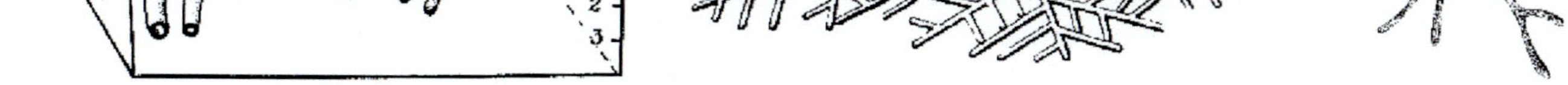


**Fig. 5.32** Different morphologies of *Chondrites*. **a** From Tauber (1949), republished with permission of the Geological Survey of Austria. **b** From Simpson (1956). **c** From Staub (1899)

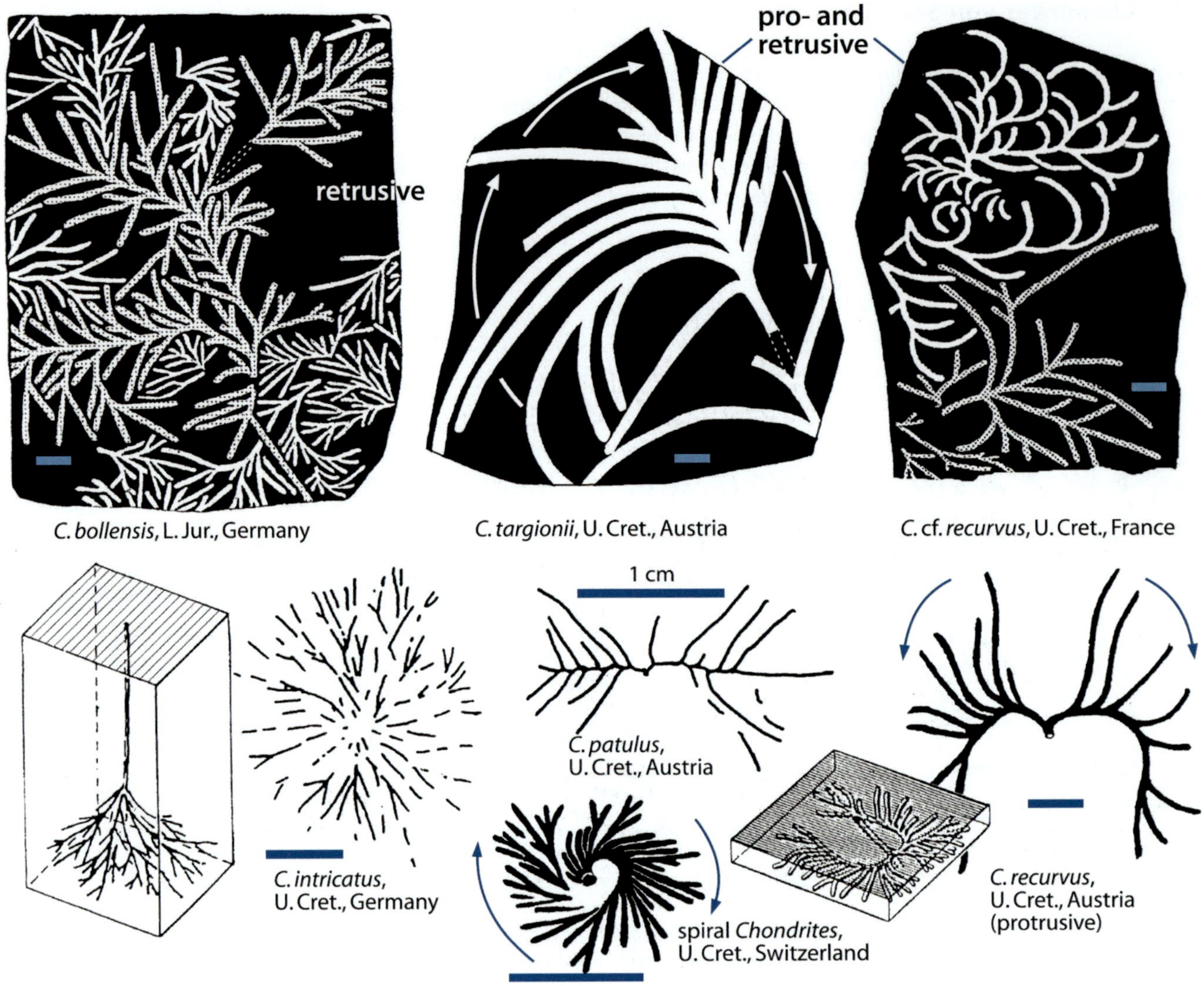


**Fig. 5.33** Characteristic ichnospecies of *Chondrites*, including the four that Fu (1991) regarded as valid. *C. bollensis* is now assigned to *Phymatoderma granulata*. From Seilacher (2007), republished with permission of Springer

extent of individual burrow systems ranges from a few centimeters to over a decimeter, although continuations between different burrow systems appear in densely bioturbated beds. The burrow systems may reach several centimeters in depth, although the true vertical extent often remains uncertain due to incomplete preservation and subsequent compaction.

*Ichnotaxonomy: Chondrites* appears with a wide range of morphologies which has given reason for establishing about 150 ichnospecies. After a revision of the ichnogenus *Chondrites*, Fu (1991) concluded that only four ichnospecies were valid (*C. targionii, C. intricatus, C. patulus* and *C. recurvus*; Figs. 5.33 and 5.34), although this view is still debated (e.g. Uchman 1999) and additional ichnospecies were subsequently introduced.

*Substrate: Chondrites* was preferably produced in fine-grained softgrounds, although the occasional occurrence of uncompacted specimens indicates relatively cohesive substrates. It occurs in siliciclastic deposits (such as mudstone, marlstone and sandstone) and carbonates (including calcilutite and chalk).

*Appearance in Core:* In core sections, *Chondrites* appears as clusters of burrows of a similar diameter. Some burrows show the typical branching at an acute angle. Depending on

**Fig. 5.34** *Chondrites* ichnospecies in outcrop. *Scale bars* = 1 cm. **a** *C. intricatus* (*i*) and *C. targionii* (*t*). Triassic part of the Hamrat Duri Group (deep marine), Hajar Mountains, Oman. **b** *C. intricatus.* Cretaceous marlstone (flysch), Ligurian Alps (east of Albenga), northern Italy. **c** *C. patulus*. Same locality as in (**b**). **d** *C. intricatus* (*i*) and *C. recurvus* (*r*). Same locality as in (**b**). **e** *C. intricatus* (*i*) and *C. targionii* (*t*). Cretaceous marlstone (flysch), Salzburg Alps (Muntigl), Austria. Coll. Stadtmuseum Berlin. Image courtesy of Beate Witzel (Berlin). **f** *C. intricatus.* Middle Triassic (Anisian) limestone (shallow marine), Udelfangen Formation near Trier, western Germany

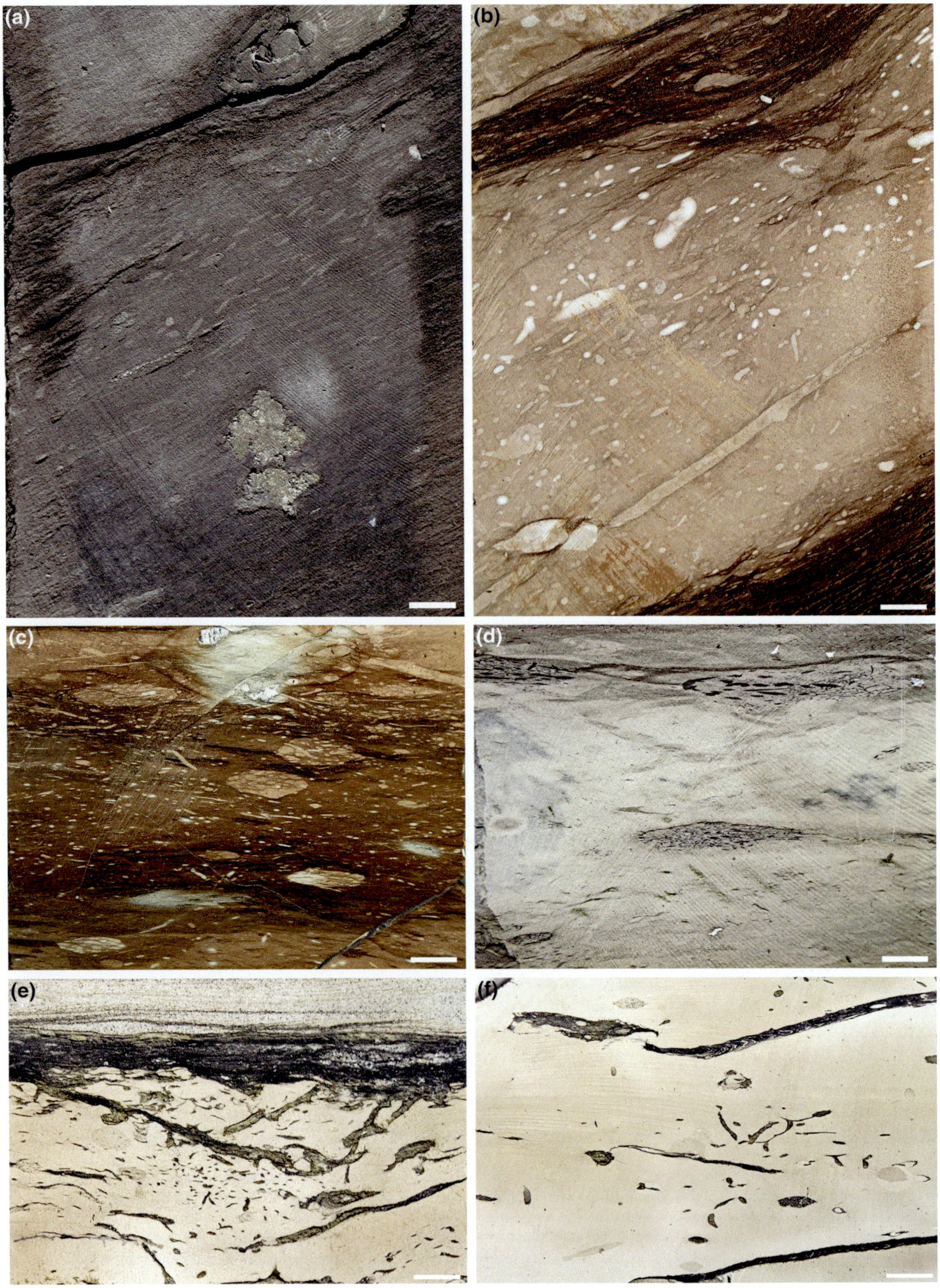
(a)
(b)
(c)
(d)
(e)
(f)

◀ **Fig. 5.35** *Chondrites* in sectioned core. *Scale bars* = 1 cm. **a** Laminated, dark-gray mudstone with sand-filled *Chondrites* as the only trace fossil, plus pyrite nodules. Note the branching of some burrows at an acute angle. Middle Jurassic (Callovian) Vestland Group (shallow marine, restricted basin), Johan Sverdrup Field, Norwegian North Sea (well 16/2-16AT2, ca. 2341.5 m). **b** Marly limestone with *Chondrites* of different sizes cross-cutting a *Zoophycos* spreite (*lower part*). Lower Cretaceous (Berriasian) Åsgard Formation (carbonate shelf), Johan Sverdrup Field, Norwegian North Sea (well 16/2-11A, ca. 2169.5 m). **c** Marly limestone (slightly faulted) with *Chondrites* of different size classes, some of which are restricted to the sediment fill of large *Thalassinoides* and *Zoophycos* burrows. Lower Cretaceous (Berriasian) Åsgard Formation (carbonate shelf), Johan Sverdrup Field, Norwegian North Sea (well 16/5-2S, ca. 1946 m). **d** Marly limestone, thoroughly bioturbated, showing indistinct *Zoophycos* spreite burrows, some of which are intensively bioturbated with *Chondrites* of different size classes. Note the dark-gray mud fill of *Chondrites*. Lower Cretaceous (Berriasian) Åsgard Formation (carbonate shelf), Johan Sverdrup Field, Norwegian North Sea (well 16/5-2S, ca. 1948.5 m). **e** Chalk with a *Zoophycos-Chondrites* ichnofabric, consisting of a colonization surface, stripe-like *Zoophycos* spreite burrows and spotty *Chondrites*. Upper Cretaceous Shetland Group (carbonate shelf, allochthonous), Oseberg Field, Norwegian North Sea (well 30/9-B-46A, ca. 3362.5 m). **f** As in (**e**), detailed view

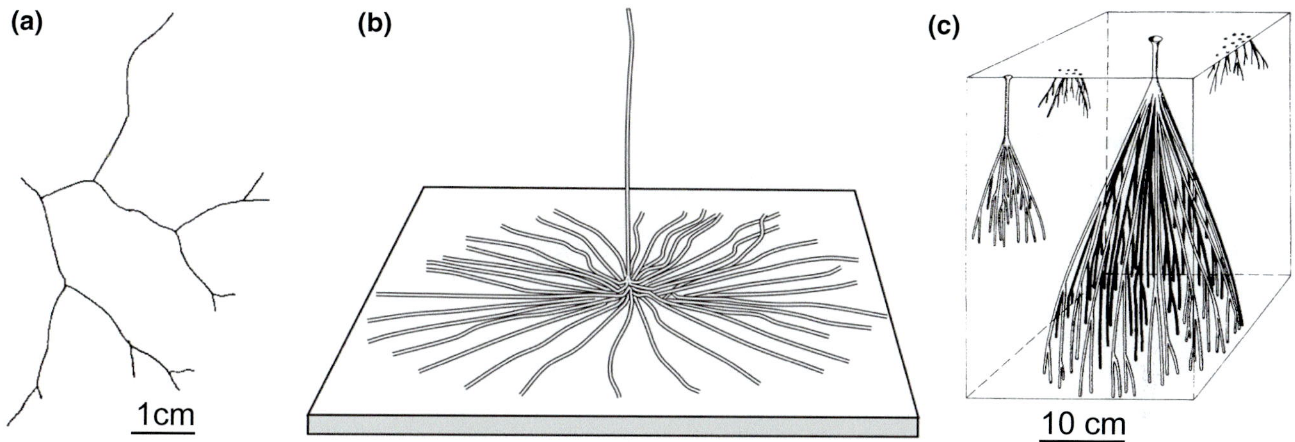


**Fig. 5.36** Schematic diagrams of ichnogenera included in the *Chondrites* group (other than *Chondrites*). **a** *Pilichnus dichotomus* (plan view), from Uchman (1999). **b** *Skolichnus hoernesii*, from Uchman (2010), republished with permission of Elsevier; permission conveyed through Copyright Clearance Center, Inc. **c** *Pragichnus fascis*, from Mikuláš (1997), republished with permission of Schweizerbart (www.schweizerbart.de/journals/njgpa)

**Fig. 5.37** *Phymatoderma granulata*, originally assigned to "*Chondrites bollensis*", has a morphology consistent with that in *Chondrites* but differs from it by its internal composition of fecel pellets. Lower Jurassic (Toarcian) black shale (bedding plane) of Holzmaden, southern Germany. Coll. Palaeontological Institute and Museum, University of Zurich

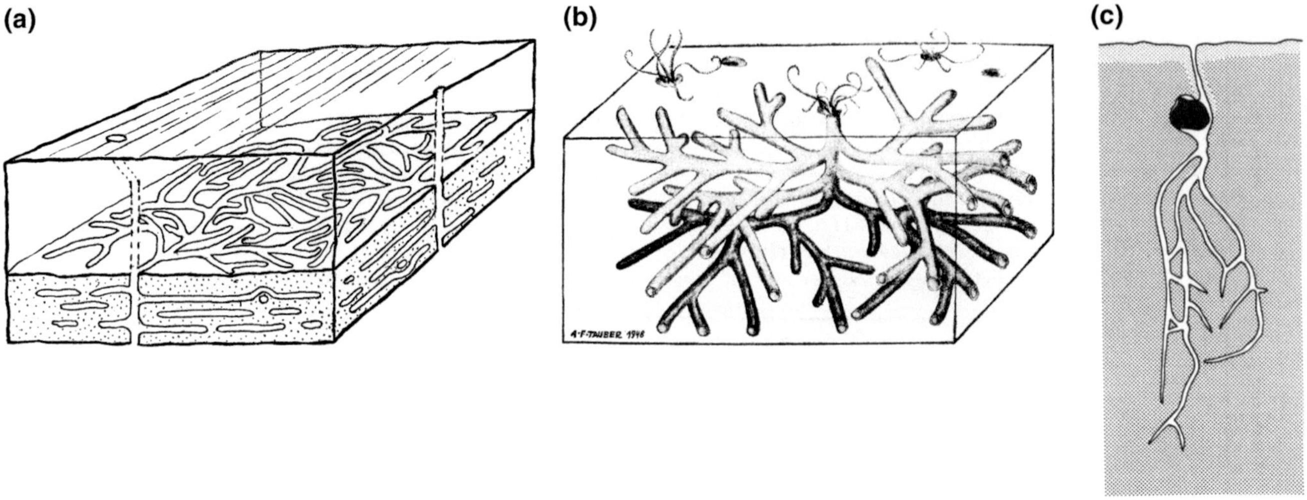


**Fig. 5.38** Different interpretations of *Chondrites* behavior and producer. **a** Deposit-feeding system (fodinichnion) within the sediment by a vermiform animal. From Richter (1931), republished with permission of Schweizerbart (www.schweizerbart.de/home/senckenberg). **b** Dwelling (domichnion) of a suspension-feeding annelid. From Tauber (1949), republished with permission of the Geological Survey of Austria. **c** Chemosymbiosis of the bivalve *Thyasira flexuosa* by means of its extensible vermiform foot. From Dando and Southward (1986), © The Marine Biological Association, published by Cambridge University Press, reproduced with permission

the direction of the section with individual burrows, circular (transverse), elliptical (oblique) and elongate (longitudinal) figures become visible (Fig. 5.35; see also Figs. 5.50a, 5.121a and 5.161f). A subhorizontal burrow orientation is most common. In composite ichnofabrics, *Chondrites* occupies the deepest tier while cross-cutting older traces such as *Zoophycos*, *Phycosiphon* and *Thalassinoides* (Ekdale and Bromley 1991; Fig. 3.1b), although in modern sediments *Thalassinoides* and *Zoophycos* can penetrate deeper than *Chondrites*. Because of this relationship, *Chondrites* often occurs within preexisting burrows (e.g. *Thalassinoides, Zoophycos, Planolites*), which the tracemaker has reburrowed.

*Similar Trace Fossils:* The *Chondrites* group contains some ichnotaxa that were previously regarded as part of the ichnogenus *Chondrites* but are now excluded from it because of their different morphology. In core, those ichnotaxa may be easily confused with *Chondrites* sensu stricto. They include irregularly winding and dichotomously branched *Pilichnus* Uchman, 1999; the radial trace fossil *Skolichnus* Uchman, 2010; and the root-like trace fossil *Pragichnus* Chlupáč, 1987 (Mikuláš 1997; Fig. 5.36). Horizontally branched tunnels filled with fecal pellets were occasionally included in *Chondrites* (e.g. Kotake 1991) but belong to the ichnogenus *Phymatoderma* (Fu 1991; Miller 2011; Izumi 2012; Fig. 5.37). The dendritic trace fossil *Hartsellea* Rindsberg, 1994 is another candidate for confusion but differs from *Chondrites* by branching upward and lining. *Rutichnus* D'Alessandro et al., 1987 branches in a similar manner as *Chondrites* but is thickly walled and has a meniscate fill and external rugosity. *Planolites* is a common trace fossil with a horizontal course, active fill and no lining, and thus may be confused with *Chondrites* in core sections, from which it differs by its unbranched nature. Finally, the complex, three-dimensional, irregular burrow system *Virgaichnus* has similarities with *Chondrites* but differs from it by a higher degree of irregularity, varying burrow diameter and passive fill (Knaust 2010a).

*Producers: Chondrites* is a heterogeneous and polygenetic trace-fossil group with many ichnospecies, reflecting its wide morphological variation. Still the burrow architecture itself remains relatively simple comprising only a few significant features. Therefore, together with its wide range in age and environments, it is very likely that animals of different groups were producers of *Chondrites*. Annelids (e.g. polychaetes; Tauber 1949; Hertweck et al. 2007) and sipunculans (Simpson 1956) are likely candidates. In homology to the peristaltic movement of a vermiform organism, the extended and branched foot of some bivalves which thrive with chemosymbiosis (e.g. thyasirid and lucinid bivalves) are known to produce *Chondrites*-like burrows in sediments (e.g. Dando and Southward 1986; Seilacher 1990, 2007; Fu 1991; Dufour and Feldbeck 2003).

*Ethology:* Due to its wide morphological variability, different behavior models have been derived for *Chondrites* (Fig. 5.38). Most commonly applied is the interpretation of

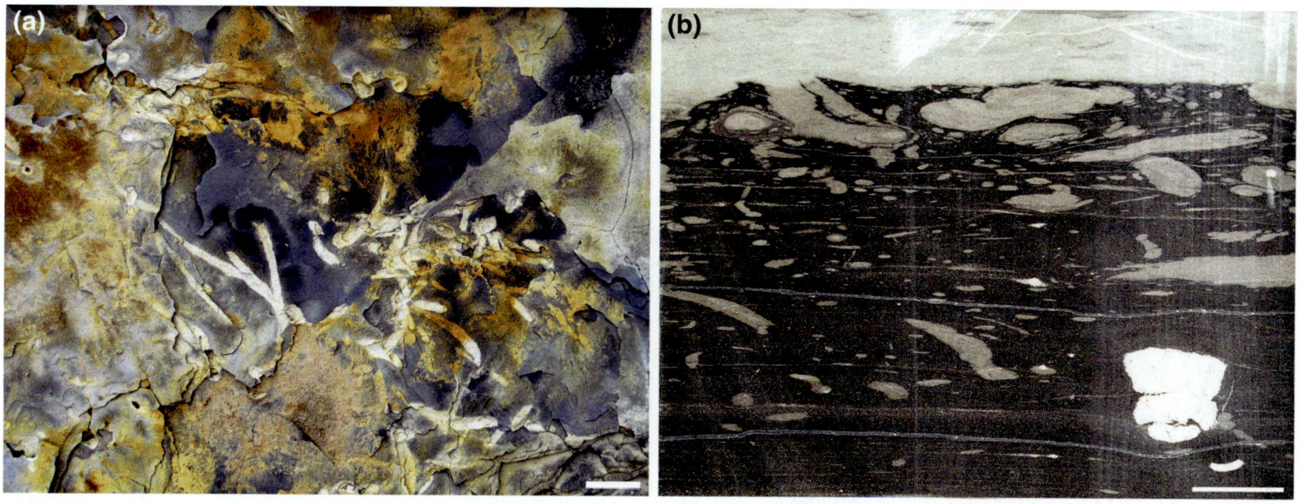


**Fig. 5.39** *Chondrites* in Devonian shale of North America. *Scale bars* = 1 cm. **a** *Chondrites* in black shale (outcrop, bedding plane) filled with contrasting material from the overlying gray shale bed. Huron Shale, Upper Devonian (Famennian), Clay City, Kentucky. **b** Laminated black shale in vertical core section, overlain by a gray shale bed from where *Chondrites* and *Zoophycos* penetrate the underlying substrate. The large bright spot in the lower right corner of the image is a pyrite nodule. Well core 57, New Albany Shale, Upper Devonian, Illinois Basin

subsurface deposit-feeding behavior (e.g. Osgood 1970), although other models include suspension-feeding (e.g. Tauber 1949), detritus-feeding on the sea floor (Simpson 1956), and chemosymbiosis (Seilacher 1990, 2007; Fu 1991).

*Depositional Environment:* Many case studies have shown that *Chondrites* can be regarded as a good indicator of dysoxic (between anaerobic and aerobic) settings, where the dissolved oxygen in bottom and pore waters of sedimentary basins is between 0.2 and 1 ml/l (Bromley and Ekdale 1984; Savrda and Bottjer 1991; Martin 2004). In existing models, such as the one developed by Savrda and Bottjer 1991), decreasing oxygenation goes hand in hand with reduction of ichnodiversity, and *Chondrites* appears to be the last trace fossil before anaerobic conditions set in as shown by lamination, high organic matter and sulfide mineralization coinciding with decreasing burrow size and decreasing depth of bioturbation. Such conditions can often be found in the deep-marine environments (e.g. basin plains, submarine fans), from which *Chondrites* is frequently reported. Stagnant conditions, however, also occur on the shelf (e.g. minibasins) and in nearshore restricted basins (e.g. estuaries, lagoons, embayments etc.). The fact that the *Chondrites*-producer tolerates low oxygen content within the sediment results in its frequent occurrence in transgressive deposits and particularly in association with maximum flooding intervals.

*Ichnofacies:* Although *Chondrites* must be regarded as a facies-crossing trace fossil, its widest distribution is documented in deep-sea (flysch) deposits belonging to the *Nereites* Ichnofacies.

*Age:* *Chondrites* is a common trace fossil from the Cambrian (Webby 1984) to Holocene (Wetzel 1981, 2008) and has a wide distribution in the Cretaceous to Paleogene alpine flysch, from which it was originally described as a plant fossil (Fu 1991).

*Reservoir Quality:* In their case study from the Cretaceous Ben Nevis Formation reservoir off Newfoundland, Tonkin et al. (2010) found that *Chondrites* in mudstone-rich facies had a net effect of permeability reduction. In other cases, however, a positive impact on porosity and permeability distribution due to the presence of *Chondrites* may be noticed, chiefly because of the introduction of sandy material into mud-dominated facies and thus an increased heterogeneity. This may be particularly important for the performance of hydrocarbon reservoirs in shale (e.g. Schieber 1999, 2003; Bednarz and McIlroy 2015; Fig. 5.39).

## 5.9 *Conichnus* Männil, 1966

*Morphology, Fill and Size: Conichnus* is a relatively large, conical burrow with subcircular cross section and subvertical orientation. Its internal fill is largely passive and a thin lining

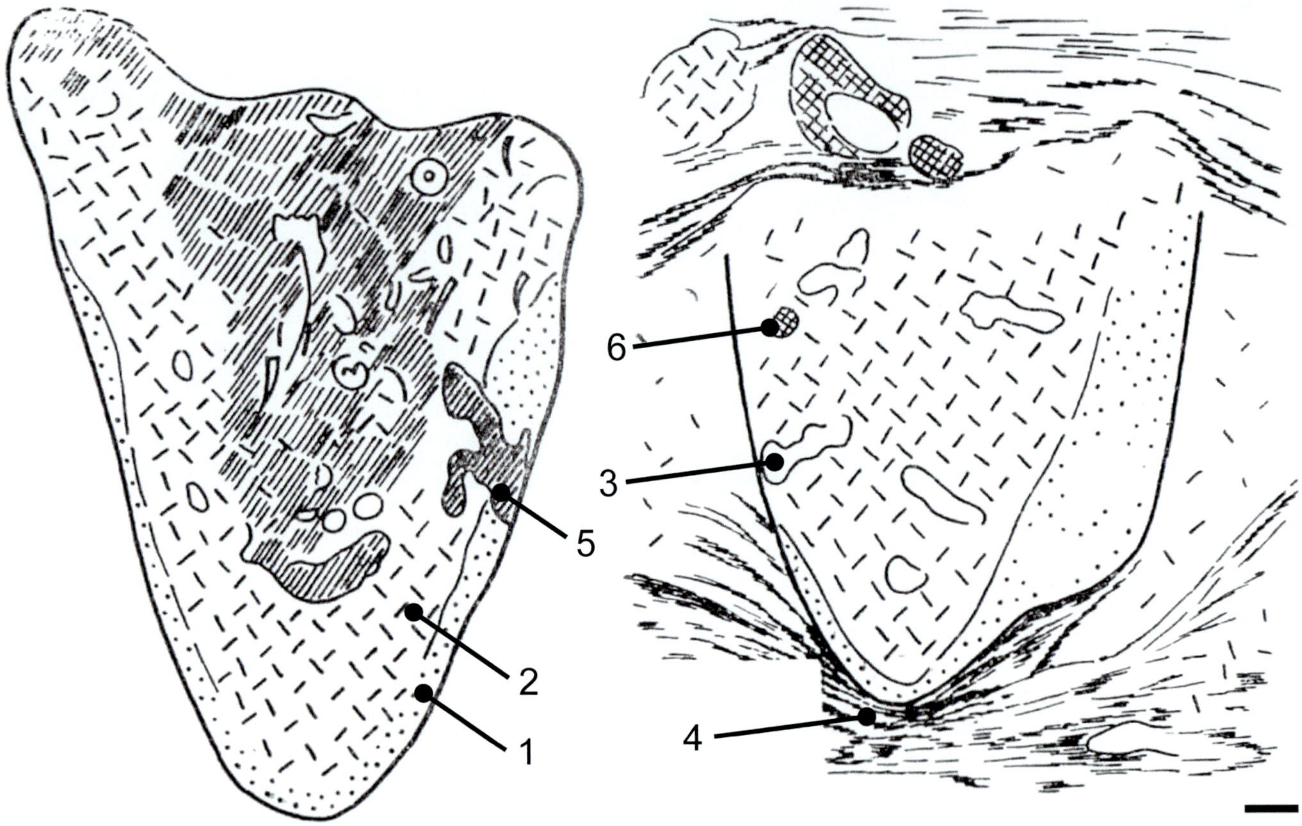


**Fig. 5.40** Vertical sections of *Conichnus conicus* type specimens with passive fill from the Ordovician of Estonia. *1* Fine clastic material, predominantly organogenic. *2* Same, relatively coarse. *3* Fine-grained limestone. *4* Marl with thin shaly layers. *5* "Bituminous" limestone. *6* Fragments of skeletons of different organisms. From Männil (1966), republished with permission of the Borissiak Paleontological Institute Moscow. *Scale bar* = ca. 1 cm

along the burrow wall may be present (Frey and Howard 1981; Pemberton et al. 1988; Figs. 5.40, 5.41 and 5.42). *Conichnus* can be related to internal, convex-down (chevron-like) structures (due to the adjustment of the producer), or it can be located above such a structure (equilibrium trace). Burrow depth typically ranges between a few centimeters and several decimeters.

*Ichnotaxonomy:* The type ichnospecies, *C. conicus*, is the most common ichnospecies of *Conichnus*. Nielsen et al. (1996) described *C. conosinus* from the Cretaceous of Bornholm, Denmark, which contains an upper dish-shaped depression above the cone-shaped lower burrow. In addition, Chen et al. (2005) introduced the questionable *C. wudangensis* from the Devonian of China. The inclusion of *Amphorichnus papillatus* in *C. conicus* (Frey and Howard 1981) remains controversial, because the former is distinguished from *C. conicus* by its amphora-like morphology and a characteristic apical, papilla-like protuberance (Männil 1966; Dronov et al. 2005).

*Substrate: Conichnus* is a characteristic trace fossil in sandy substrate (looseground) of siliciclastic and carbonate origin.

*Appearance in Core:* Complete specimens of *Conichnus* are rarely displayed in core samples because of their relatively large size. The subvertical burrows show a typically plug-shaped morphology (Fig. 5.43). The length/diameter ratio of the burrows is relatively high (about 3:1 to 10:1 or more) and is related to the upward movement (adjustment) of the producing animal during sediment aggradation. Internally, *Conichnus* burrows often show downward-deflected laminae which lead to a chevron-like appearance. Such a structure can also be present below the actual *Conichnus* burrow as a response to the vertical adjustment of the producer. In addition, some parts of the laterally adjacent sediment may be impacted by the activity of the producer

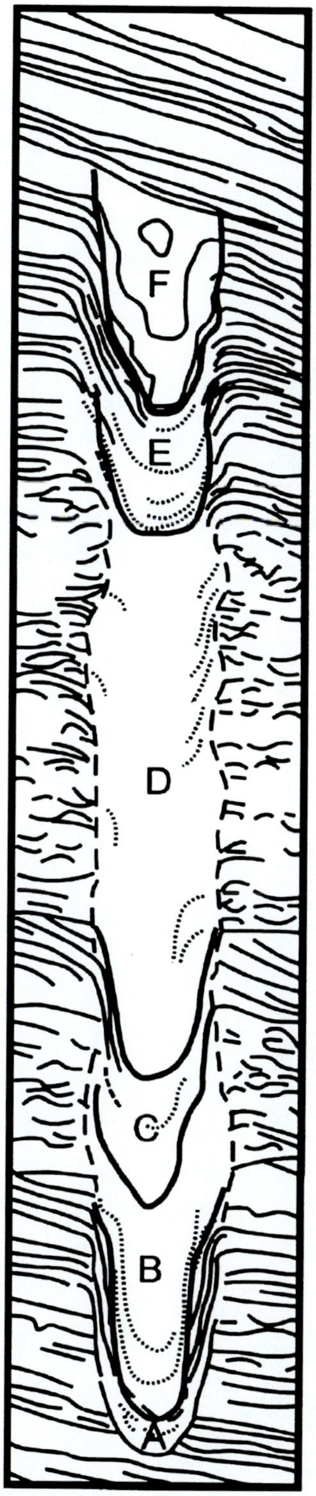


**Fig. 5.41** Vertically extensive *Conichnus* specimen occurring in Upper Cretaceous sandstone of Alabama, USA. Bold lines and letters delineate burrow segments. Burrow width is ca. 15 cm. From Savrda (2002), reprinted by permission of the publisher (Taylor & Francis Ltd., http://www.tandfonline.com)

and thus show downward-deflected laminae, listric faults or collapsed sediment. The fill of *Conichnus* can be reburrowed to various extents.

*Similar Trace Fossils:* As pointed out by Buck and Goldring (2003), *Conichnus* may be confused with a number of conical biogenic structures and sedimentary structures of different natures, such as sand collapse into a cavity, escape and equilibrium traces of organisms, and biodeformational excavations by organisms (Fig. 5.44). In addition, fluid-escape structures can also resemble *Conichnus*. *Conostichus* resembles *Conichnus*, but differs from it by having an ornamented wall, which unfortunately can hardly be seen in core slabs. Slabbed specimens of *Conichnus* with internal downward-deformed laminae may resemble *Diplocraterion*, which can be roughly in the same size range. *Diplocraterion*, however, consists of a spreite and marginal burrow, occasionally accompanied by fecal pellets. The lower, retrusive burrow part of large spreite burrows (e.g. *Teichichnus*) may look like *Conichnus* in cases where the upper burrow part has been eroded. Even the footprints of tetrapods (e.g. dinosaurs) may produce imprints superficially similar to deformed *Conichnus*.

*Producers:* Based on modern analogs, *Conichnus* can be regarded as the product of the activity of sea anemones (Actiniaria) (Schäfer 1962; Shinn 1968; Frey and Howard 1981; et al. 2008; Figs. 5.45, 5.46 and 5.47) and, to a lesser degree, bivalves (Gingras et al. 2008).

*Ethology:* The domichnion *Conichnus* usually results from the dwelling activity of anemones, which often is accompanied by an adjustment of the animal's position in order to keep pace with shifting substrate thickness (Fig. 5.41). The latter process may result in equilibrium traces associated with *Conichnus*.

*Depositional Environment: Conichnus* is a common element of nearshore to shallow-marine environments with high-energy sedimentary processes, high sediment supply and frequently shifting substrate (Abad et al. 2006). *Conichnus* also occurs in intertidal to shallow subtidal environments (e.g. tidal flats, estuaries; Curran and Frey 1977) and associated sand waves, dunes and megaripples (Shinn 1968; Savrda 2002) as well as flood-tidal deltas and lagoons (Mata et al. 2012).

*Ichnofacies: Conichnus* is a constituent of the *Skolithos* Ichnofacies.

*Age:* Anthozoa-produced trace fossils such as *Conichnus* occur from the Early Cambrian (Paczesna 2010; Mata et al. 2012) to Holocene (e.g. Gingras et al. 2008).

*Reservoir Quality:* Given their relatively large size and passive (sand) fill, deep penetration of *Conichnus* may contribute to an increased net-reservoir distribution and vertical connectivity.

**Fig. 5.42** *Conichnus conicus* in shallow-marine (nearshore) sandstone. *Scale bars* = 1 cm (**a**, **b**) and 22 cm (**c**, **d**). **a**, **b** Lower Cretaceous (Berriasian) Robbedale Formation, Arnager Bay, Bornholm, Denmark. Note the deformed coarse sand layer related to the lowermost burrow segment. **c** Lower Miocene Chenque Formation, Santa Cruz, Patagonia, Argentina. Note the overlap of several specimens on top of each other due to small adjustments of the producer. *C. conicus* is reburrowed with *Macaronichnus segregatis*. This trace resembles the feeding trace of rays similar to *Piscichnus waitemata* and was tentatively interpreted as such by Carmona et al. (2008). **d** Upper Cretaceous (Campanian) Neslen Formation, East Canyon, Book Cliffs, Colorado, USA

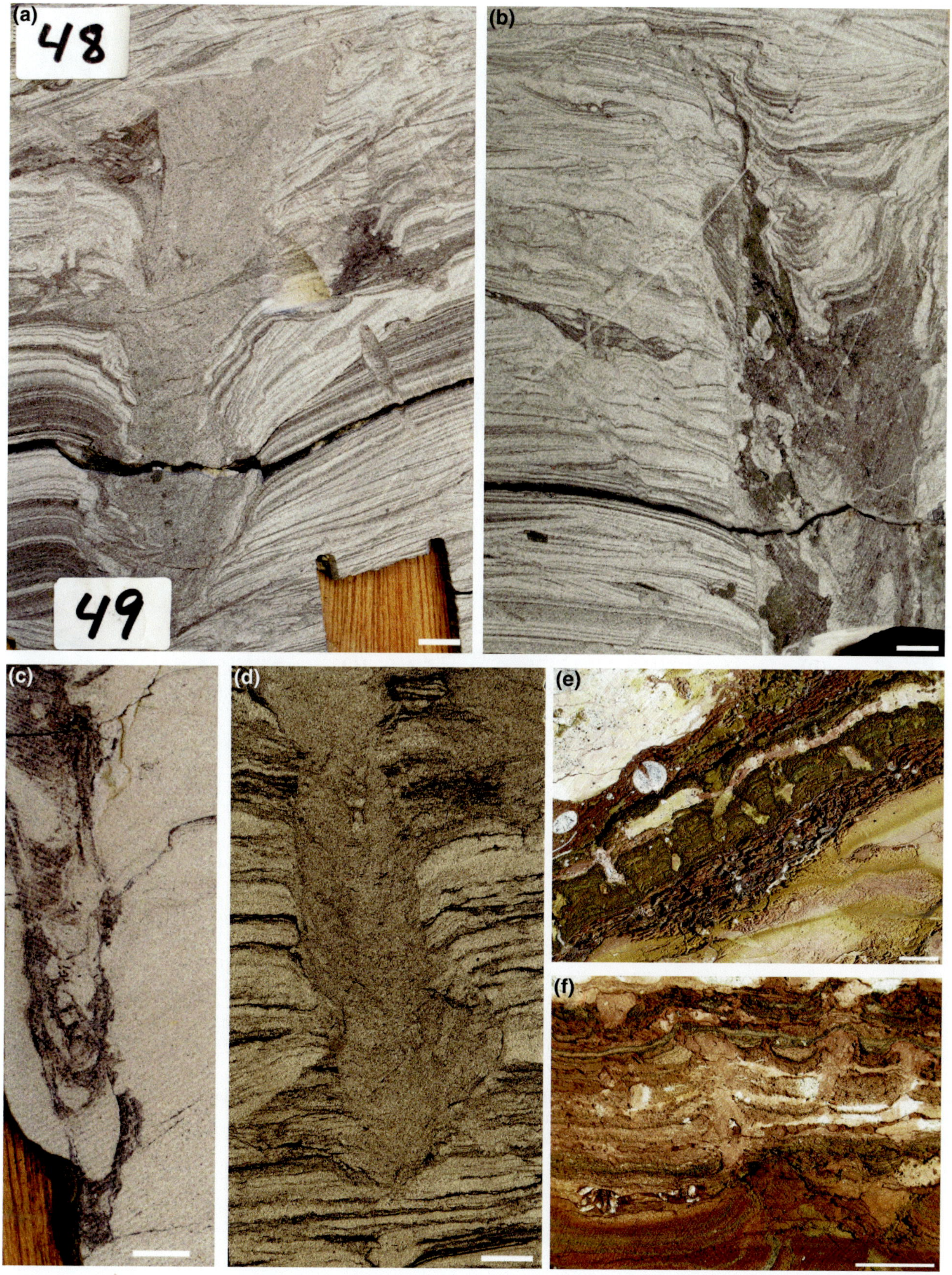


**Fig. 5.43** *Conichnus* in sectioned core. *Scale bars* = 1 cm. **a–c** Lower to Middle Jurassic (Toarcian-Aalenian) Ile Formation (tidal-influenced delta), Njord Field area, Norwegian Sea (well 6407/10-2). All specimens show significant soft-sediment deformation related to the process of burrow generation. **a** Ca. 3456.8 m. **b** Ca. 3455.4 m. **c** Ca. 3685.7 m. **d** Upper Cretaceous (Campanian) Neslen Formation (lower delta plain), East Canyon, Book Cliffs, Colorado, USA. Well HCR#1, ca. 75.6 ft. **e**, **f** Lower Cretaceous (Berriasian) Åsgard Formation (carbonate shelf), Johan Sverdrup Field, Norwegian North Sea. The relatively small burrows are located within a thin layer of biolaminite. **e** Well 16/2-11A, ca. 2175.5 m. **f** Well 16/2-21, ca. 1930.5 m

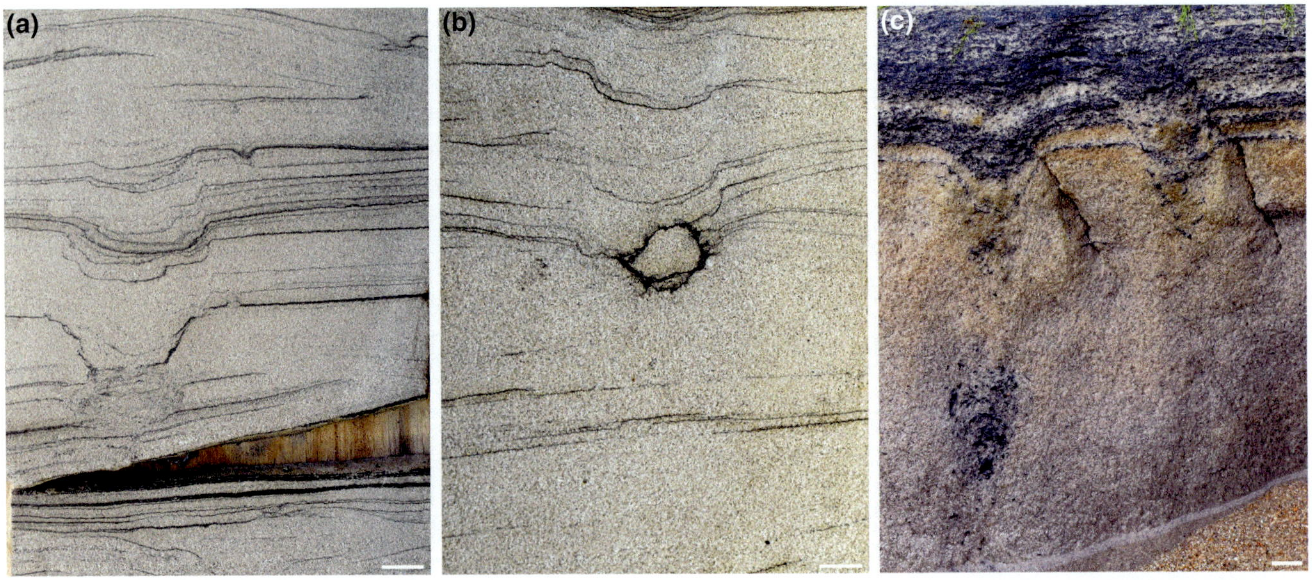


**Fig. 5.44** *Conichnus*-like sedimentary structures and traces. *Scale bars* = 1 cm. **a**, **b** Sedimentary collapse structures above *Ophiomorpha* tunnels in sectioned core. Middle Jurassic Hugin Formation (marginal marine, tidally influenced), well 15/9-8 (Sleipner Field, ca. 3483.5 m, 3481.3 m). **c** Lower parts of burrows showing retrusive spreite structures. Middle Jurassic sandstone, coastal outcrop in the Brora area, Scotland, UK

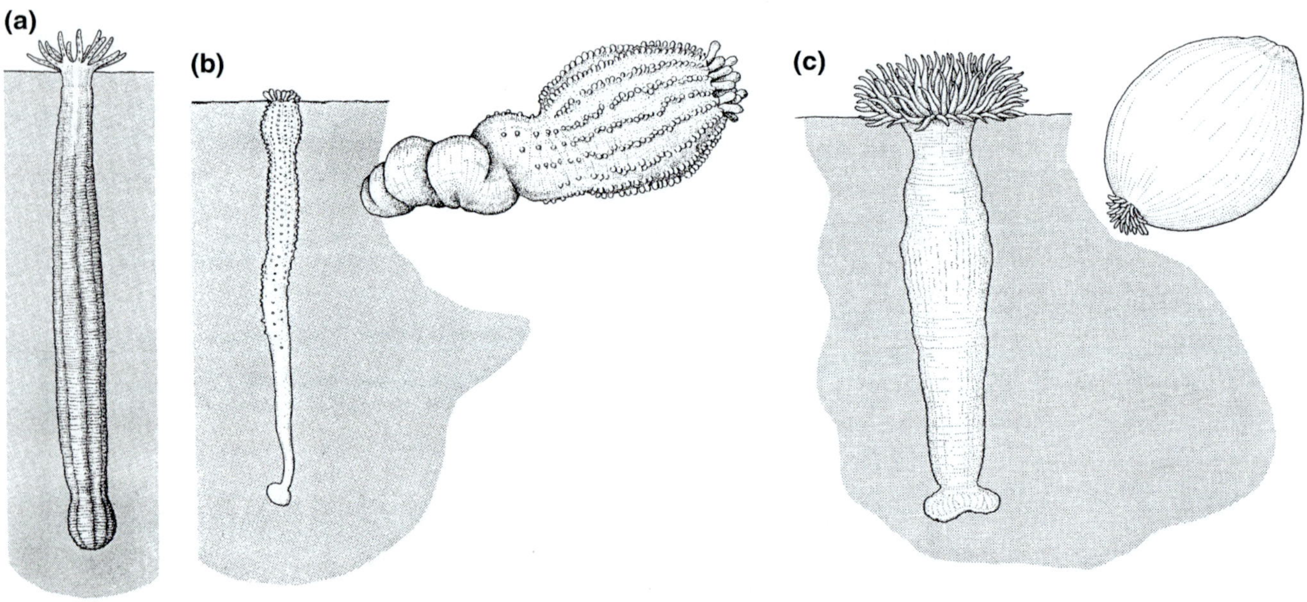


**Fig. 5.45** Schematic illustration of modern anemones within their burrows from the southeastern Atlantic coast of the United States. From Ruppert and Fox (1988). **a** Extended specimen of *Edwardsia elegans* from a muddy sand flat. **b** *Haloclava producta*. **c** The sea onion *Paranthus rapiformis*

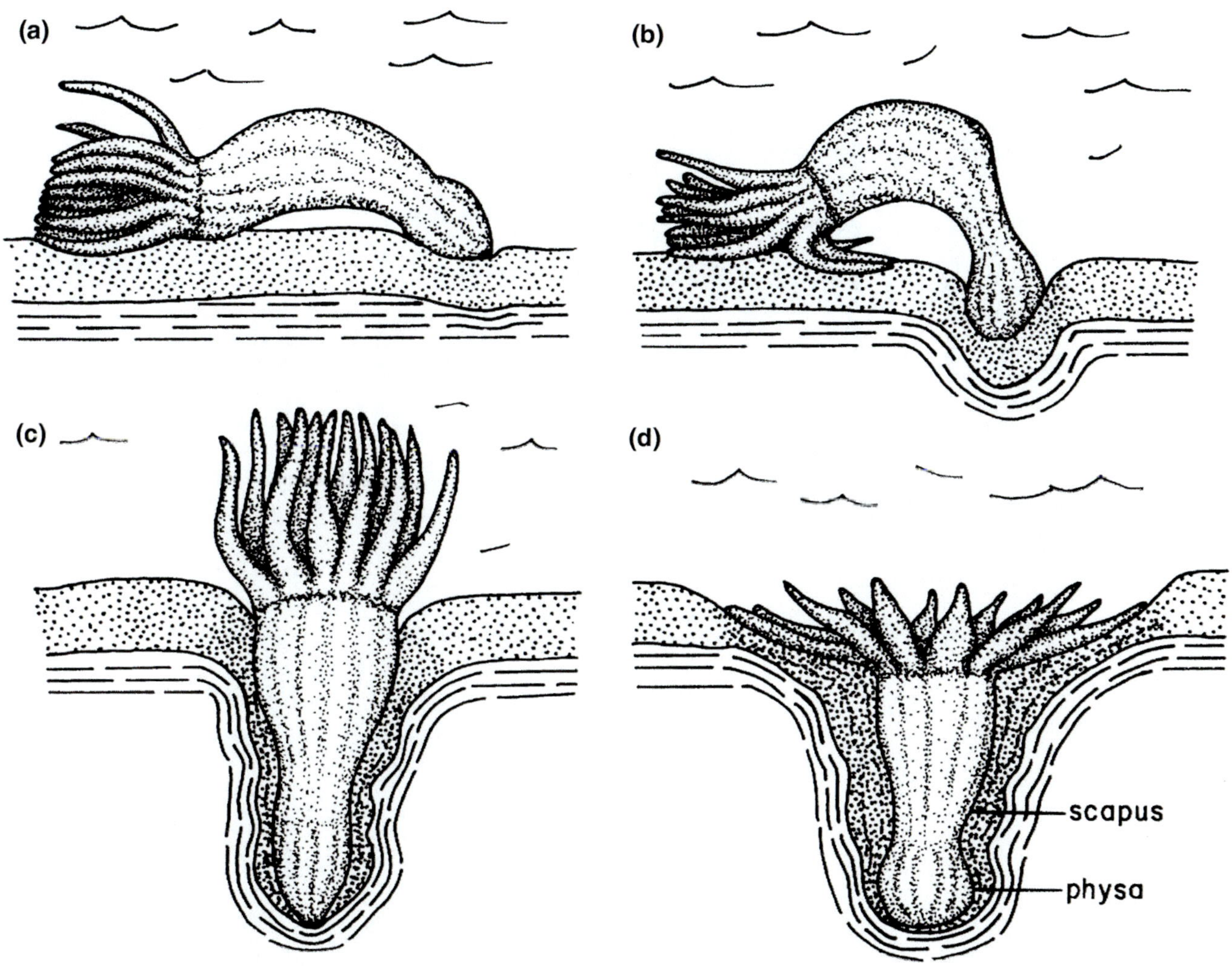


**Fig. 5.46** Development of *Conostichus*, a trace fossil similar to *Conichus*. **a** Burrowing anemone on surface. **b** Physa driving into substrate using its body weight, then drawing its body into the substrate. **c** Body column (scapus) expanded against sediment, and contracted physa driven deep. **d** Anemone in place with expanded physa as holdfast. From Chamberlain (1971), © Paleontological Society, published by Cambridge University Press, reproduced with permission

## 5.10 *Cylindrichnus* Toots in Howard, 1966

*Morphology, Fill and Size: Cylindrichnus*, originally described from the petroleum-rich Western Interior Basin (USA), includes bow-shaped to broadly U-shaped burrows with a passive fill and a concentric lining (Fig. 5.48). They are commonly unbranched, although branching has been documented (e.g. in *C. candelabrus* Głuszek, 1998). Burrow apertures may be slightly enlarged to form funnel-shaped entrances, or may be constricted. The diameter of *Cylindrichnus* is in the range of a centimeter or less, and the burrows may reach a length of several centimeters to decimeters, with a vertical extent of several centimeters.

*Ichnotaxonomy:* After some controversy over the ichnotaxonomic status of *Cylindrichnus* (Goldring 1996), it is now regarded as valid ichnogenus (Ekdale and Harding 2015). *C. concentricus* is the type ichnospecies (Fig. 5.49). *C. japonicus* Shuto and Shiraishi, 1979, *C. pustulosus* Frey and Bromley, 1985, *C. errans* D'Alessandro and Bromley, 1986, *C. operosus* Orłowski, 1989, *C. candelabrus* Głuszek, 1998, and *C. helix* de Gibert et al., 2006 were introduced subsequently, but *C. hollowus* Nilsen and Kerr, 1978 remains a nomen nudum.

*Substrate: C. concentricus* is a common element in siliciclastic sedimentary rocks, where it occurs in a deep-tier position and is typically associated with sandy substrates (Goldring 1996; Fig. 5.50a–c, f–h). It also occurs in carbonates, including chalk (e.g. Frey and Bromley 1985; Goldring et al. 2005; Knaust 2009a; Fig. 5.50d, e). *Cylindrichnus* was built in softground sediment but can show

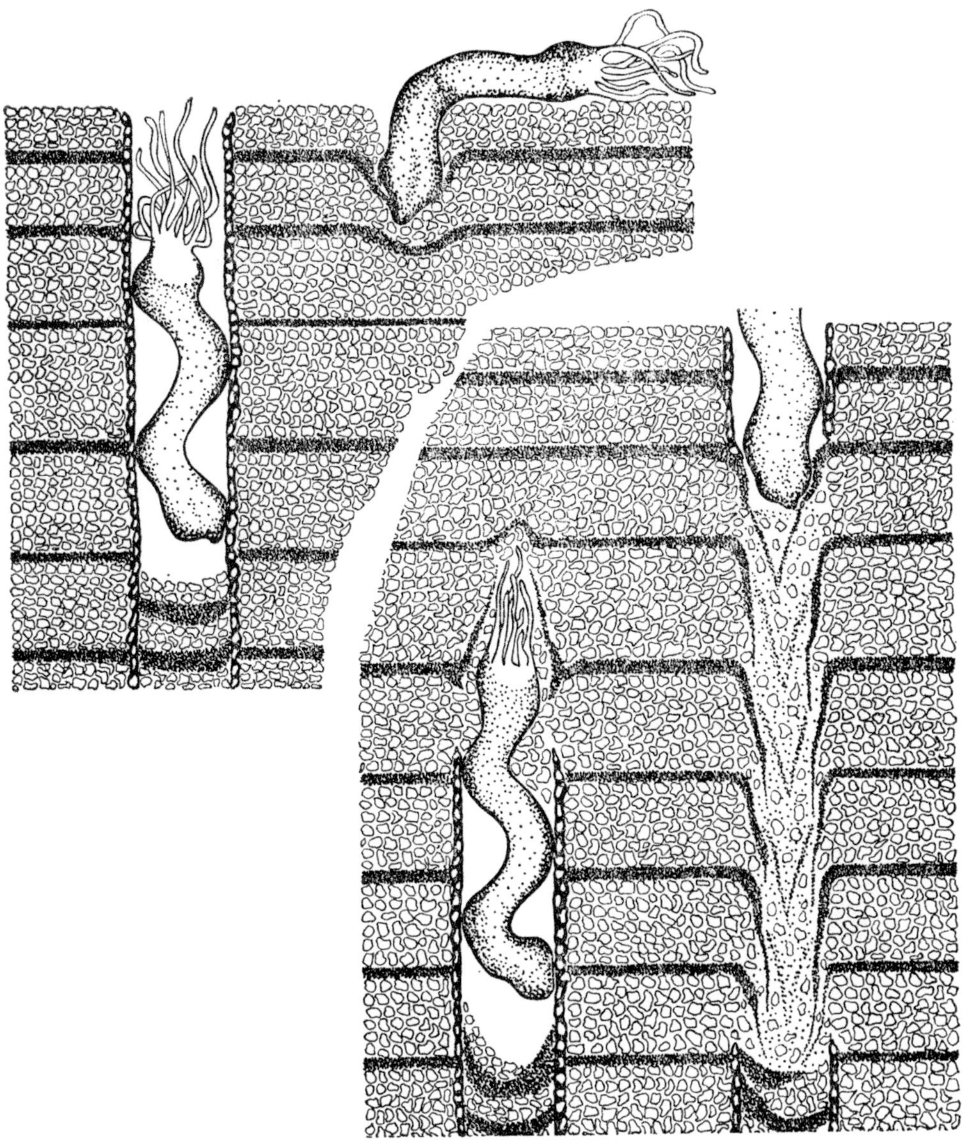

**Fig. 5.47** Modern anemone *Cerianthus* producing *Conichnus*-like burrows. *Left* Dwelling structure with mucus-lined wall and sediment fill below. Second animal burrowing from the sediment surface into the sediment. *Right* Equilibrium trace of a sediment-buried animal adjusting its burrow while moving towards the sediment surface. From Schäfer (1962), republished with permission of Schweizerbart (www.schweizerbart.de/home/senckenberg)

transitions to such firmground burrows as *Glyphichnus* (Goldring et al. 2002).

*Appearance in Core:* Because of the limited size of core samples, only parts of *Cylindrichnus* burrows are exposed (Fig. 5.50; see also Figs. 4.3 and 5.129a, e). Funnel-shaped openings of the overall bow-shaped burrows are common features of *Cylindrichnus* in core, which continue with tapering and inclining tubes in their deeper parts. Vertical sections of such burrow parts appear with a V-shaped structure consisting of a steeply inclined central tube with passive fill, which is surrounded by a wall of predominantly finer-grained sediment that thickens upwards. Most frequent are random cross sections through the burrow system, displaying the diagnostic thick lining enclosing a thin central to eccentric tube with passive fill. Clustering of burrow sections may occur in dense *Cylindrichnus* ichnofabrics.

*Similar Trace Fossils: Cylindrichnus* shares its internal composition, such as the thick lining enclosing a passively filled tube, with few other trace fossils, particularly *Asterosoma*, *Rosselia* and *Artichnus*. So far, no consensus has been reached whether or not *Cylindrichnus* must be regarded as a transitional form to the last mentioned ichnogenera.

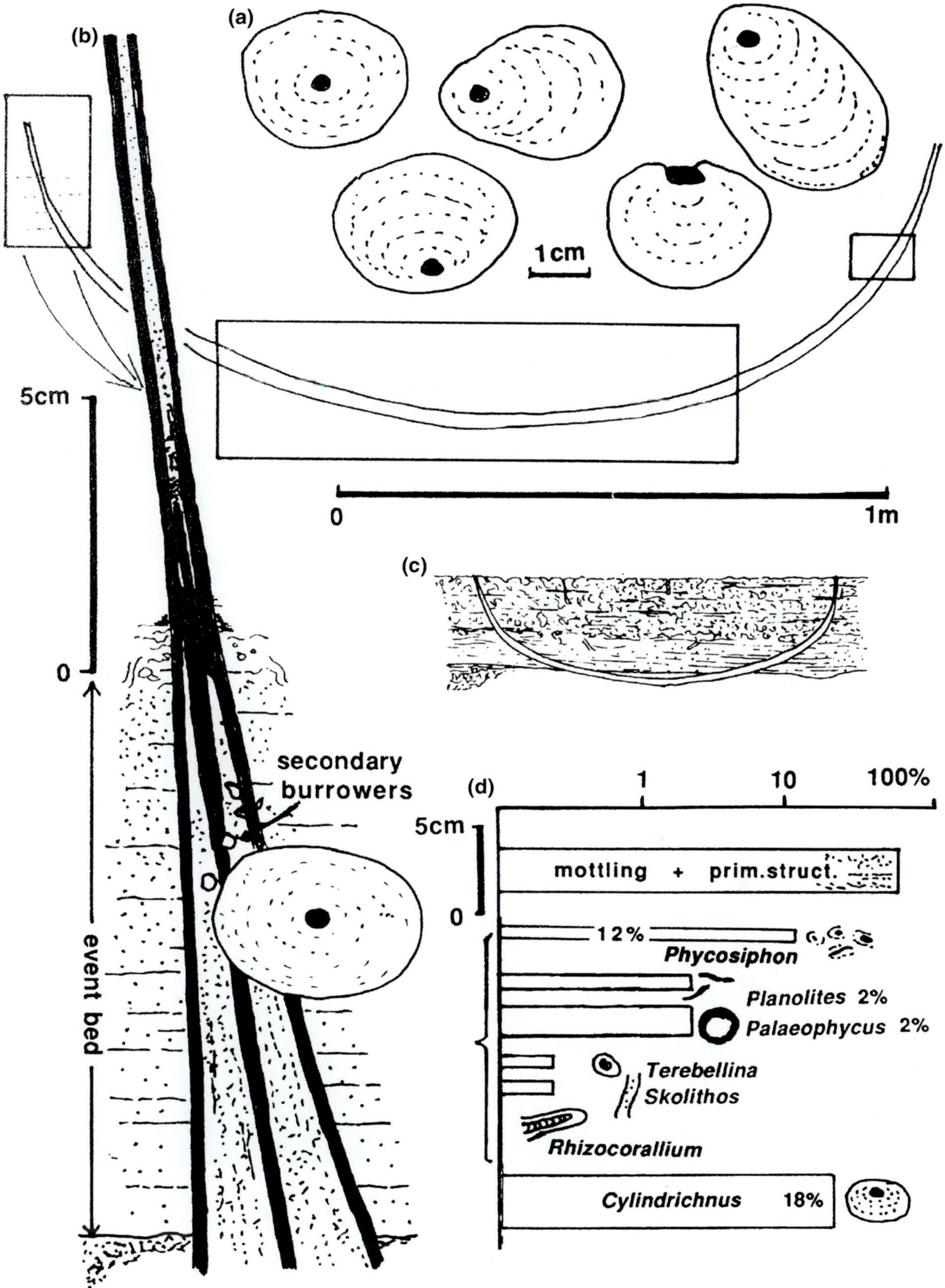


**Fig. 5.48** *Cylindrichnus concentricus* (bow-shaped burrows) and their infill (from Goldring 1996). **a** Typical cross sections of five burrows. **b** Reconstruction as a shallow, bow-like burrow with constricted apertural neck (enlarged) leading down to main section of burrow. **c** Reconstruction of a complete burrow. **d** Ichnofabric constituent diagram constructed from the interval with *C. concentricus*

**Fig. 5.49** *Cylindrichnus concentricus* in outcrop. *Scale bars* = 1 cm. **a** Cross-laminated sandstone with a bow-shaped burrow in longitudinal section and many oblique and cross sections. Miocene Mount Messenger Formation (deep marine, overbank), sea cliffs of the Taranaki Peninsula, North Island, New Zealand. **b** Cross-laminated sandstone with numerous burrow cross sections. Same locality as in (**a**). **c** Cross-bedded sandstone with numerous bow-shaped burrows displayed in various sections and enhanced by brownish iron mineral staining. Eocene Battfjellet Formation (deltaic), Brongniart Fjellet, Svalbard. **d** Sandstone beds with burrow penetration from the top surface (mainly oblique and cross sections), and a longitudinal section in the middle. Weakly developed scratches in the latter indicate partly firmground conditions and suggest transition to the ichnogenus *Glyphichnus*. Same locality as in (**c**). **e** Heterolithic sandstone with a thickly mud-lined and branched burrow. Campanian Neslen Formation (marginal marine), Jim Canyon, Book Cliffs, Utah, USA. **f** Cross-bedded sandstone with an elongate and thickly sand-lined burrow. Campanian Sego Formation (tide-dominated river delta), San Arroyo, Book Cliffs, Colorado, USA. **g** Sandstone bedding plane showing the lower part of a bow-shaped burrow. Campanian Sego Formation (tide-dominated river delta), Book Cliffs, Colorado, USA. **h** Sandstone with an oblique burrow section. Same locality as in (**g**)

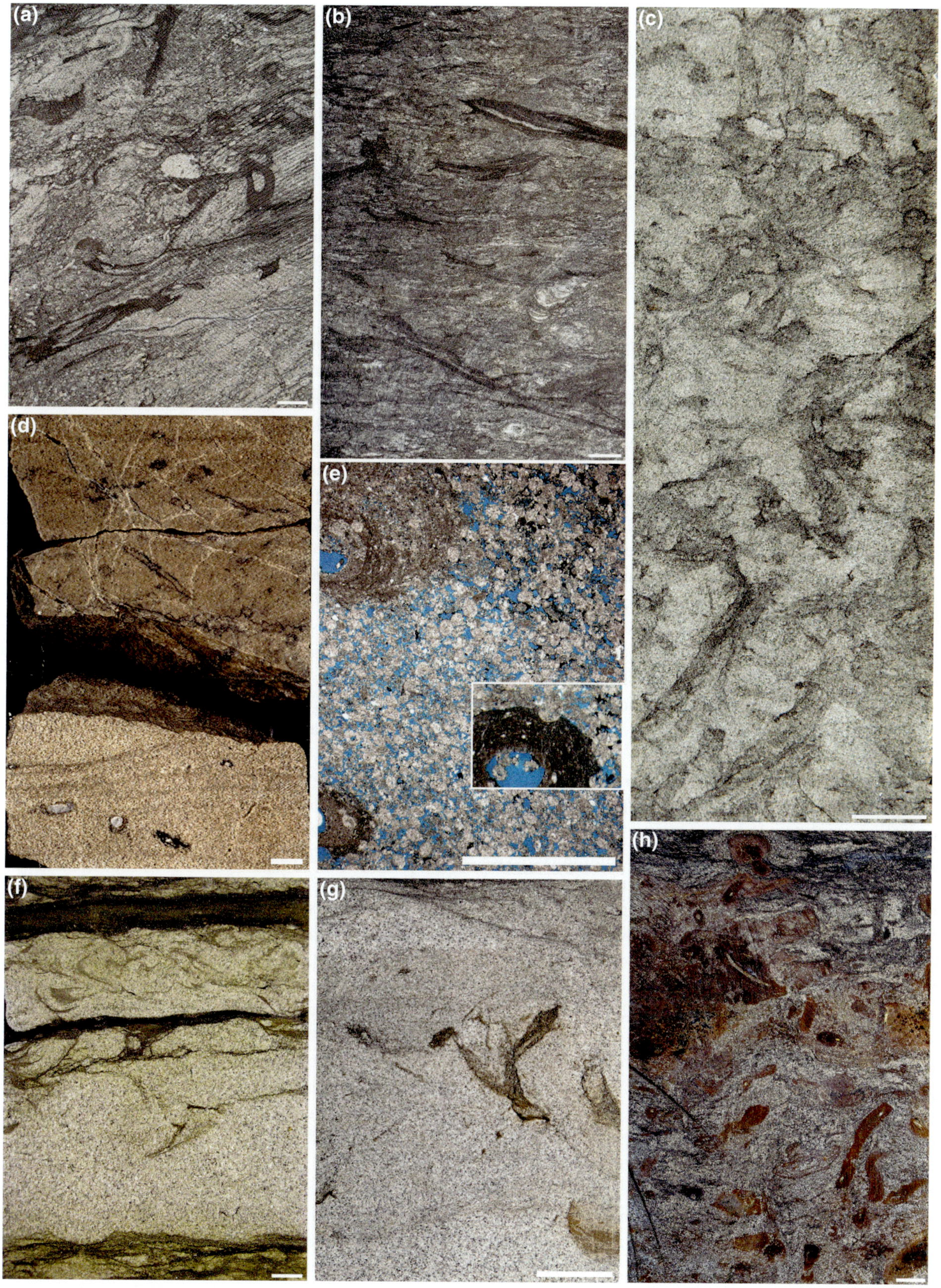
(a)
(b)
(c)
(d)
(e)
(f)
(g)
(h)

◀ **Fig. 5.50** *Cylindrichnus* in sectioned core. *Scale bars* = 1 cm. **a** Heterolithic sandstone with high degree of bioturbation, resulting in an ichnofabric composed of mud-lined *Cylindrichnus* overprinting *Chondrites*. Middle Jurassic (Bathonian-Oxfordian) Hugin Formation (restricted lower shoreface), Gina Krog Field, Norwegian North Sea (well 15/5-7, ca. 3908.5 m). **b** Silty sandstone with several bow-shaped burrows in longitudinal sections. Early Jurassic (Pliensbachian-Toarcian) Ror Formation (offshore), Lavrans Discovery, Norwegian Sea (well 6406/2-1, ca. 4853.1 m). **c** Ichnofabric with multiple burrows overlying each other. Paleocene Grumantbyen Formation (lower shoreface to offshore transition), Sysselmannbreen well, Svalbard (well BH 10-2008, ca. 825.5 m). **d** *Cylindrichnus* in cross-bedded oolitic limestone (grainstone and wackestone). Upper Permian Khuff Formation (storm-reworked sand shoals and sand waves, inner to outer carbonate ramp), South Pars Field, Persian Gulf, Iran (well SP9, ca. 3173.9 m). From Knaust (2009a), republished with permission of GulfPetroLink. **e** As in (**d**), thin section showing *Cylindrichnus* in cross section and significant mud lining. **f** Heterolithic sandstone with *Cylindrichnus* burrows (partly with funnel-shaped aperture) on several colonization surfaces. Upper Jurassic (Oxfordian) Heather Formation (shelf turbidites), Fram Field, Norwegian North Sea (well 35/11-11, ca. 2715.5 m). **g** Sandstone with funnel-shaped burrow aperture. Eocene Battfjellet Formation (deltaic), Sysselmannbreen well, Svalbard (well BH 10-2008). **h** Sandstone with numerous reworked parts of *Cylindrichnus*. These ichnoclasts became resistant due to early diagenetic cementation by siderite along an omission surface prior to reworking. Middle Jurassic (Bathonian-Oxfordian) Hugin Formation (shoreface), Sleipner Vest Field, Norwegian North Sea (well 15/9-5, ca. 3554 m)

In general, the bow-shaped bauplan of *Cylindrichnus* differs from the one in *Asterosoma* (with multiply branched or radiating horizontal parts), *Rosselia* (vertical and conical termination) and *Artichnus* (thickly lined, predominantly horizontal lumen). Subvertical parts of *Cylindrichnus*, displaying a funnel-shaped aperture, may lead to confusion with other vertical trace fossils such as *Laevicyclus* (formerly *Monocraterion* sensu lato; Stanley and Pickerill 1998; Knaust 2015a), while cylindrical parts resemble *Skolithos* (Frey and Howard 1985; Fig. 5.51), or have been mistaken as root traces (e.g. Frébourg et al. 2010; see Fig. 5.50d). *Cylindrichnus* can be associated with omission (hiatus) surfaces and may show intergradations with the firmground burrow *Glyphichnus* (Goldring et al. 2002). *Catenarichnus* Bradshaw, 2002 shares its overall bow-shaped morphology and passive fill with *Cylindrichnus* but only sometimes has a thin lining.

*Producers:* Polychaete worms (such as terebellids) are good candidates for producing *Cylindrichnus*-like burrows (Dashtgard et al. 2008; Belaústegui and de Gibert 2013; Fig. 5.52). In estuarine environments, maldanid polychaetes have been observed in *Cylindrichnus*-like tubes that are used for head-down mining as well as head-up suspension and interface deposit-feeding (MacEachern and Gingras 2007). Funnel-feeding holothurians (sea cucumbers) also produce bow-shaped burrows with a thick lining around a central lumen (e.g. Ayranci and Dashtgard 2013) similar to *Artichnus*.

*Ethology:* A suspension-feeding behavior of polychaetes living within the sediment (domichnion) is inferred for *Cylindrichnus*.

*Depositional Environment: Cylindrichnus* preferably occurs in shelf settings up to the lower shoreface with moderate- to low-energy regime, where more horizontal burrow components and forms slightly oblique to bedding occur (e.g. Fürsich 1974a). In addition, it is a characteristic trace fossil in high-energy deposits, including storm deposits, sand dunes and shoals, where vertical and steeply inclined forms predominate (e.g. Howard 1966; McCarthy 1979; Pemberton and Frey 1984; Frey and Howard 1985; Frey 1990; Olariu et al. 2012). *Cylindrichnus* is a common constituent of marginal-marine and estuarine environments with brackish conditions (Jurassic and younger; Netto and Rossetti 2003; MacEachern and Gingras 2007; Buatois and Mángano 2011; Gingras and MacEachern 2012) and occurs in association with delta-front and prodelta deposits (e.g. Tonkin 2012). Monoichnogeneric occurrences in such settings are a good indicator for stressed environments with reduced salinities. *Cylindrichnus* is occasionally reported from deep-marine deposits (e.g. Nilsen and Kerr 1978; Fig. 5.49a).

*Ichnofacies: Cylindrichnus* preferably occurs in the distal *Skolithos* Ichnofacies and is typical within the *Cruziana* Ichnofacies.

*Age: Cylindrichnus* is common in Mesozoic and Cenozoic deposits (Goldring 1996) but also occurs in Paleozoic strata (e.g. Głuszek 1998; Buatois et al. 2002). Orłowski (1989), Gámez Vintaned et al. (2006) and Desai et al. (2010) described *Cylindrichnus* from Early Cambrian deposits. Modern examples are given by Dashtgard et al. (2008).

*Reservoir Quality:* Suspension-feeding animals such as the producers of *Cylindrichnus* introduce a certain amount of mud when colonizing the top surface of sandy waves or shoals. In this way, dwelling burrows commonly reduce the sand/mud ratio within the affected interval. As a result, horizons with a considerable decrease in porosity and permeability can occur within highly permeable units and act as relatively thin baffles or barriers (Knaust 2009a). Those horizons with reduced porosity and permeability can either occur locally or follow flooding surfaces at the base of shallowing-upward cycles and then are widespread. In contrast to this, La Croix et al. (2013) noticed that vertical connections were generated by *Cylindrichnus* within fine-grained sandstone of a gas reservoir.

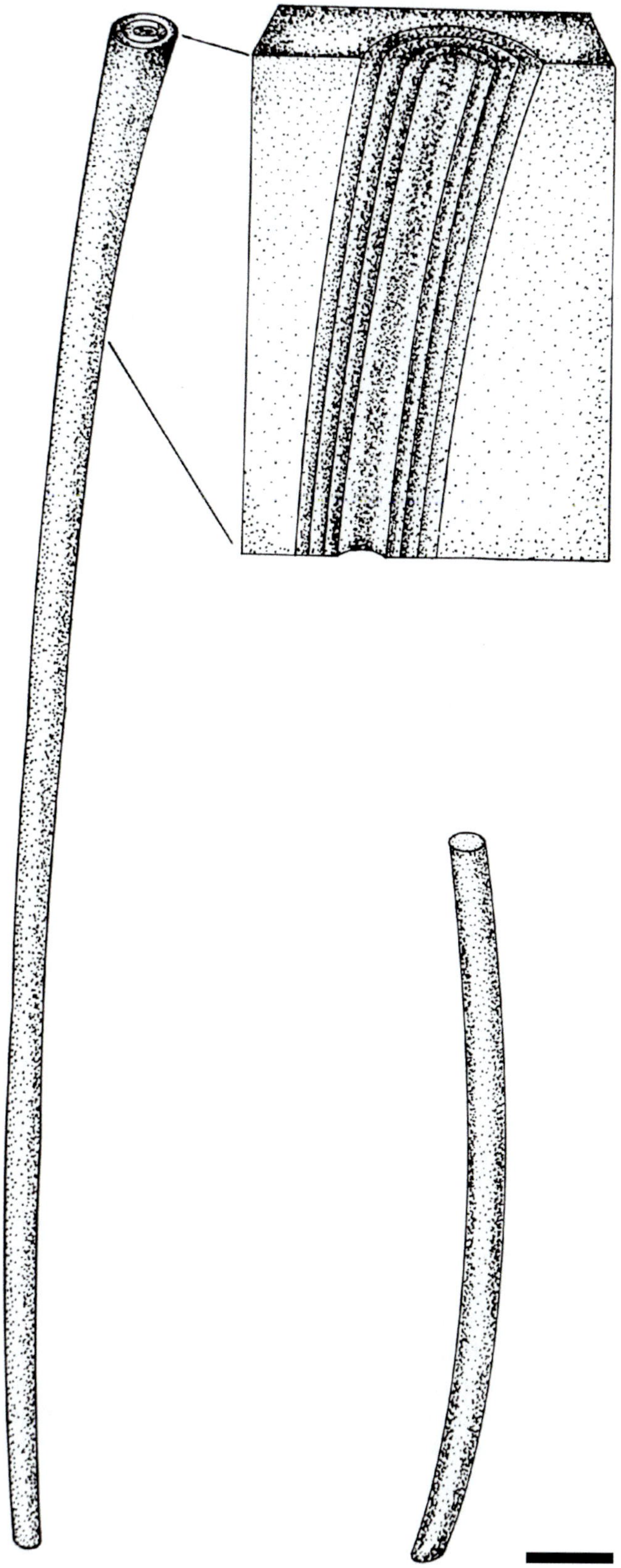

**Fig. 5.51** *Cylindrichnus concentricus* (*left*) and *Skolithos linearis* (*right*). *Scale bar* = 1 cm. After Howard and Frey (1984), republished with permission of Canadian Science Publishing; permission conveyed through Copyright Clearance Center, Inc.

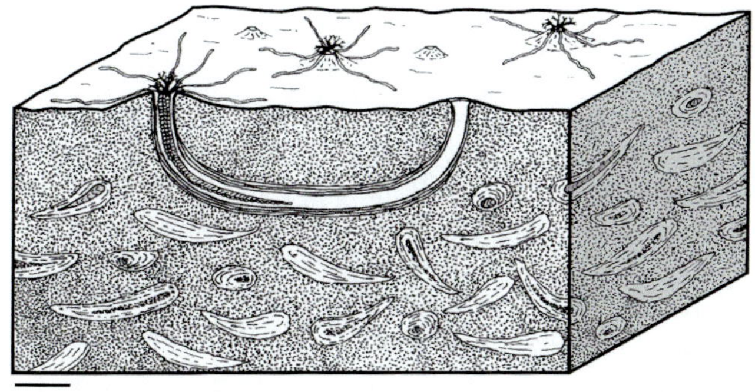

**Fig. 5.52** Diagram showing the reconstruction of *Cylindrichnus concentricus* as produced by a terebellid polychaete, and the ichnofabric resulting from its activity. *Scale bar* = 1 cm. Reproduced from Belaústegui et al. (2011). Copyright © 2011 Elsevier Masson SAS. All rights reserved

## 5.11 *Diplocraterion* Torell, 1870

*Morphology, Fill and Size:* The ichnogenus *Diplocraterion* consists of vertical U-shaped spreite burrows (Fürsich 1974b; Fig. 5.53). The spreite can be retrusive or protrusive, or both (e.g. Goldring 1962, 1964; Fig. 5.54). It ranges in size from a few millimeters in length to several decimeters (Fig. 5.55), the larger size group being more obvious in cores.

*Ichnotaxonomy:* Morphological differences in *Diplocraterion* have resulted in the erection of several ichnospecies, of which the type, *D. parallelum*, remains most important. In addition to the five ichnospecies distinguished by Fürsich (1974b), three ichnospecies have subsequently been introduced, resulting in a total of eight ichnospecies. Schlirf (2011) also includes burrows with an oblique orientation into *Diplocraterion*, a procedure which is not followed here (cf. Knaust 2013).

*Substrate: Diplocraterion* is produced in softgrounds and firmgrounds in siliciclastic and carbonate settings.

*Appearance in Core:* The spreite bounded by the U-shaped tube ("limbs") is conspicuous and relatively easy to recognize in core. It can be sectioned at various angles and thus produces different projections (Fig. 5.56; see also Chakraborty and Bhattacharya 2013; Fig. 5.57). Longitudinal sections along the plane of the spreite burrow produce the most complete expression, while oblique sections across the spreite are more common and result in a thin but elongate spreite combined with the passively filled tube. Depending on the protrusive or retrusive character of the burrow, the tube may be located either at the base of the spreite or higher up within it. Oblique sections are also common and show a "feathering-out" of the spreite. The marginal tube can be filled with sand or mud, in the latter case often containing high content of organic material or, more rarely, plant debris. The tube can be also laterally enlarged and display features

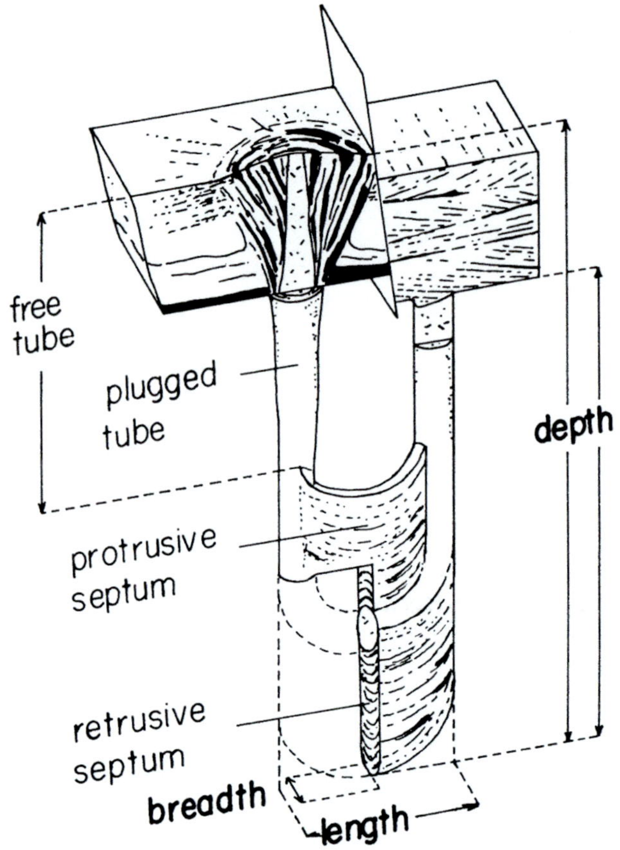


**Fig. 5.53** Idealized sketch of *Diplocraterion* with its elements, in which the spreite (=septum) between the tube results from the protrusive and retrusive movement of the tracemaker. One free tube is shown opening to a normal aperture; the other tube is shown as having been plugged before erosion and sedimentation took place. From Goldring (1962), republished with permission of Springer

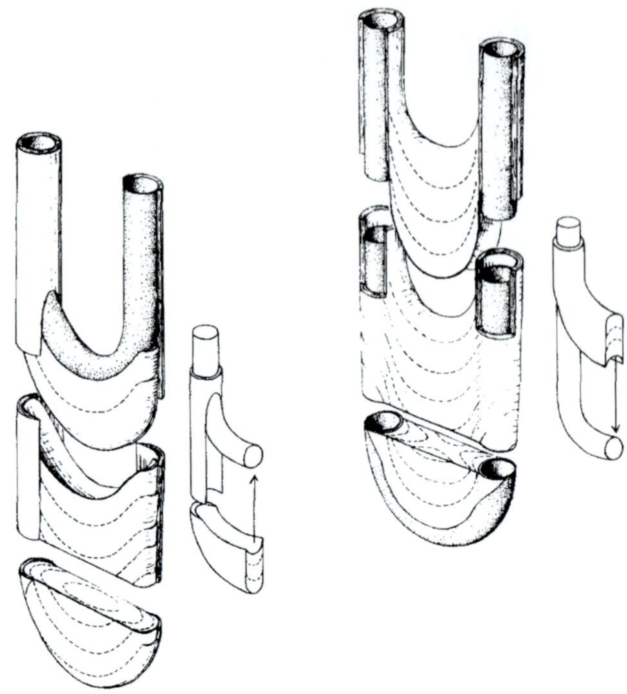

**Fig. 5.54** Retrusive (*left*) and protrusive (*right*) *Diplocraterion*. From Seilacher (1967), republished with permission of Elsevier; permission conveyed through Copyright Clearance Center, Inc.

of lateral accretion, resulting from the adjustment of the producer in response to loose and shifting substrate. The spreite may incorporate granulae and pebbles and may show diagenetic modifications such as sideritic cementation. A conspicuous feature of some burrows is the concentration of millimetric elliptical mud pellets (fecal pellets, *Coprulus oblongus*) within the spreite.

*Similar Trace Fossils: Diplocraterion* is part of the ichnofamily Rhizocoralliidae and shows close affinity to the ichnogenus *Rhizocorallium*, which includes horizontal to oblique spreite burrows (Schlirf 2011; Knaust 2013). Both ichnogenera may partly originate from the same kind of producer. Incomplete core sections of *Diplocraterion* may be confused with the spreite burrow *Teichichnus* (Fig. 5.58). *Catenichnus* McCarthy, 1979 is a bow-shaped spreite burrow similar to *Diplocraterion* and *Teichichnus*. Some specimens of the slender and very narrow U-shaped burrow *D. habichi* resemble *Tisoa siphonalis* and may better be attributed to it, particularly when lacking a spreite. Collapsed U-burrows (e.g. *Arenicolites*) can resemble *Diplocraterion*, the collapsed structure mimicking a spreite. Large *Ophiomorpha* shafts with a thick mud lining and active fill can actually look alike small *Diplocraterion* (Fig. 5.59).

*Producers:* Two main groups are generally regarded as producers of *Diplocraterion*: crustaceans and polychaete worms. The amphipod *Corophium*, for instance, is known to produce incipient *Diplocraterion* (e.g. Dashtgard and Gingras 2012), although fossorial crustaceans seem to be absent in the Paleozoic (e.g. Carmona et al. 2004) and thus alternative interpretations become necessary. Boxcores taken in firm substrate of the Baltic Sea have revealed oblique spreite burrows akin to *Diplocraterion* and *Rhizocorallium* (Winn 2006; Knaust 2013), where they were produced by

**Fig. 5.55** *Diplocraterion parallelum* in outcrop. *Scale bars* = 1 cm. **a**, **b** Sandstone with a dense *D. parallelum* assemblage seen as slit-like traces on the bedding plane (**a**) and as vertical spreite burrow in vertical section (**b**). Lower Cambrian Hardeberga Formation (shallow marine), Snogebæk, Bornholm, Denmark. See Clausen and Vilhjálmsson (1986). **c** Vertical spreite burrow. Lower Cambrian Hardeberga Formation (shallow marine), Due Odde, Bornholm, Denmark. **d** Sandstone bedding surface with slit-like burrow apertures. Lower Cambrian erratic boulder, northeastern Germany. Coll. Natural History Museum Berlin. **e**, **f** Vertical spreite burrows in massive (micritic) limestone. Triassic (Rhaetian)/Jurassic (Hettangian) boundary, Western Bergamasc Alps, Lombardy, Italy

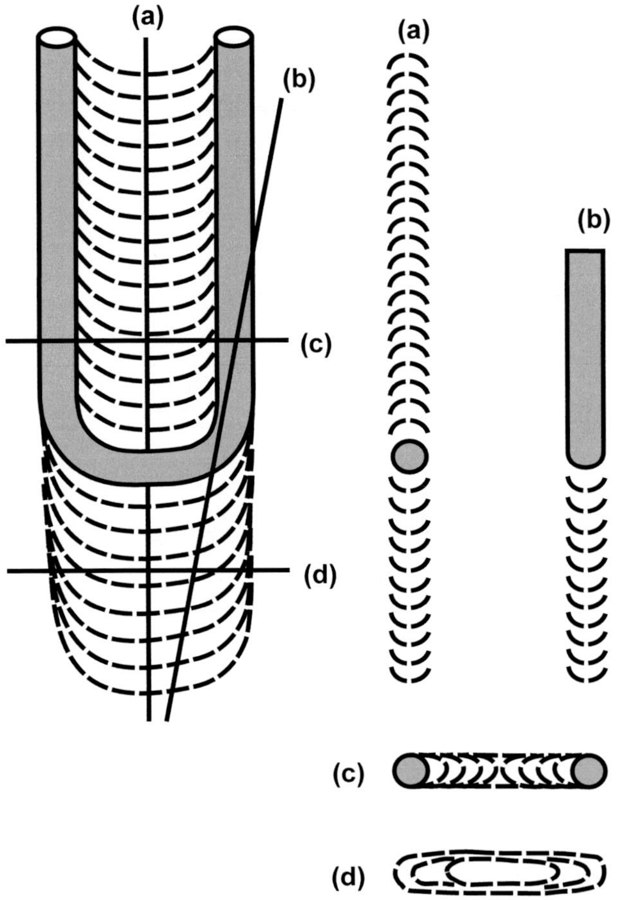


**Fig. 5.56** Schematic diagram showing different sections through *Diplocraterion parallelum* with retrusive (*lower part*) and protrusive (*upper part*) spreite burrows as they would appear in sectioned core. **a** Longitudinal section perpendicular to burrow plane. **b** Oblique section through lower part of marginal tube and protrusive spreite part. **c** Cross section through retrusive spreite burrow. **d** Cross section through protrusive spreite burrow

terebellid polychaetes. The polychaete *Polydora* is another candidate for producing *Diplocraterion* in firmgrounds (Seilacher 1967; Gingras et al. 2001, 2012a; Fig. 5.60). The fecal pellets *Coprulus oblongus* are similar to those produced on modern tidal flats by the polychaete *Heteromastus*, which makes polychaetes likely producers for *Diplocraterion*. More subordinately, holothurians (sea cucumbers) and echiurans are known to produce U-shaped burrows with minimal spreite (Bromley 1996; Dashtgard and Gingras 2012; Smilek and Hembree 2012) similar to *Diplocraterion* and *Teichichnus*.

*Ethology:* Suspension-feeding behavior within their dwelling (domichnion) can be assumed for the producers of most *Diplocraterion* (e.g. Goldring 1962; Fürsich 1974b), although deposit-feeding has been also considered (e.g. Leaman and McIlroy, 2016).

*Depositional Environment:* Ichnospecies of *Diplocraterion* have been commonly recorded in marginal-marine environments such as tidal settings and estuaries (e.g. bars and sand flats; Buatois and Mángano 2011; Gingras et al. 2012a; Desjardins et al. 2012; Higgs and Higgs 2015). This implies that their producers are able to tolerate reduction in salinity (brackish conditions) and are often subject to high-energy sedimentation with repeated erosion and deposition (e.g. storm deposition; Figs. 5.61 and 5.62). Measurements of high-density occurrences of *Diplocraterion* have shown that the burrow planes may show a preferred orientation along the bedding plane as a reaction to waves and currents (e.g. Clausen and Vilhjálmsson 1986; Buckman 1992; Gaillard and Racheboeuf 2006; Rodríguez-Tovar and Pérez-Valera 2013). Brackish conditions with fluctuating salinity are also common in the glaciomarine and deltaic settings from which *Diplocraterion* has been recorded (Netto et al. 2012; Tonkin 2012), as well as hyperpycnal flow deposits (Buatois et al. 2011). *Diplocraterion* may also occur in shoreface and offshore environments, where it is often related to transgressive surfaces (ravinement surfaces, omission surfaces) due to the action of waves or tides (e.g. Mason and Christie 1986; Dam 1990; Taylor and Gawthorpe 1993; Goldring et al. 1998; Rodríguez-Tovar et al. 2007). Such surfaces can be utilized for stratigraphical correlation and serve as markers (e.g. sequence boundaries; see Olóriz and Rodríguez-Tovar 2000). Monoichnospecific *Diplocraterion* produced in laminated rocks may indicate oxygen-deficient sediment (Leszczynski et al. 1996). Furthermore, *Diplocraterion* is one of those ichnotaxa that first appear after mass extinctions such as at the end Permian (e.g. Knaust 2010b; Chen et al. 2011). Isolated occurrences of *Diplocraterion* in continental (e.g. Kim and Paik 1997; Xing et al. 2016) and deep-marine deposits (e.g. Crimes et al. 1981) are sporadically reported, although Martin et al. (2016) have shown that shallow-marine *Diplocraterion* may occur within continental deposits. Nevertheless, mayflies are known to produce spreite-like burrows into continental firmgrounds (see Knaust 2013), which could resemble *Diplocraterion*.

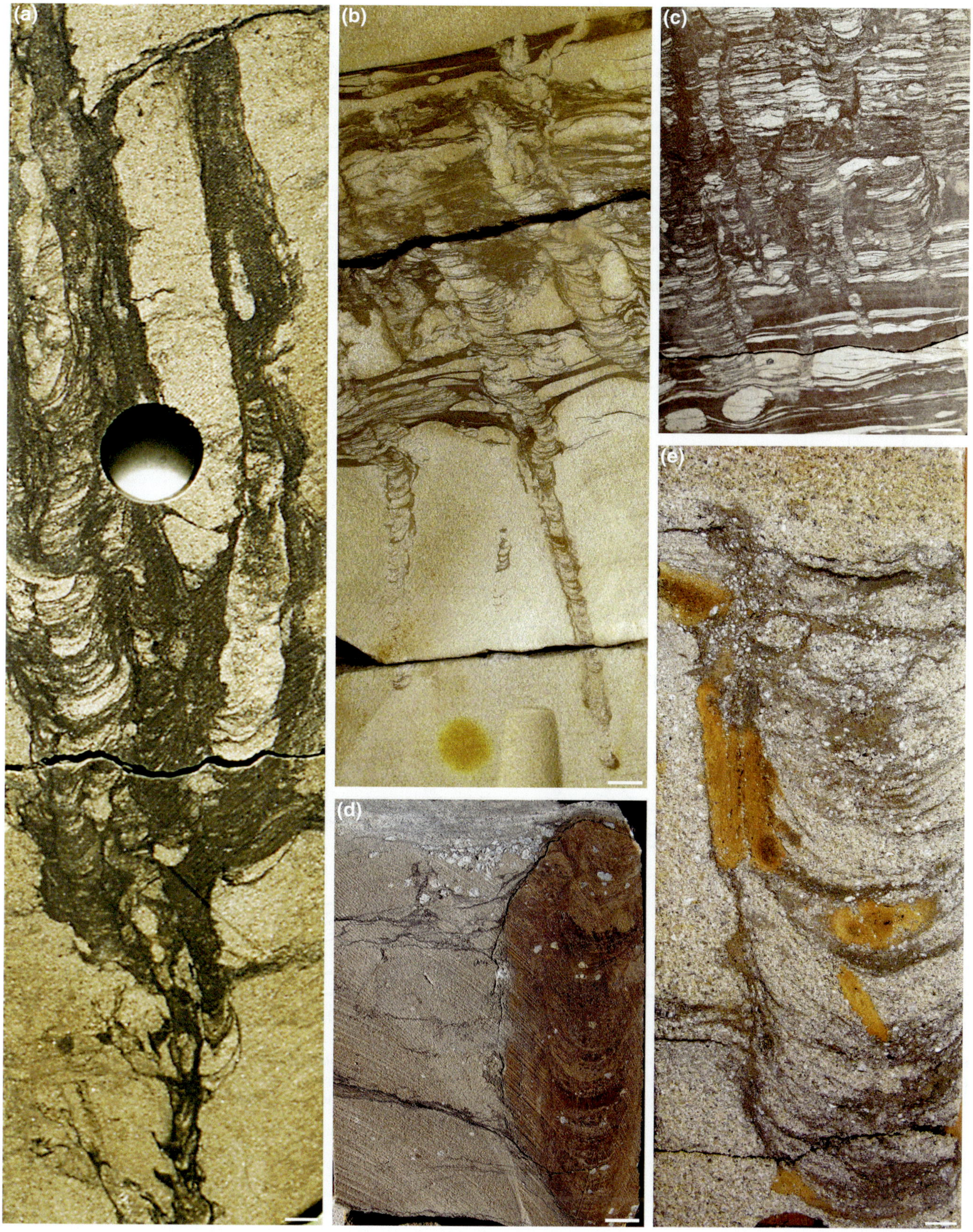
(a)
(b)
(c)
(d)
(e)

◀**Fig. 5.57** *Diplocraterion parallelum* in sectioned core. *Scale bars* = 1 cm. **a** Highly bioturbated cross-bedded sandstone with large mud- and sand-filled burrows (partly retrusive) sectioned roughly longitudinal to the burrow plane. Lower Jurassic (Sinemurian-Pliensbachian) Tilje Formation (nearshore, tidal-influenced), Skarv Field, Norwegian North Sea (well 6507/5-1, ca. 3580.8 m). **b** Heterolithic sandstone with wavy bedding and large *D. parallelum* (retrusive) sectioned mostly oblique to the burrow plane. Lower Jurassic (Pliensbachian) Tilje Formation (mixed tidal flat), Njord Field area, Norwegian North Sea (well 6407/10-1, ca. 2990.85 m). **c** Heterolithic, ripple-laminated sandstone with a *D. parallelum* ichnofabric. The spreite burrows (partly retrusive) were sectioned at various angles (transverse, oblique and longitudinal to the burrow plane). Lower Jurassic (Sinemurian-Pliensbachian) Tilje Formation (mixed tidal flat), Njord Field area, Norwegian North Sea (well 6407/10-1, ca. 2995.1 m). **d** Heterolithic sandstone with a ravinement surface, from which a muddy spreite burrow with a thick U-shaped tube penetrates the underlying sediment. The spreite contains millimetric elliptical fecal pellets (*Coprulus oblongus*) and crystalline granulae incorporated from the ravinement (omission) surface. Preferred cementation by iron-bearing minerals (siderite?) has led to differential compaction between burrow and surrounding sediment. Middle Jurassic (Bajocian) Hugin Formation (transition between sandy tidal flat below the ravinement surface and deeper-marine environment above it), Sleipner Field, Norwegian North Sea (well 15/9-1, 3660.35 m). **e** Medium-grained sandstone with *D. parallelum* developed below an erosion surface. The fill of the spreite burrow is heterolithic (sand and mud) and includes sideritic mud clasts (ocher color). Middle Jurassic (Callovian) Fensfjord Formation (upper shoreface), Gjøa Field, Norwegian North Sea (well 36/7-1, ca. 2341.8 m)

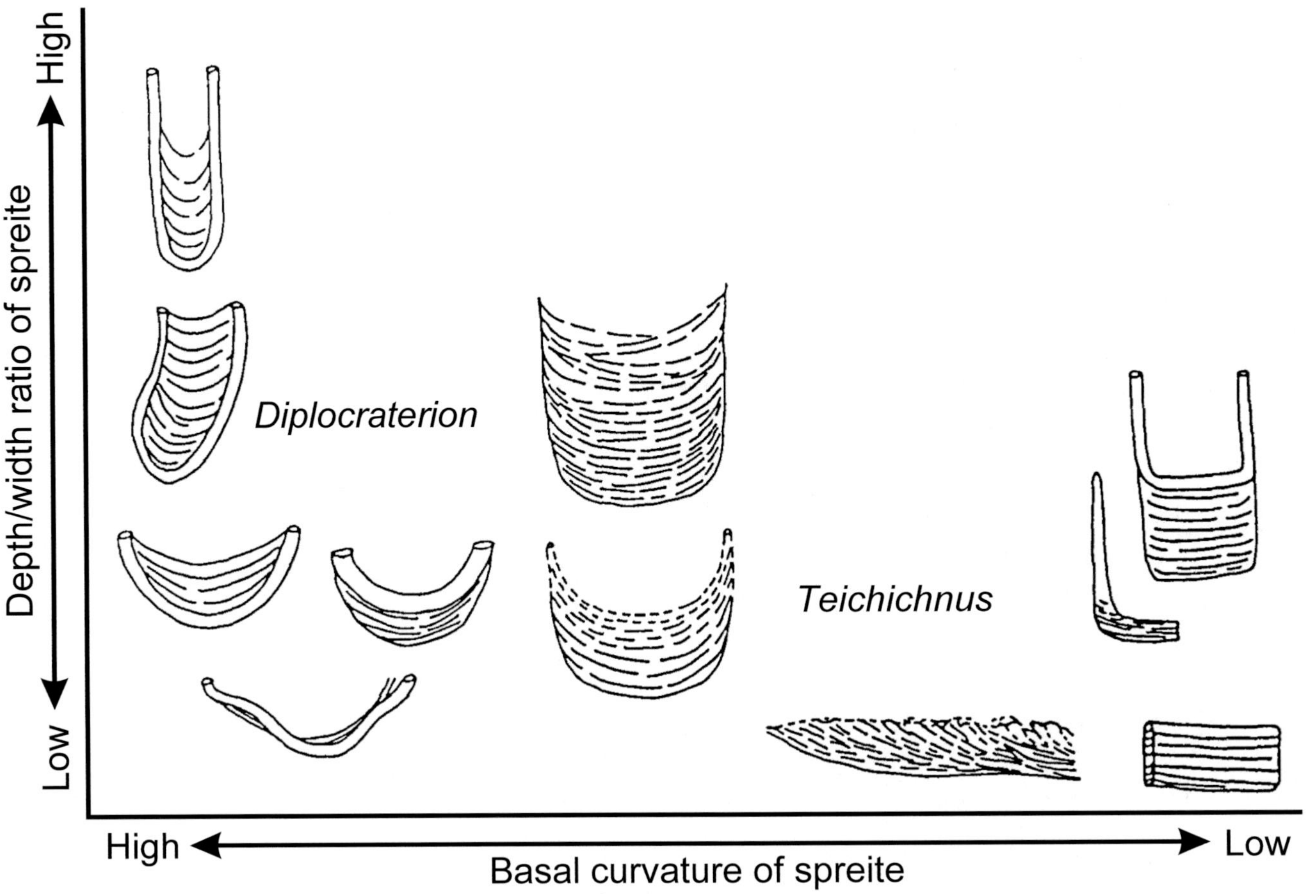


**Fig. 5.58** Various forms of vertical spreite trace fossils showing the transition field between *Diplocraterion* and *Teichichnus*. The forms are arranged roughly according to the depth/width ratio and basal curvature of the spreite. Variable scale. Different sources, modified after Corner and Fjalstad (1993), reprinted by permission of the publisher (Taylor & Francis Ltd., http://www.tandfonline.com)

**Fig. 5.59** Dense *Ophiomorpha* ichnofabric in Upper Cretaceous (Maastrichtian) Springar Formation (deep marine, turbidites) from the Gro Discovery (well 6604/10-1, ca. 3545.5 m). Some of the thickly lined and actively filled vertical shafts (e.g. *upper left*) resemble *Diplocraterion* and may be confused with it. *Scale bar* = 1 cm

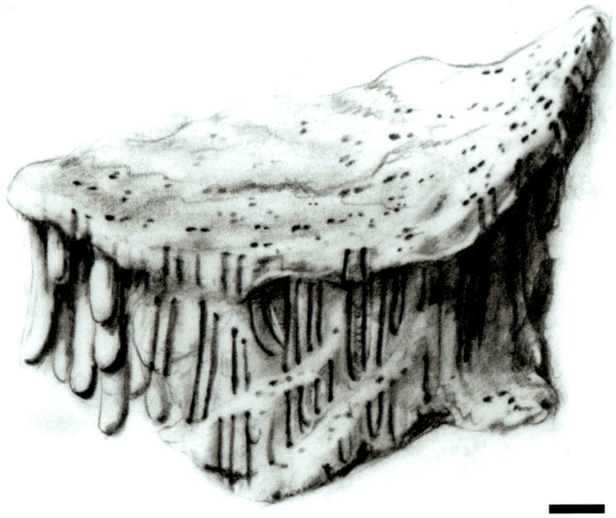

**Fig. 5.60** Diagram illustrating a *Polydora* (polychaete worm) association and the resulting *Diplocraterion*-like traces from modern firmground at Willapa Bay, Washington. *Scale bar* = ca. 1 cm. Illustration by Tom Saunders in Gingras et al. (2001), republished with permission of Elsevier; permission conveyed through Copyright Clearance Center, Inc.

*Ichnofacies: Diplocraterion* is a major constituent of the *Skolithos* Ichnofacies (softgrounds), but also occurs within the substrate-controlled *Glossifungites* Ichnofacies (firmgrounds).

*Age: Diplocraterion* is a long-ranging trace fossil known from the Cambrian (e.g. Cornish 1986; Bromley and Hanken 1991) to Holocene (e.g. Corner and Fjalstad 1993; Dashtgard and Gingras 2012).

*Reservoir Quality: Diplocraterion* often occurs at a relatively large size and high concentration along omission surfaces (e.g. ravinement surfaces, sequence boundaries) and therefore may have an influence on the producibility of a reservoir. For instance, high-density occurrences of *Diplocraterion* with a passive sand fill may promote vertical connectivity, while the incorporation of mud may act as a baffle. Improved reservoir quality due to sediment cleaning and the occurrence of highly permeable vertical conduits has been documented by Leaman and McIlroy (2016).

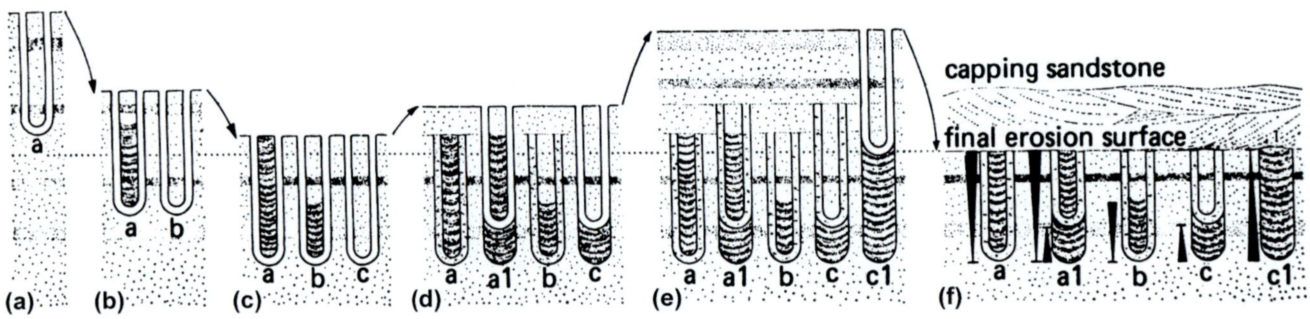


**Fig. 5.61** Movement pattern of *Diplocraterion parallelum* in response to the amount of sedimentation or erosion as indicated by the adjustment to depth and mode of preservation. Upper Devonian Baggy Beds, England, UK. After Goldring (1964), republished with permission of Elsevier; permission conveyed through Copyright Clearance Center, Inc. Heights of solid arrows show amount of sedimentation or erosion. *D. parallelum* occurs in the various types shown in (**f**), where all have been truncated to a common erosion surface. It is considered that repeated phases of erosion and sedimentation led to the development of the various types. Stage (**a**), development of burrow (*1*). With degradation of the surface, this tube migrates downwards, and at intervals, new tubes (*2* and *3*) are constructed (**e** and **c**). Sedimentation follows (**d** and **e**) but some of the tubes are abandoned. Stage (**f**), all tubes are abandoned and erosion reduces them to a common base

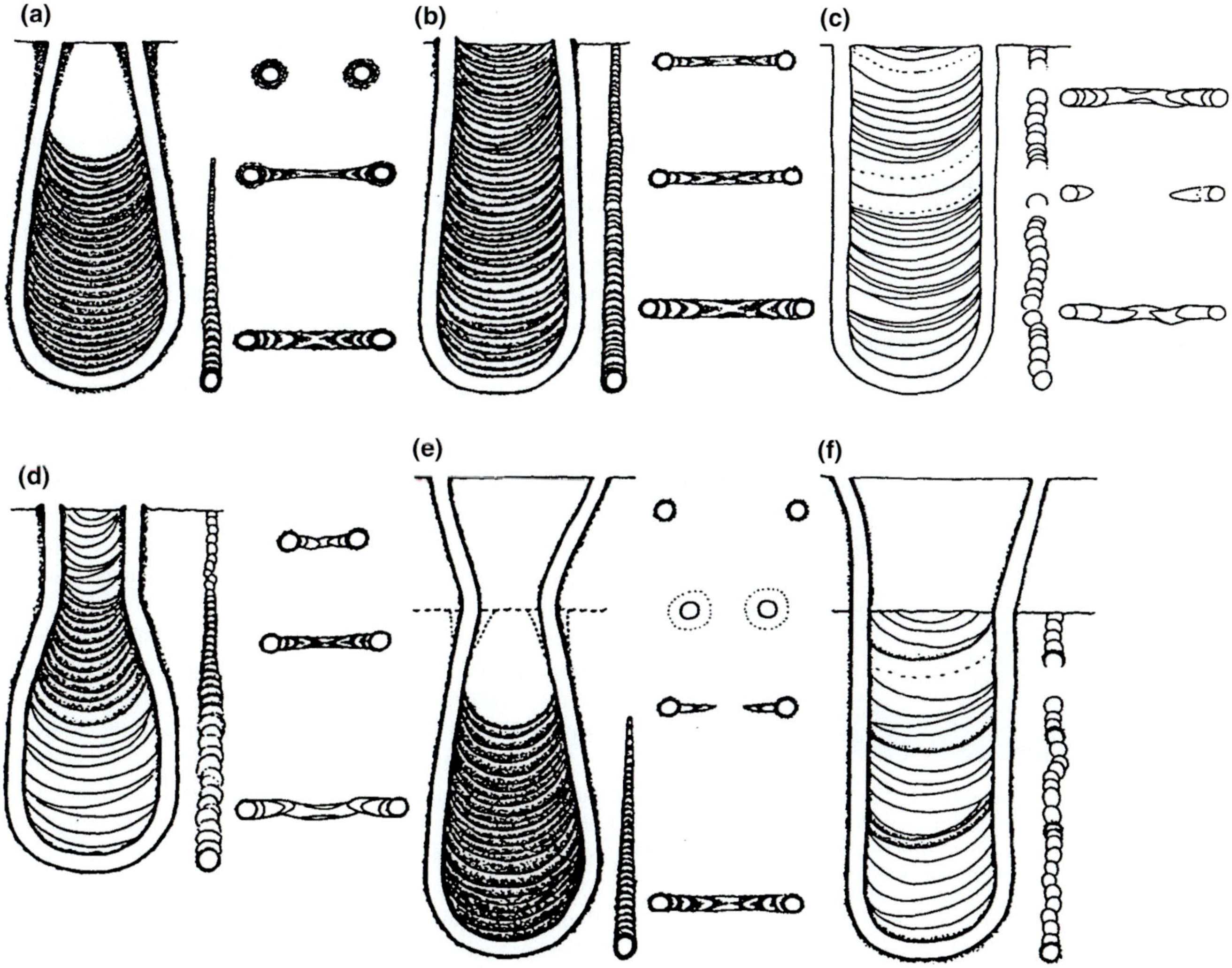


**Fig. 5.62** Different models of *Diplocraterion parallelum*, lacking the retrusive vector. These represent variation in three factors: ontogenetic growth, rate of deposition or erosion, and amount of deposition or erosion. **a** Stable sedimentary surface. **b** Slow, steady erosion. **c** Rapid erosion. **d** Erosion—stability—erosion. **e** Stability followed by deposition. **f** Rapid erosion followed by deposition. After Bromley and Hanken (1991), reprinted by permission of the publisher (Taylor & Francis Ltd., http://www.tandfonline.com)

## 5.12 *Hillichnus* Bromley et al., 2003

*Morphology, Fill and Size: Hillichnus* is a highly complex trace fossil comprising various elements occurring at different levels, which could be assigned to different ichnotaxa if occurring in isolation. A basal axial tube complex is accompanied on either side by feather- and spreite-like structures. The upper part contains an array of upward-curving, linear to undulating tubes that can be mantled (Bromley et al. 2003; Fig. 5.63). The entire trace fossil has a rather large size, ranging from 10 to 20 cm in width and vertical extension. This disparity in diameter within the same trace fossil is one of its most diagnostic features.

*Ichnotaxonomy:* Beside the type ichnospecies *H. lobosensis, H. agrioensis* was introduced by Pazos and Fernández (2010) for more regular forms with lateral and vertical spreite-like components. Due to the occurrence of rising tubes, *Hillichnus* can be regarded as part of the ichnofamily Siphonichnidae (Knaust 2015a; Fig. 5.141).

*Substrate: Hillichnus* preferably occurs in thin-bedded sandstone and mudstone facies, or ripple-laminated sandstone with a slightly heterogeneous appearance due to the admixture of clay material and organic matter.

*Appearance in Core:* The complex nature of *Hillichnus* makes this trace fossil difficult to recognize, particularly in core. The individual elements of this trace fossils resemble

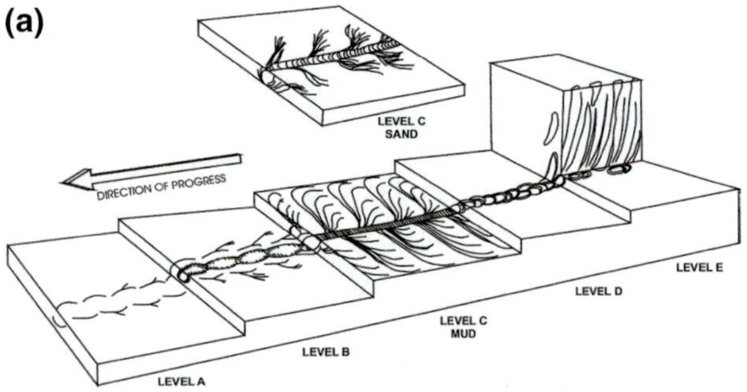

**Fig. 5.63** *Hillichnus* architecture. **a** The structure of *H. lobosensis* dissected as a five-level model. Structures show a range of variation, in particular those at level C. From Bromley et al. (2003), republished with permission of Elsevier; permission conveyed through Copyright Clearance Center, Inc. **b** Sandstone bedding plane with numerous burrow systems aff. *Hillichnus* isp. from the Barremian to Aptian Almargem Formation (fluvial to marginal marine?) at Quinta do Grajal, Belas, Portugal. The twig-like, palmate branching pattern is accompanied by numerous short ascending tubes which represent the probing traces of the siphon. The dense occurrence of those probing traces led to the interpretation of this structure as algae or meniscate backfill of a different ichnotaxon and recently described as *Cladichnus lusitanicum* (Neto de Carvalho et al. 2016). Original specimen in the Geological Museum of Lisbon, Portugal. *Scale bar* = 5 cm

different ichnotaxa and could be assigned to those if found in isolation. Thus, the recognition of *Hillichnus* depends on the combined occurrence of various diagnostic elements (Fig. 5.64). In addition, association with other bivalve-produced trace fossils facilitates the identification of *Hillichnus* in core. Following the terminology proposed by Bromley et al. (2003; Fig. 5.63a), the basal segmented structure "appears as a series of discontinuous arcs arranged in line so as to suggest a chain of barrels". Slightly higher up in the section, the basal tube is a thinly mud-lined and segmented structure. Lateral to this basal tube, alternating structures "are now seen to expand in sand into distinct bunches of curved lamellae, black, somewhat resembling spreiten" (referred to as lateral lamellae). "Within mud laminae, however, these structures become very extended, feather-like and curving and become more slender distally", called lateral tubules. In the upper part, rising tubes (inclined, arcuate) are developed, typically having a distinct mud lining.

*Similar Trace Fossils:* Due to the nature of core material, only a small and perhaps unrepresentative part of *Hillichnus* can be ordinarily be exposed, which in turn may result in its identification as different trace fossils. Lateral lamellae of the basal part can be clustered and thus be reminiscent of *Asterosoma*, although no passively filled central tube is present. The curved lamellae and lateral tubules (feathered serpent structures) seem to be the trickiest part of *Hillichnus*, because they may be easily confused with *Lophoctenium* or crowded *Palaeophycus*. Individual rising tubes, although inclined and arcuate, may resemble *Skolithos*. In addition, funnel-shaped construction of rising tubes probably results from stationary feeding and conforms to *Parahaentzschelinia*. Finally, spreite structures are common elements of both, *H. lobosensis* and *H. agrioensis*, and may result in *Teichichnus*-like burrows.

*Producers:* Functional interpretation of individual burrow parts and comparison with modern analogs allowed Bromley et al. (2003) to interpret *Hillichnus* as the product of deposit-feeding bivalves (Paleotaxodonta or Protobranchia).

*Ethology:* A subsurface deposit-feeding (fodinichnial) mode of life of bivalves (e.g. tellinacean) is inferred for *Hillichnus*. The bivalve moved through the sediment and utilized its palpal tentacles to exploit adjacent sediment for food (Figs. 5.65 and 5.66). Siphonal excursions to the sea floor result in the preservation of an array of upward-curving tubes. Association with pyrite may indicate chemosymbiosis with the farming of sulfide-oxidizing bacteria (Bromley et al. 2003).

*Depositional Environment: H. lobosensis* was originally described from deep-marine inner fan environments of submarine canyons, where it occurs in overbank/levee deposits (Bromley et al. 2003). Pazos and Fernández (2010) described *H. agrioensis* from marginal-marine deposits with tidal influence. Similar marginal-marine deposits with sandy tidal flats can also be assumed for the above presented data from core as well as for the example from Portugal (Fig. 5.63b). Given the wide distribution of its producers, *Hillichnus* should be present with a very broad distribution.

*Ichnofacies:* The few records of *Hillichnus* belong to the deep-marine *Nereites* Ichnofacies as well as the marginal-marine *Skolithos* and *Cruziana* ichnofacies.

*Age:* Only recently, a decent understanding of this complex trace fossil has been achieved, which explains its poor record so far, including Early Jurassic (Ekdale et al. 2012), Early Cretaceous (Pazos and Fernández 2010; Neto de Carvalho et al. 2016) and Paleocene (Bromley et al. 2003) examples. Further research will probably extend the stratigraphical range of *Hillichnus*.

**Fig. 5.64** Various burrow elements potentially belonging to *Hillichnus lobosensis* in sectioned core. *Scale bars* = 1 cm. **a** Clusters of wrinkled structures composed of dark lining material, which are interpreted as lateral lamellae (resembling *Asterosoma*, *arrow heads*) and associated with tube sections, which are interpreted as parts of the basal tube (*t*). Lower Jurassic (Toarcian) Stø Formation (lower shoreface), Snøhvit Field, Norwegian Barents Sea (well 7120/8-2, ca. 2109.8 m). **b** Feathered serpent structures, which consist of clustered sand-filled tubes surrounded by a thick mud lining (resembling *Palaeophycus* or *Lophoctenium*). Lower Jurassic (Pliensbachian) Tilje Formation (marginal marine), Norwegian Sea (well 6607/12-3, ca. 4222.85 m)

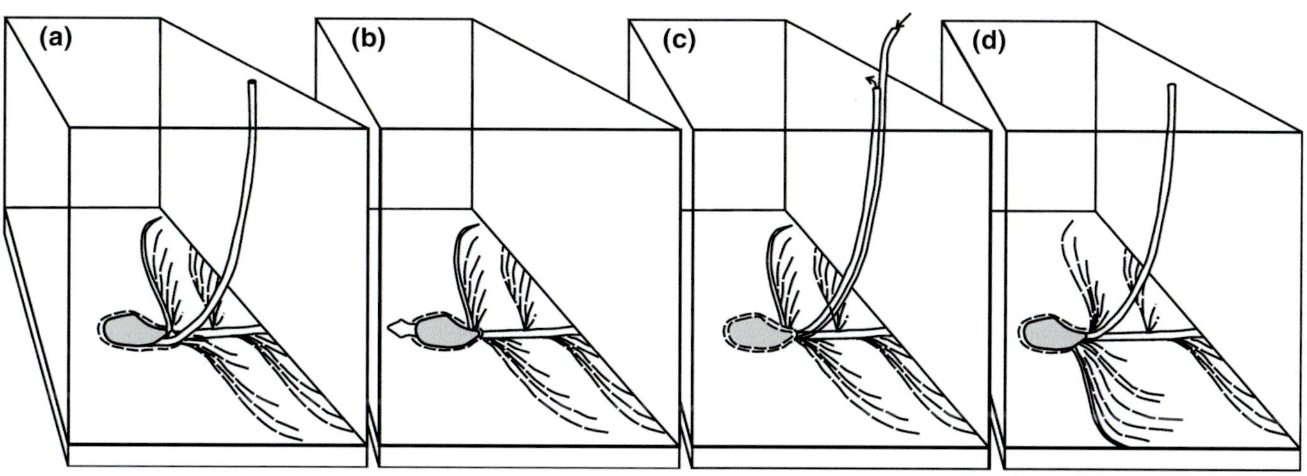


**Fig. 5.65** Successive activities of a deposit-feeding tellinacean bivalve that could have produced the trace fossil *Hillichnus lobosensis*. From Bromley et al. (2003), republished with permission of Elsevier; permission conveyed through Copyright Clearance Center, Inc. **a** The inhalant siphon probes successively at a horizon of exploitation. **b** Having withdrawn its siphons, the bivalve digs a short distance forwards. **c** The siphons extend along a new path to the surface for respiration. **d** A new bunch of probes is undertaken by the inhalant siphon at the feeding level

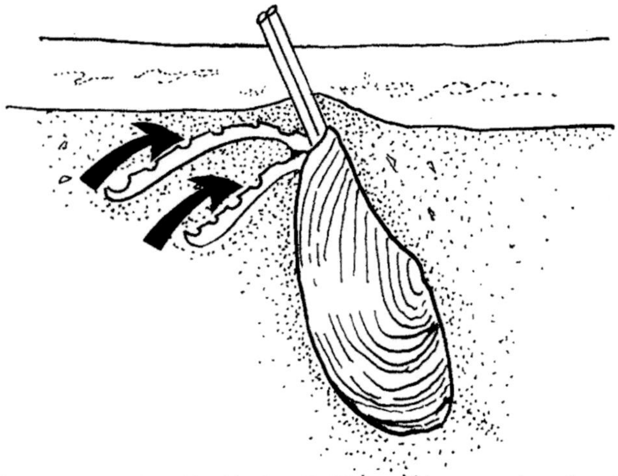

**Fig. 5.66** The benthic bivalve *Yoldia* sp., demonstrating deposit-feeding by means of palp proboscides. From Ward and Shumway (2004), republished with permission of Elsevier; permission conveyed through Copyright Clearance Center, Inc.

*Reservoir Quality:* Overall, a slight reduction of reservoir properties can be assumed because of intermingling of mud-rich sediment into sandy facies by deposit-feeding of the *Hillichnus* producer.

## 5.13 *Lingulichnus* Hakes, 1976

*Morphology, Fill and Size: Lingulichnus* consists of vertically or obliquely oriented burrows having a straight, sinuous or J- to U-shaped morphology (Zonneveld and Pemberton 2003; Fig. 5.67). Cross sections vary with respect to their position within the burrow and range from elliptical to subcircular. Portions of the burrows are filled with concentric laminae with a spreite-like appearance, which may be penetrated by a passively filled central lumen. The size of *Lingulichnus* is typically in the range of a few centimeters with burrow diameters around 1 cm.

*Ichnotaxonomy:* Depending on their shape and orientation within the substrate, three ichnospecies are differentiated (Zonneveld and Pemberton 2003): *L. verticalis* (straight, vertical; Fig. 5.68), *L. inclinatus* (straight, inclined), and *L. hamatus* (J- and U-shaped, vertical).

*Substrate:* The distribution of lingulide brachiopods (the producers of *Lingulichnus*) is controlled by the grain size of the substrate (Zonneveld et al. 2007). Accordingly, the abundance of *Lingulichnus* continuously decreases with increasing mud content, and is lacking in muddy substrates. *Lingulichnus* preferably occurs in very fine- to medium-grained sandstone but has been reported rarely from arenitic limestone with initial firmground conditions (Knaust et al. 2012; Fig. 5.68).

*Appearance in Core:* The overall morphology of the burrows (vertical, inclined, J- or U-shaped) gives a first indication on *Lingulichnus* in core (Fig. 5.69). The pedicle trace is a relatively small (few millimeters), elongate burrow with circular cross section and passive fill (Fig. 5.69a). In *L. verticalis*, the pedicle trace is commonly located in the lower part of the burrow, but often it cross-cuts the upper, laminated portion of the burrow. The valve trace is larger and may have an extended spade shape (Fig. 5.69b). It may be passively filled or show internal lamination (active fill) that can be extended as a long funnel (Fig. 5.69c, d). Proximal parts of *Lingulichnus* are typically elliptical in cross section (Fig. 5.69e–g).

*Similar Trace Fossils: Siphonichnus* is probably the most similar trace fossil to *Lingulichnus*. It differs from *Lingulichnus* by its circular instead of elliptical cross section, and the internal lamination is penetrated by the passively filled core in a more consistent manner. The internal trace (core) is generally wider than the pedicle trace in *Lingulichnus*, and only preserves a relatively small portion of the outer laminated burrow. Some of the adjusted *Lingulichnus* resemble *Rosselia* but are more irregularly laminated and lack the passively filled terminal burrow in their upper part. They could also be mistaken as *Conichnus*, which may occur as stacked funnel-shaped burrows. However, *Conichnus* does not have a passively filled burrow continuing to the apex of the funnel-shaped burrow. In horizontal section, *Lingulichnus* can resemble *Lockeia*, which differs by its almond-shaped cross section and the lack of a pedicle trace (Rindsberg 1994). Finally, the pedicle trace could be mistaken as a root trace or as *Skolithos*, if occurring in isolation (e.g. due to erosion).

*Producers:* On the basis of modern analogs (Emig et al. 1978) and direct evidence in fossil *Lingulichnus* (Zonneveld et al. 2007), lingulide brachiopods (e.g. *Lingula* and *Glottidia*) can be assumed as the producers of this trace fossil (Figs. 5.69 and 5.70).

*Ethology: Lingulichnus* mainly represents a dwelling trace (domichnion) of lingulide brachiopods, which have a suspension-feeding behavior. The tracemaker is able to cope with increased sedimentation rates and adjusts its burrow accordingly (equilibrichnion; see Zonneveld et al. 2007).

*Depositional Environment: Lingulichnus* occurs in many marine environments but commonly dominates in shallow- and marginal-marine successions (Zonneveld et al. 2007).

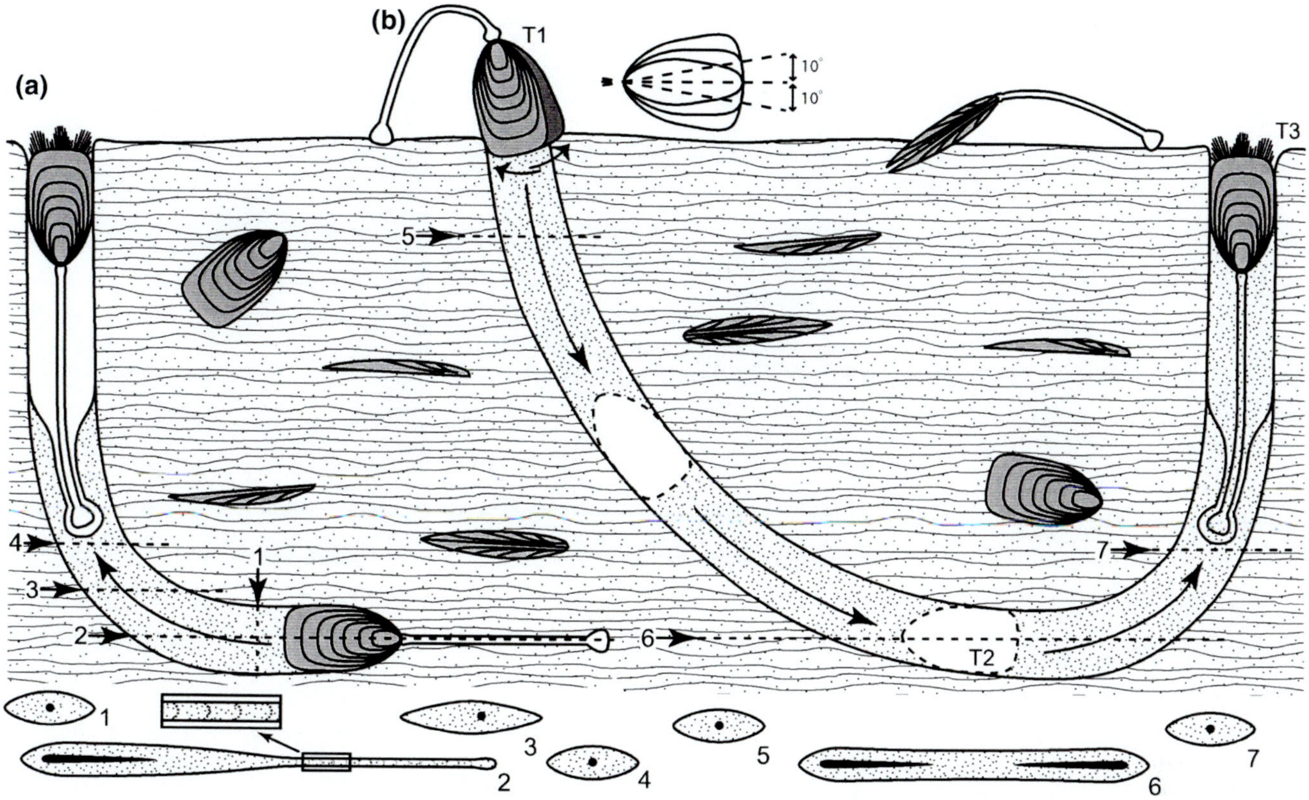


**Fig. 5.67** Lingulide infaunal behavior and resulting ichnospecies of *Lingulichnus* in response to exhumation and sudden burial. Cross sections showing preservation in horizontal aspect are displayed at base. After Thayer and Steele-Petrović (1975), from Zonneveld and Pemberton (2003), reprinted by permission of the publisher (Taylor & Francis Ltd., http://www.tandfonline.com). **a** Lingulide brachiopod in normal dwelling position (*L. verticalis*) and in oblique orientation (*L. inclinatus*). **b** A lingulide in process of arching its pedicle to initiate reburrowing after exhumation and the formation of a U-shaped burrow (*L. hamatus*). The brachiopod props itself up with its pedicle, orienting its shell downward (*T1*). The brachiopod burrows using scissorlike motion of valves. The brachiopod typically burrows deeply enough (*T2*) to return to the surface in a vertical orientation (*T3*)

**Fig. 5.68** *Lingulichnus verticalis* in grainy packstone (arenite) with incipient firmground conditions. Middle Triassic (Anisian) Jena Formation (Muschelkalk), Thuringia, Germany. *Scale bars* = 1 cm. **a** Slightly abraded bedding plane. **b** Vertical section

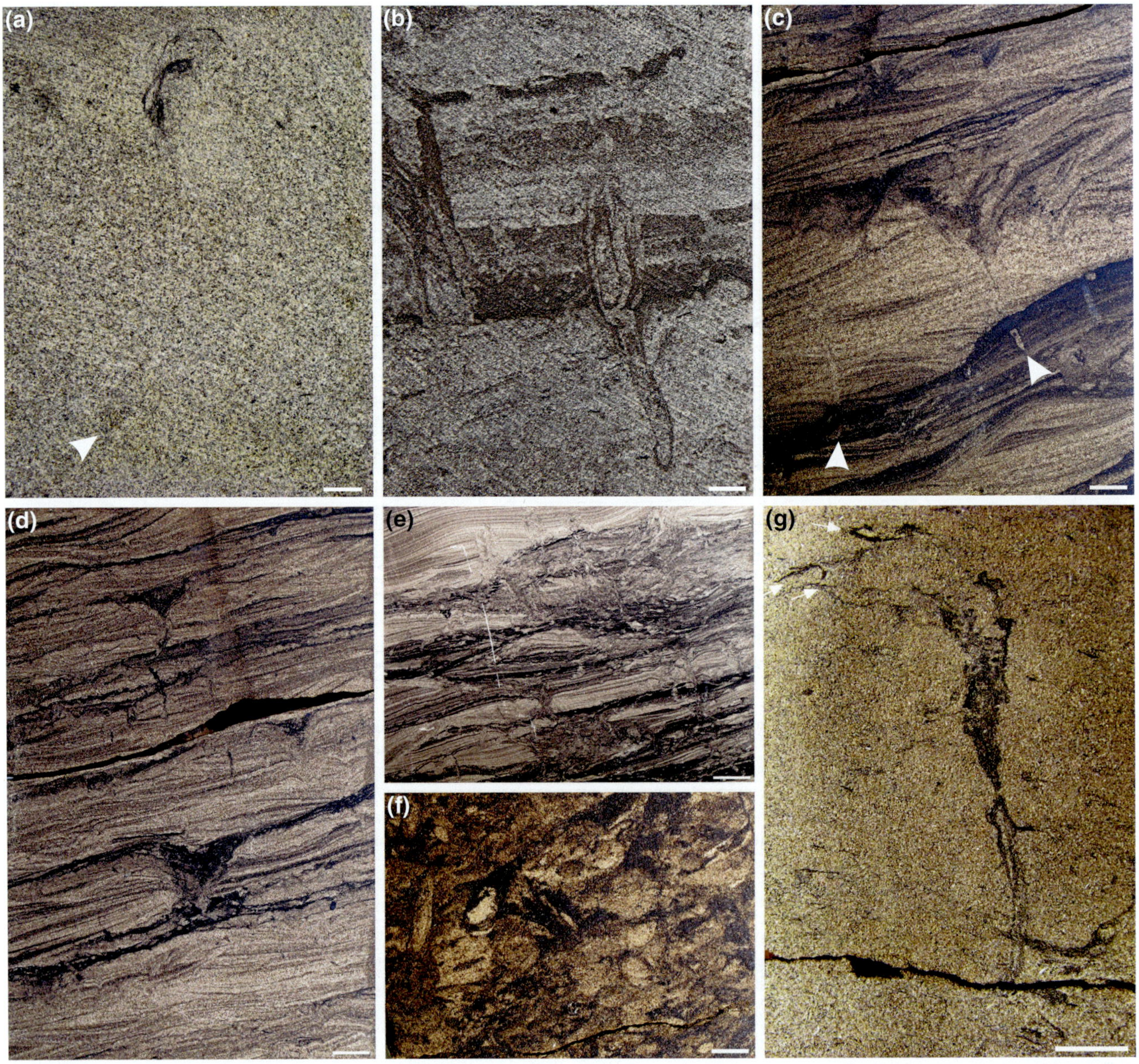


**Fig. 5.69** *Lingulichnus verticalis* in sectioned core. *Scale bars* = 1 cm. **a** Burrow in homogeneous sandstone, with a long pedicle trace (*arrow head*) and a mud-laminated upper part. Note the adjustment of the burrow in response to rapid deposition. Upper Jurassic (Oxfordian) Sognefjord Formation (shallow marine), Vega Field, Norwegian North Sea (well 35/11-6, ca. 3187.65 m). **b** Individual specimens showing the extended pedicle trace in the lower part of the burrow, and the spade-shaped laminated trace in the upper part of the burrows. Middle Jurassic (Bathonian) Hugin Formation (marginal marine, tidal flat), Gudrun Field, Norwegian North Sea (well 15/3-9T2, ca. 4503.5 m). **c** Ripple-laminated sandstone with several burrows showing the extended pedicle traces in the lower part (*arrow heads*) and a wide funnel with lamination in the upper part. Lower Jurassic (Pliensbachian-Toarcian) Tofte Formation (fan delta, tidal flat), Skuld Field, Norwegian North Sea (well 6608/10-14S, ca. 2591.95 m). **d** Ripple-laminated sandstone with several colonization surfaces bearing *L. verticalis*. Lower Jurassic (Pliensbachian) Åre Formation (delta plain, tidal flat), Skuld Field, Norwegian North Sea (well 6608/10-14S, ca. 2629.35 m). **e** Same interval as in (**d**), but with increased bioturbation resulting in a dense *L. verticalis* ichnofabric. Individual burrows (in addition to many pedicle-trace cross-sections) can be recognized in the upper left of the image. **f** Dense ichnofabric of *L. verticalis* due to total bioturbation. Some discrete pedicle traces and laminated burrow parts are still recognizable. Middle Jurassic (Aalenian) Ile Formation (delta plain, tidal flat), Skuld Field, Norwegian North Sea (well 6608/10-14S, ca. 2550.65 m). **g** Well-preserved burrow with a pedicle trace in the lower part and a spade-shaped burrow above, the latter strongly inclined to the left, reflecting adjustment of the tracemaker to continuous sedimentation. Several casts of linguloids are preserved at the burrows aperture (*arrows*). Lower Jurassic (Toarcian-Pliensbachian) Cook Formation (marginal marine), Norwegian North Sea (well 35/10-1, ca. 3653.0 m)

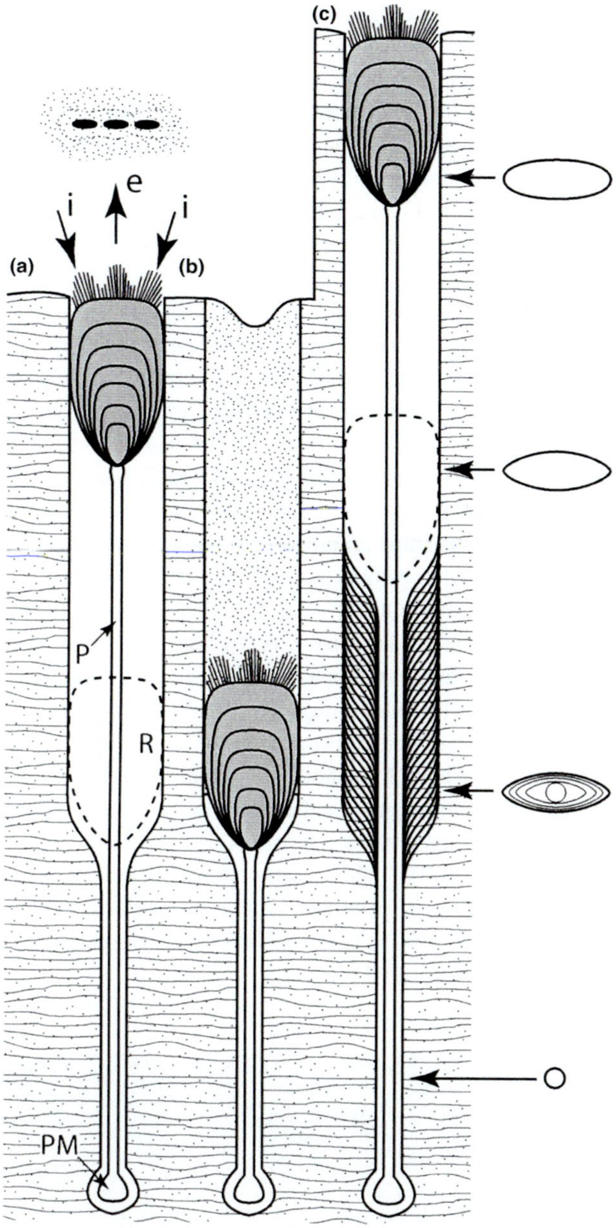


**Fig. 5.70** Sedimentary structures constructed by modern lingulide brachiopods that resulted in *Lingulichnus verticalis*. After Emig et al. (1978), from Zonneveld and Pemberton (2003), reprinted by permission of the publisher (Taylor & Francis Ltd., http://www.tandfonline.com). **a** Longitudinal section of a lingulide within its dwelling structure (*L. verticalis*). *P* pedicle; *PM* pedicle mass; *R* retraction position; *i* inhalant pseudosiphon; *e* exhalant pseudosiphon projecting above the sediment-water interface. **b** Retraction of a lingulide brachiopod into its burrow. **c** Formation of lingulide equilibrium trace (*L. verticalis*). Cross sections through different levels of a lingulide equilibrium trace are shown at *right*

It can be related to storm deposits in proximal offshore to lower shoreface settings and is common in intertidal environments. Due to the tracemaker's preference for grainy substrates, *Lingulichnus* is often associated with sandy event beds such as tempestites. Rapid or sudden burial is considered as a requirement to preserve *Lingulichnus*, whereas specimens in soft substrates with extensive bioturbation only have a low preservation potential (Kowalewski and Demko 1997). Because lingulide brachiopods can tolerate fluctuations and temporarily lowered salinity, *Lingulichnus* can be found in marginal-marine settings dominated by brackish conditions (e.g. estuaries). In such environments, burrow size is commonly reduced (Buatois et al. 2005; Gingras et al. 2012a).

*Ichnofacies: Lingulichnus* belongs to the *Skolithos* Ichnofacies but may occur in transitions to the *Cruziana* Ichnofacies.

*Age: Lingulichnus* is recorded from the Early Cambrian to the Holocene. It is most commonly reported from Paleozoic strata, which is probably related to a preservation bias (Zonneveld and Pemberton 2003).

*Reservoir Quality:* The influence of *Lingulichnus* on reservoir quality has not been studied so far. Judging from its morphological features such as subvertical orientation and the nature of its (partly) passive fill, *Lingulichnus* is capable of slightly increasing reservoir quality and vertical connectivity.

## 5.14 *Macaronichnus* Clifton and Thompson, 1978

*Morphology, Fill and Size: Macaronichnus* includes predominantly horizontal, cylindrical burrows of indefinite length with a straight, winding, meandering or spiral-like course (Fig. 5.71), although oblique and vertical burrows can occur too (e.g. Uchman et al. 2016; see Fig. 5.42c). The burrows are characterized by an active fill of pale sand, and an outer mantle composed of dark mineral grains (Bromley et al. 2009; Figs. 5.72 and 5.73; see also 5.178c). The burrows are unbranched (although occasional branching has been described by Rodríguez-Tovar and Aguirre 2014) and commonly occur in high density. Their diameter ranges between 0.2 and 2.0 cm, and their length can be at least several centimeters (Savrda and Uddin 2005).

*Ichnotaxonomy: M. segregatis* is the only ichnospecies of *Macaronichnus*. It includes *M. segregatis segregatis* and three other morphological forms described as ichnosubspecies by Bromley et al. (2009), *M. s. lineiformis, M. s. maeandriformis* and *M. s. spiriformis* (Fig. 5.71). Rodríguez-Tovar and Aguirre (2014) added a fourth, ichnosubspecies *M. s. degiberti*, for occasionally branched and partly oblique to vertically oriented burrows.

*Substrate: Macaronichnus* is typically associated with sandy substrate (looseground).

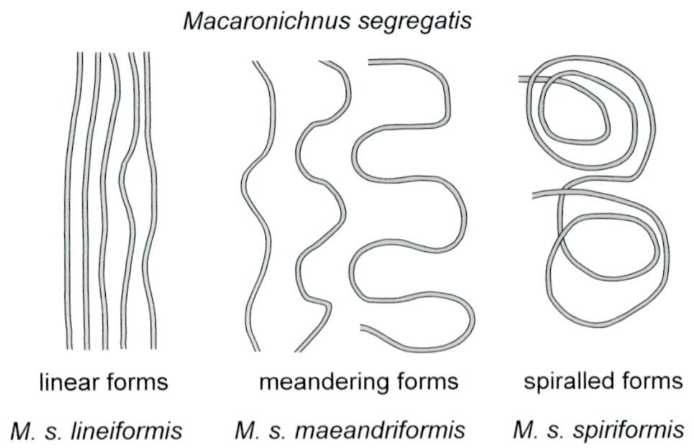


**Fig. 5.71** Morphological variability of *Macaronichnus segregatis*. From Bromley et al. (2009), reprinted by permission of the publisher (Taylor & Francis Ltd., http://www.tandfonline.com)

*Appearance in Core:* In core, *Macaronichnus* typically appears as crowds of more or less horizontal cylindrical burrows with a winding course, which can be recognized as elongate to elliptical and circular burrow sections (Fig. 5.74, see also Fig. 2.6c). Burrow fill contrasts from the surrounding sediment by its pale color and the diagnostically dark mantle (consisting of mica or heavy minerals) surrounding that core.

*Similar Trace Fossils: Macaronichnus* is a simple horizontal burrow that shares similarities with several other trace fossils. The active fill of *Planolites* differs from *Macaronichnus* by lacking a mantle, which also applies to *Gordia*. *Chondrites* may also create a dense ichnofabric similar to crowded *Macaronichnus* but the latter having no bifurcation. This also applies to *Nereites*, in cases where the actively filled central core is surrounded by a poorly lobed mantle.

*Producers:* By neoichnological comparison, opheliid polychaetes have been identified as producers of modern *Macaronichnus*, such as *Ophelia limacina* (Clifton and Thompson 1978), *Euzonus* sp. (Seike 2007, 2008; Dafoe et al. 2008a, b) and *Travisia japonica* (Seike et al. 2011) (Fig. 5.75).

*Ethology: M. segregatis* is produced by intrastratal deposit-feeding of opheliid polychaetes (Clifton and Thompson 1978; Seike 2008). The polychaetes feed on the microbes on the surface of the (quartz) grains and process these grains through their gut, while segregating dark-colored grains with their bristles around their bodies.

*Depositional Environment: M. segregatis* is a shallow-marine trace fossil most common in foreshore, shoreface and delta-front deposits (Nara and Seike 2004; Seike 2007; Bromley et al. 2009; Quiroz et al. 2010), as well as intertidal and shallow subtidal deposits (Clifton and Thompson 1978; Seike 2008). Rodríguez-Tovar and Aguirre (2014) described *M. segregatis* from shelf deposits. Under certain circumstances, it also can occur in upper slope environments (e.g. Figure 5.74d) where longshore or upwelling currents provide favorable conditions for the tracemaker. The polychaete *Euzonus* responds with its burrows to beach morphodynamics (Seike 2008). The worms burrow horizontally in various directions in relatively stable conditions, whereas they move preferentially landward after heavy erosion of the beach face by a storm (Fig. 5.76).

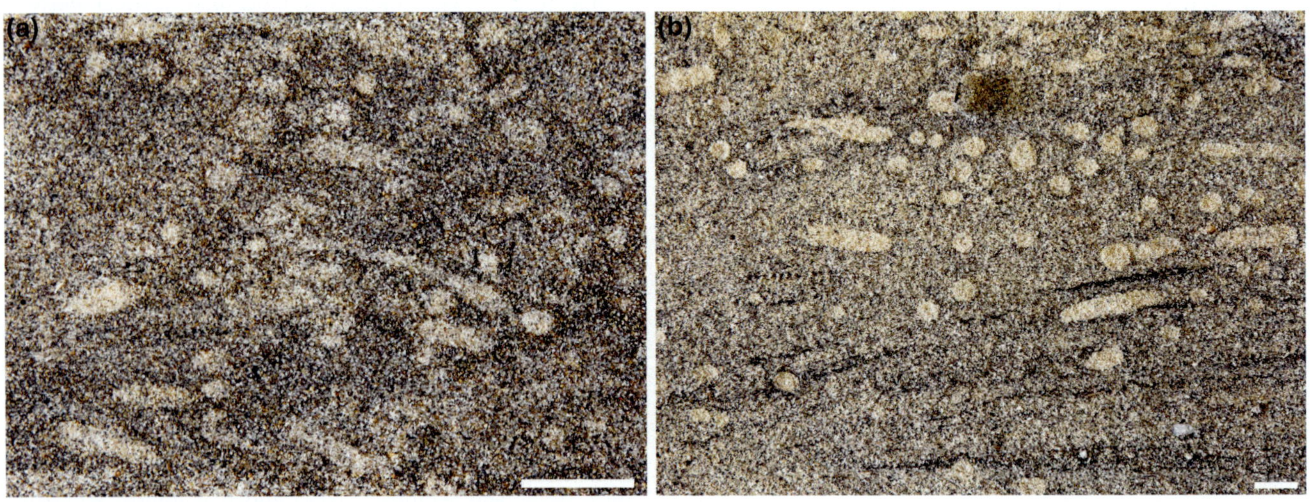


**Fig. 5.72** Burrow architecture of *Macaronichnus segregatis*-like traces produced by the polychaete worm *Euzonus* in modern beach sand of the Hasaki coast, central Japan (original resin peels from Seike 2008). Note the pale core surrounded by a mantle with dark mineral grains. *Scale bars* = 1 cm. **a** Bedding-plane view. **b** Vertical section

**Fig. 5.73** *Macaronichnus segregatis* in outcrop. *Scale bars* = 1 cm. **a** *M. s. maeandriformis* on a sandstone bedding. Eocene Battfjellet Formation (lower delta plain), Brongniartfjellet (Van Keulenfjorden), Svalbard. **b** Sandstone with a dense *M. s. maeandriformis* ichnofabric on the bedding plane. Paleocene Firkanten Formation (shallow marine), Longyearbyen, Svalbard. **c** *M. s. maeandriformis* in bioturbated siltstone. Paleocene Grumantbyen Formation (shelf), near Longyearbyen, Svalbard. **d** Sandstone with a dense *M. s. lineiformis* ichnofabric on the bedding plane. Lower Cambrian Hardeberga Formation (sandy tidal flat), Snogebæk, Bornholm, Denmark. **e** Sectioned vertical surface of a glauconite-rich cross-bedded sand within a fault zone (overturned section) displaying a cluster of *M. segregatis*. Lower Cretaceous Arnager Greensand Formation (storm-dominated shoreface), near Rønne, Bornholm, Denmark

**Fig. 5.74** *Macaronichnus segregatis* in sectioned core. *Scale bars* = 1 cm. **a** Heterolithic (silty) sandstone with ripple lamination and discrete burrows consisting of a pale core surrounded by a thin dark mantle. Middle Jurassic (Callovian) Fensfjord Formation (shallow marine, lower shoreface), Gjøa Field, Norwegian North Sea (well 36/7-1, ca. 2384.5 m). **b** Heterolithic (silty) sandstone with ripple lamination and small burrows. Middle Jurassic (Bathonian-Callovian) Hugin Formation (shallow marine), Gina Krog Field, Norwegian North Sea (well 15/5-7, ca. 3894.0 m). **c** Cross-bedded sandstone with an intensively burrowed layer chiefly containing *Macaronichnus*. Associated equilibrium trace fossils (upper layer) are consistent with a high-energy environment and rapid sedimentation. Middle Jurassic (Bathonian) Tarbert Formation (sandy tidal flat), Oseberg Sør Field, Norwegian North Sea (well 30/9-14, ca. 3133.35 m). **d** Dense *Macaronichnus* ichnofabric on top of a thick turbiditic sandstone bed (light quartzitic sandstone). The overlain dark silty material is partly incorporated in some burrows of the shallower tier. The burrows consist of an actively filled core (quartz sand) and a reworked sandy mantle (darker sand). Lower Cretaceous (Albian, deep-marine channel system), off Tanzania

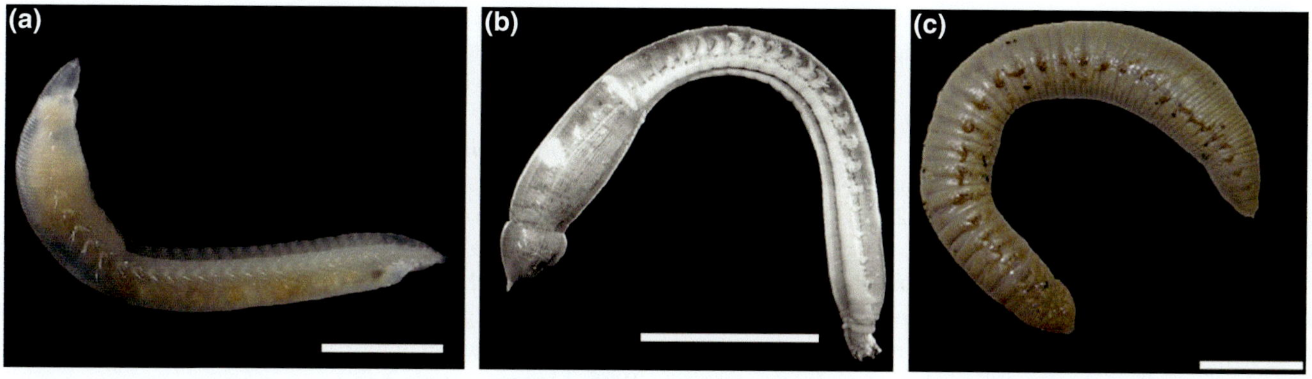


**Fig. 5.75** Opheliid polychaetes as the producers of modern *Macaronichnus*-like traces. *Scale bars* = 1 cm. **a** *Ophelia limacine*. From www.marinespecies.org. **b** *Euzonus mucronata*. From Nara and Seike (2004). **c** *Travisia japonica*. From Seike et al. (2011), republished with permission of Elsevier; permission conveyed through Copyright Clearance Center, Inc.

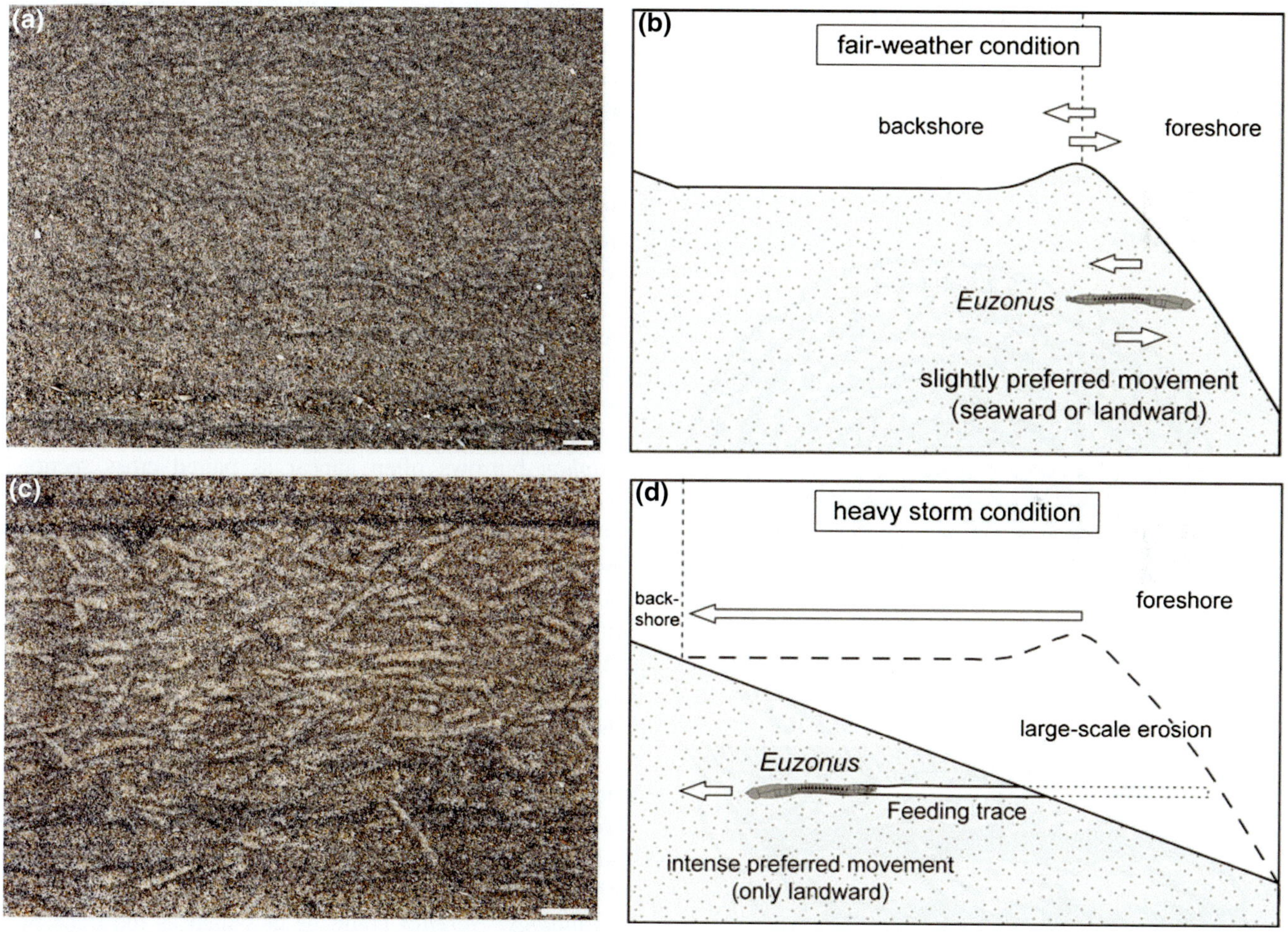


**Fig. 5.76** Resin peels of vertical sections of modern beach deposits from the Hasaki coast, central Japan (**a** and **c**) together with schematic diagrams showing the corresponding burrowing behavior inferred from the traces (**b** and **d**). Under fair-weather conditions, the worms burrow in various directions (**a** and **b**), whereas under storm conditions, the tracemakers are forced landward, which results in that preferred orientation (**c** and **d**). Original resin peels and drawings from Seike (2008), republished with permission of Springer. *Scale bars* = 1 cm

*Ichnofacies: Macaronichnus* assemblages are common constituents of the *Skolithos* Ichnofacies (e.g. Pemberton et al. 2012).

*Age: M. segregatis* has frequently been reported from Mesozoic and Cenozoic deposits (e.g. Clifton and Thompson 1978; Quiroz et al. 2010), occasionally also from the Paleozoic (e.g. Bromley 1996; Knaust 2004a) as old as Early Cambrian (Fig. 5.73d).

*Reservoir Quality:* Several studies have demonstrated the subtle but positive effect of *Macaronichnus* ichnofabrics on enhanced reservoir quality, which is related to the sorting and cleaning effect due to sediment feeding by the worms (Gingras et al. 2002; Pemberton and Gingras 2005; Pemberton et al. 2008; Dafoe et al. 2008b; Knaust 2009a, 2014a; Gordon et al. 2010).

## 5.15 *Nereites* MacLeay in Murchison, 1839

*Morphology, Fill and Size: Nereites* is defined as a predominantly horizontal, unbranched, meandering to winding burrow or trail, consisting of an actively filled central core and a thick lobed mantle (Uchman 1995; Fig. 5.77). The backfilled core can show a meniscate fill, consisting of either muddy substrate (e.g. fecal material) or sandy sediment. *Nereites* ranges in size from few millimeters to over 1 cm in width/diameter.

**Fig. 5.77** *Nereites* from the Lower Devonian of Germany in plan view (*top*) and oblique view (*bottom*). From Seilacher (2007), republished with permission of Springer

*Ichnotaxonomy:* About 30 ichnospecies of *Nereites* have been described, many of them now regarded as preservational variants of a few ichnospecies (Uchman 1995; Mangano et al. 2000; Fig. 5.78). *Helminthoida*, a junior synonym of *Nereites* (Uchman 1995), can still sometimes be found in the literature (e.g. Pemberton et al. 2001), as it is the case for *Scalarituba*, another junior synonym of *Nereites* (Uchman 1995; Mangano et al. 2000).

*Substrate: Nereites* preferably occurs in silty to fine-grained sandy substrate with a certain amount of mud admixture. In siltstone, the *Scalarituba* preservational aspect is more usual.

*Appearance in Core: Nereites* typically appears in mass occurrences with burrows displaying an actively filled tunnel enveloped by a halo of reworked sediment (Uchman 1995). *Nereites* with a muddy core and sandy halo are common, although sandy burrows also occur (Fig. 5.79; see also Figs. 4.3, 5.121b, 5.129b, 5.132d and 5.156d). Depending on the degree of winding and meandering of the burrows, a wide range of sections can occur in core, including semicircular cross sections and more or less elongate longitudinal sections.

*Similar Trace Fossils:* Particularly in core, *Nereites* occurs with *Phycosiphon*, and both trace fossils can be quite similar to each other, which weakens their distinction. However, *Phycosiphon* is a winding burrow with a muddy or silty core and a sandy spreite around it. As a result, *Phycosiphon* burrows often appear as paired cross sections. *Nereites* is generally larger than *Phycosiphon* and has a concentric or bilobate halo around the mud-rich core (Callow et al. 2013). In contrast to *Phycosiphon*, vertical loops and twists are not developed. *Nereites* shares similarities with the trace fossil *Macaronichnus*, which consists of an actively sand-filled core with surrounding mantle. *Nereites* of low contrast could be also mistaken for *Chondrites*, which, however, shows branching in three dimensions.

*Producers:* The producer of *Nereites* remains uncertain, but a kind of vermiform animal, probably an enteropneust, is most likely (Mangano et al. 2000). Rindsberg and Martin (2003) and Martin and Rindsberg (2007) also consider arthropods as the producer of *Nereites*.

*Ethology: Nereites* could be classified as the trace of a deposit-feeder (fodinichnion), although the combined locomotion and feeding activities would be consistent with an interpretation as grazing trace (pascichnion; Mangano et al. 2000).

**Fig. 5.78** *Nereites* from outcrop. *Scale bars* = 1 cm. **a** Tidly meandering *N. irregularis* in Eocene shale (flysch), Bregenzerwald, Austria. Senckenberg coll., Frankfurt Main (original of Richer 1928). **b** Detail of *Nereites* isp. from the Middle Devonian of Thuringia, Germany. Senckenberg coll., Frankfurt Main (original of Richer 1928). **c** Another specimen from the Middle Devonian of Thuringia, Germany (coll. University of Greifswald). **d** *Nereites* (formerly *Helminthoida*) in Cretaceous helminthoid flysch of the Ligurian Alps east of Albenga, Italy

*Depositional Environment: Nereites* is a typical element of deep-sea deposits, where it preferably occurs in sediments originating from deposition under moderate energy (Wetzel 2002, and references therein). It also occurs in slope deposits (*Zoophycos* Ichnofacies, e.g. channel-levee deposits; e.g. Callow et al. 2013), and is common in shelf deposits (*Cruziana* Ichnofacies). Finally, *Nereites* has been reported from sandy estuarine deposits and tidal flats (Martin and Rindsberg 2007; Neto de Carvalho and Baucon 2010) and even from lacustrine (Hu et al. 1998) and glacial (Netto et al. 2012) deposits. *Nereites* is a shallow-tier component and occurs as postdepositional trace fossil just below the surface layer within oxygenated sediment (Fig. 5.80).

*Ichnofacies: Nereites* is the namesake of the *Nereites* Ichnofacies (Seilacher 1967) and is typical of basin-floor (flysch) deposits (Uchman 1995). It also occurs in other ichnofacies, for instance in the *Zoophycos* and *Cruziana* ichnofacies.

*Age: Nereites* is reported from the Cambrian (e.g. Aceñolaza and Alonso 2001) to Holocene (Wetzel 2002).

*Reservoir Quality:* The impact of *Nereites* on reservoir quality depends on the composition of the burrows and their contrast with the hosting sediment. Sand-dominated burrows occurring in high density, for instance, have the potential to increase connectivity in a reservoir (Bednarz and McIlroy 2015; Fig. 3.6).

**Fig. 5.79** *Nereites* in sectioned core (**a**–**d**) and full core (**e**). *Scale bars* = 1 cm. **a** Heterolithic (silty) sandstone with *Nereites* consisting of a dark (silty) inner zone and a light (sandy) mantle. In silty (dark) layers, burrow density is higher than in sandy (light) layers. Lower Jurassic (Pliensbachian) Amundsen Formation (shallow marine, shelf), Norwegian North Sea (well 35/10-1, ca. 3657.2 m). **b** Cluster of large burrows with dark muddy core and light sandy mantle in cross section. Upper Cretaceous (Campanian) Neslen Formation (lower delta plain), East Canyon, Book Cliffs, Colorado, USA. Well HCR#1, ca. 266 ft. **c** Silty sandstone with complete bioturbation due to *Nereites*, in which the original stratification seems to remain intact because of reworking at a small scale (small burrow size, cryptic bioturbate texture). Upper Jurassic (Oxfordian) Heather Formation (offshore), Fram Field, Norwegian North Sea (well 35/11-9, ca. 2650.25 m). **d** Completely bioturbated silty sandstone with a dense *Nereites* ichnofabric and discrete *Schaubcylindrichnus*. Upper Jurassic (Oxfordian) Heather Formation (offshore), Fram Field, Norwegian North Sea (well 35/11-9, ca. 2645.3 m). **e** Large *Nereites* with complex muddy core surrounded by sandy mantle. Paleocene Grumantbyen Formation (lower shoreface to offshore transition), Svalbard (well BH 9-2006, ca. 389 m)

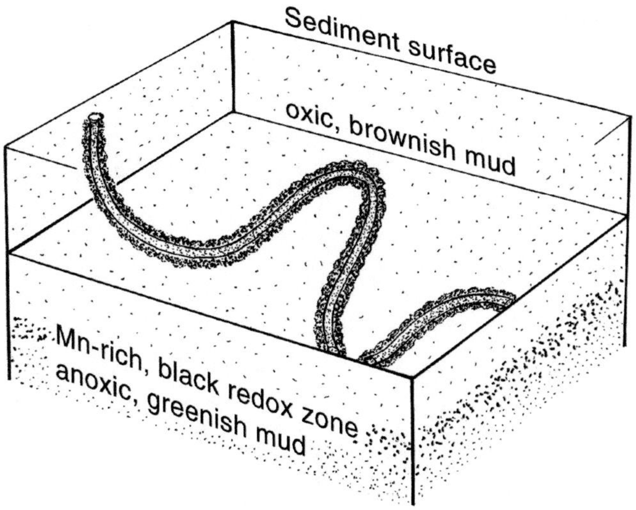


**Fig. 5.80** *Nereites* as shallow-tier trace fossil in deep-marine sediments. From Wetzel (2002)

## 5.16 *Ophiomorpha* Lundgren, 1891

*Morphology, Fill and Size: Ophiomorpha* consists of boxworks constituting a horizontal maze with vertical shafts (Fig. 5.81). The burrows are circular to elliptical in cross section. Branching is Y- and T-shaped, typically with enlargement of the junctions. Passive fill is common, although some burrow segments may have active meniscate fill (Figs. 2.6b and 5.84e–g). Burrow lining is diagnostic of *Ophiomorpha*, consisting of pellets of sand and/or mud along the wall. The burrow diameter varies from one ichnospecies to another and ranges between 3 and 30 mm, while complete burrow systems are reported to be several meters in extent and more than 1 m in depth.

*Ichnotaxonomy:* Several ichnospecies are distinguished on the basis of overall morphology and the shape and composition of the pellets (Uchman 2009):

- *O. nodosa*—regularly distributed, knobby sand pellets (Figs. 5.22c, 5.82, 5.83c, d, 5.84d and 5.178d)
- *O. borneensis*—regularly distributed, bilobate sand pellets (Fig. 5.83a, b)
- *O. irregulaire*—contorted, sand-cored mud pellets (Fig. 5.84a–c)
- *O. annulata*—elongated sand pellets perpendicular to burrow axis
- *O. recta*—small mud pellets
- *O. rudis*—irregularly distributed, knobby sand pellets (Figs. 5.83e, f and 5.84e–g)
- *O. puerilis*—cylindrical, rod-shaped pellets with rounded ends
- *O. ashiyaensis*—granular ornamentation.

*O. isabeli* Mayoral (1986) is based on the occurrence of a pseudopelletal structure due to diagenetic processes (e.g. compaction or differential dissolution) and thus does not warrant its own ichnotaxonomical name. In contrast to all other *Ophiomorpha* ichnospecies, *O. puerilis* de Gibert et al. 2006 contains fecal pellets assignable to *Coprulus oblongus*, which today is produced by polychaetes (Knaust 2008).

*Substrate: Ophiomorpha* is a typical constituent of clean sandstone, both homogeneous and cross-bedded. Due to the burrowing activity of its producer, a relatively large amount of organic matter and fine-grained sediment can be introduced into the burrow and may lead to a heterogenization of the host sediment. The *Ophiomorpha* tracemaker is capable of penetrating sandy beds 1 m thick or more (e.g. storm or turbidite deposits) before turning horizontally at the mud/sand interface. Thus, thin mudstone interlayers can be completely penetrated due to the producer's ability of searching for food (Fig. 2.6b). *Ophiomorpha* also occurs in chalk.

*Appearance in Core:* Given their diagnostic appearance with the pelleted wall, identification of *Ophiomorpha* in core is comparatively straightforward (Fig. 5.84). Burrows are commonly larger than accompanying trace fossils and the pelleted wall is quite distinct, although a more homogeneous mud lining can give reason for confusion with *Palaeophycus*. Horizontal burrow parts typically dominate (Fig. 5.146e) but vertical shafts are sometimes exposed in core sections (Fig. 5.146f). It must be stressed that *Ophiomorpha* can be passively or actively filled. Active fill typically is meniscate (Figs. 5.84e, g, 5.146f and 5.161d).

*Similar Trace Fossils:* The local discontinuation of pellets may give reason for confusion with other (supposed)

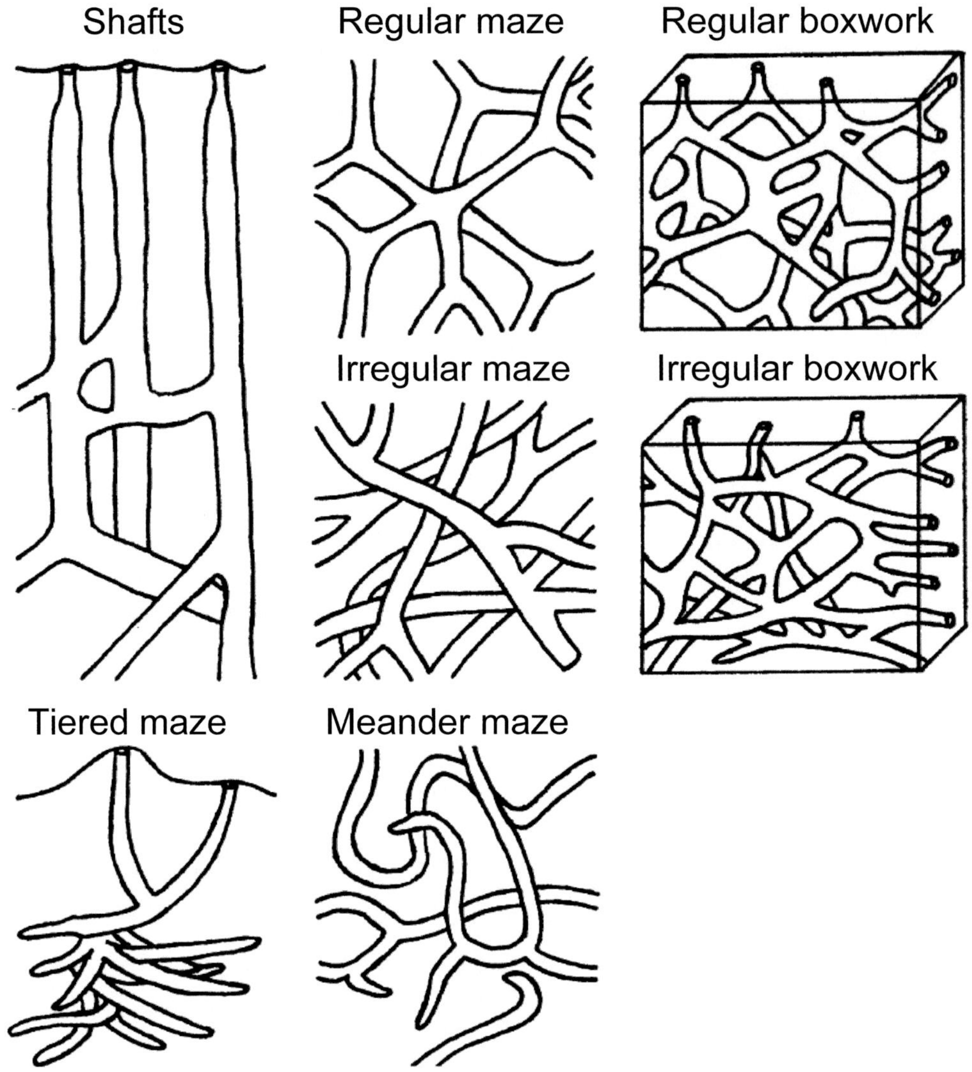


**Fig. 5.81** Morphological variability of *Ophiomorpha* and modern analogs. After Chamberlain and Baer (1973), Frey et al. (1978), adapted from Anderson and Droser (1998), republished with permission of Wiley; permission conveyed through Copyright Clearance Center, Inc.

crustacean burrows such as *Thalassinoides, Spongeliomorpha, Psilonichnus*, *Pholeus* and *Scoyenia* (Fig. 5.85). Burrows with compact linings may resemble *Palaeophycus* and gradational transitions to other crustacean burrows can be found in dependence on the substrate properties. However, *Palaeophycus* differs from *Ophiomorpha* by preferred horizontal orientation and its usually unbranched nature. Actively filled *Ophiomorpha* tunnels are superficially similar to the backfilled burrows *Taenidium* and *Scolicia*, which are unbranched and lack pellets or significant vertical components, while actively filled shafts may resemble *Diplocraterion* (Fig. 5.58). Small *Ophiomorpha* produced by juvenile tracemakers may resemble *Chondrites* but differ from it by their boxwork architecture and the occurrence of pellets. Open *Ophiomorpha* tunnel elements may be subject to collaps, which then may lead to dawnwards deflected laminae in the overburden sediment, mimicking *Conichnus* (Fig. 5.44a, b).

*Producers:* By comparison with modern analogs, there is good confidence that the producers of *Ophiomorpha* belong to thalassinidean shrimp, particularly callianassids (Figs. 5.25, 5.86 and 5.87). In fact, several hundred fossorial species of the extant thalassinideans (families Callianassidae and Upogebiidae) are known (Knaust et al. 2012).

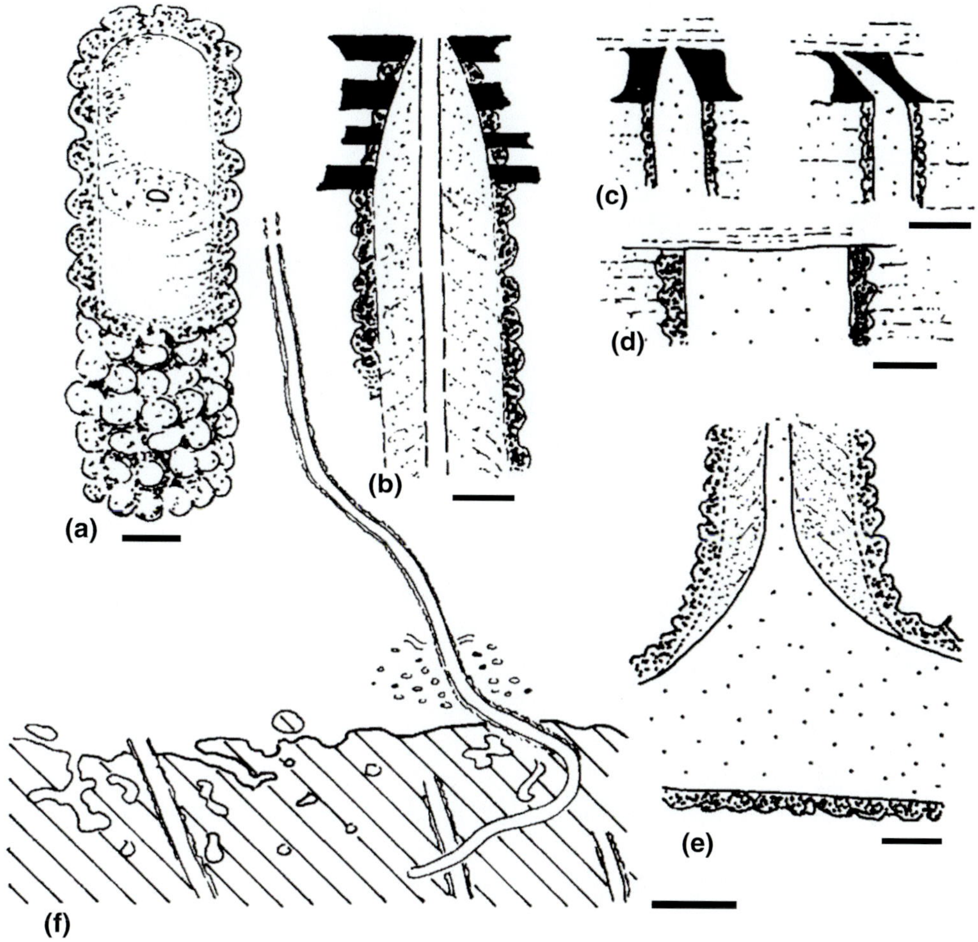


**Fig. 5.82** General characteristics of *O. nodosa*. After Pollard et al. (1993). *Scale bars* = 1 cm except (**f**) = 10 cm. **a** Dissected burrow with sandy wall pellets, smooth inner surface to lining and muddy sand restriction. **b** Taper of restricted burrow. **c** Two shafts with taper into muddy layer, assumed to be colonization surface. **d** Shaft truncation. **e** Base of restricted shaft leading into *T*-junction with another gallery. **f** Lined shaft passing down into humic rich sand without lining. Humic sand truncated and burrowed by *Thalassinoides* cf. *suevicus*

*Ethology: Ophiomorpha*-producing shrimp can be deposit- and suspension-feeders (domichnial or fodinichnial), depending on the involved species but often also within the same species (Nickell and Atkinson 1995). Interconnected, highly organized and systematic burrow elements (such as polygons and sinusoidal segments), particularly along the sandstone/mudstone interface, indicate more advanced behavior and have been interpreted as agrichnia (Cummings and Hodgson 2011). The pellets of *Ophiomorpha* often originated from the mixture of mucus and sediment. Because of their occurrence along the interface between open, oxygenized burrow and anoxic host sediment, these pellets have a high potential for diagenetic modification.

*Depositional Environment: Ophiomorpha* is one of the commonest trace fossils and occurs in a wide range of paleoenvironments (Leaman et al. 2015). Originally regarded as a significant component of shallow-marine facies (Frey et al. 1978; Pollard et al. 1993), several ichnospecies of *Ophiomorpha* are characteristic of deep-marine deposits (Tchoumatchenco and Uchman 2001; Uchman 2009), and its cross-facies relationship is discussed by Monaco et al. (2009). Few reports exist where *Ophiomorpha* is documented from continental settings, although those burrows can be also produced by other means and assigned to different ichnotaxa (Goldring and Pollard 1995; cf. Baucon et al. 2014). Although not exclusively, *Ophiomorpha* is a typical component related to high-energy environments. Estimations of the time available for colonization and burrow construction, referred to the colonization window, allowed Pollard et al. (1993) to differentiate sandy shoreline sedimentary environments such as shoreface, offshore tidal shelf sand-wave facies, and estuarine facies (Fig. 5.88). On

**Fig. 5.83** *Ophiomorpha* in outcrop. *Scale bars* = 1 cm except (**f**) = 10 cm. **a** *O. borneensis* in sandstone. Lower Cretaceous (Berriasian) Robbedale Formation (shallow marine, nearshore), Arnager Bay, Bornholm, Denmark. **b** Part of *O. borneensis*. Upper Cretaceous (Campanian) Bearpaw-Horseshoe Canyon Formation (marginal marine), near Drumheller, Alberta, Canada. **c**, **d** *O. nodosa* in limestone. Note the adjacent occurrence of passive and active burrow fill. Erratic boulder, Cretaceous, Greifswalder Oie, northeastern Germany. **e** *O. rudis* on the bedding plane of a sandstone. Eocene Grès d'Annot Formation (deep marine, turbiditic), southeastern France. From Knaust et al. (2014), republished with permission of Wiley; permission conveyed through Copyright Clearance Center, Inc. **f** A cluster of *O. rudis* on the bedding plane of a sandstone. Miocene Mount Messenger Formation (deep marine, channel-levee system), sea cliff of the Taranaki Peninsula, North Island, New Zealand

**Fig. 5.84** *Ophiomorpha* in sectioned core. *Scale bars* = 1 cm. **a** *O. irregulaire* shaft. Lower Jurassic (Sinemurian-Pliensbachian) Tilje Formation (nearshore, tidal-influenced), Skarv Field, Norwegian North Sea (well 6507/5-1, 3581.2 m). **b** *O. irregulaire* shaft. Middle Jurassic (Bajocian) Ile Formation (shoreface, tidal-influenced), Norwegian Sea (well 6406/8-1, ca. 4421.5 m). **c** *O. irregulaire* shaft. Middle Jurassic (Bathonian-Callovian) Hugin Formation (shoreface, tidal-influenced), Sleipner Vest Field, Norwegian North Sea (well 15/9-5, ca. 3627.2 m). **d** *O.* cf. *nodosa* ichnofabric in sandstone. Upper Jurassic (Callovian) Heather Formation (offshore), Fram Field area (well 35/11-11, ca. 2724.5 m). **e** *O. rudis* in thin-bedded heterolithics. Upper Cretaceous (Maastrichtian) Springar Formation (deep marine), Norwegian Sea (well 6604/10-1, ca. 3648.5 m). **f** *O. rudis* shaft and tunnels filled with glauconitic sediment. Lower Cretaceous (Albian, deep marine, channel system), off Tanzania. **g** *O. rudis* ichnofabric with total bioturbation in the upper part and discrete, actively filled burrows in the lower part. Upper Cretaceous (Santonian) Kvitnos Formation (deep marine, turbiditic), Aasta Hansteen Field (well 6707/10-2A, 4477.0-4477.5 m). See Fig. 3.4e for porosity/permeability changes due to bioturbation

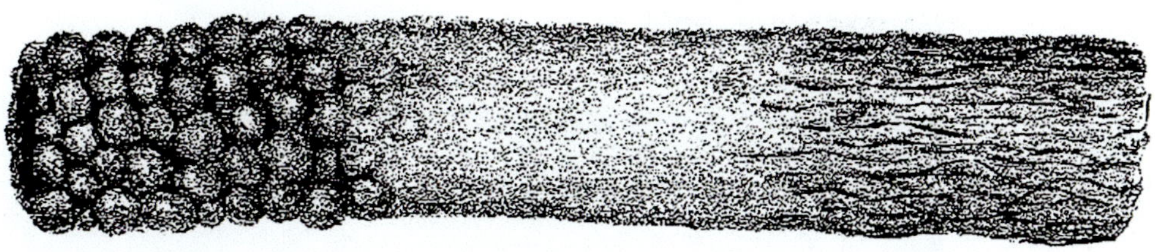


**Fig. 5.85** Burrow part showing three different components, which result from the activity of the same kind of organism (a supposed thalassinidean crustacean) in different substrate. Due to this contrasting features, three different names are applied on the ichnogenus level. Modified after Schlirf (2000)

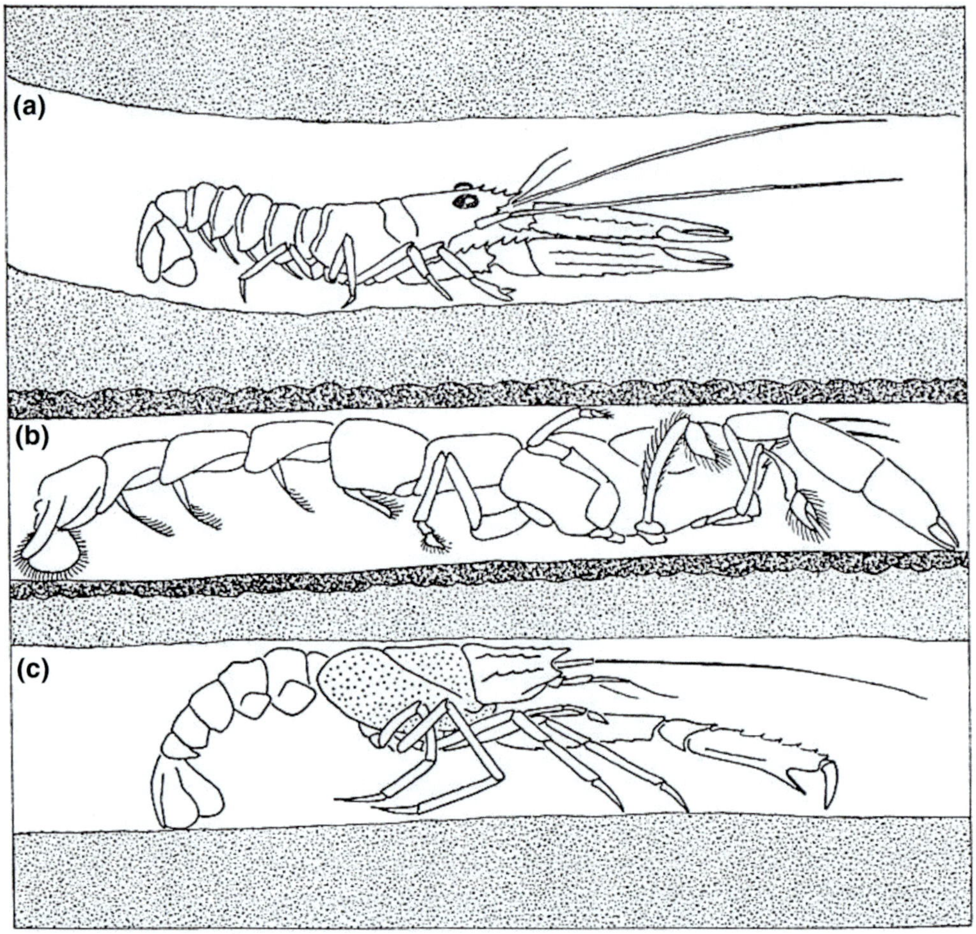


**Fig. 5.86** Three examples of burrowing crustaceans. From Bromley and Asgaard (1972), republished with permission of GEUS. **a** *Nephrops norvegicus*, **b** *Callianassa major*, and **c** *Glyphea rosenkrantzi*

**Fig. 5.87** Mounded topography created by the deep-burrowing callianassid *Glypturus acanthochirus* on a wide tidal flat in the Bahamas. From Knaust et al. (2012), republished with permission of Elsevier; permission conveyed through Copyright Clearance Center, Inc.

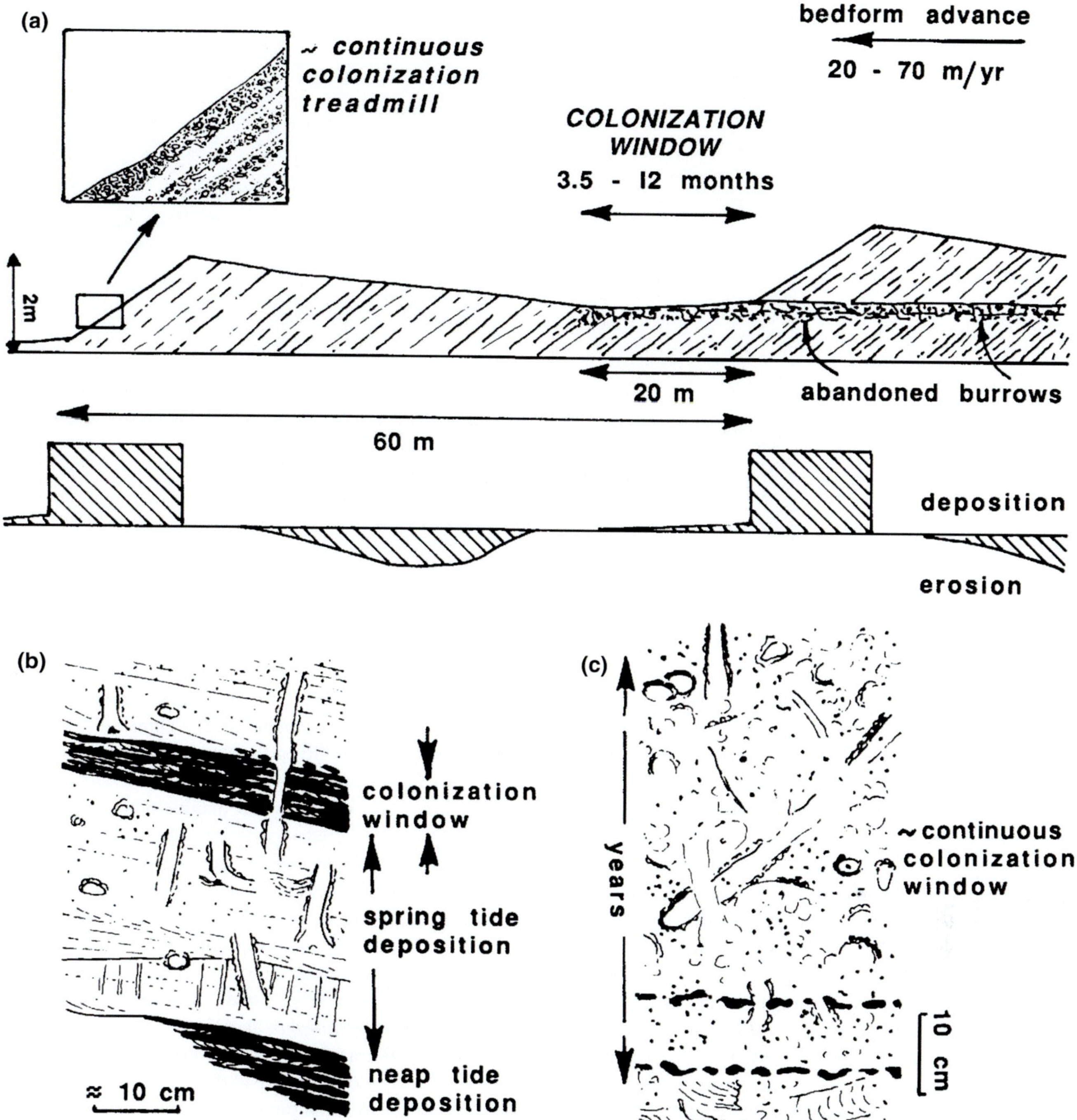


**Fig. 5.88** Diagram to illustrate colonization window. From Pollard et al. (1993). **a** Sand-wave migration and opportunity for colonization of the trough area. **b** Schematic diagram to illustrate depositional processes on estuarine point bar where opportunity for colonization was limited to periods of mud deposition. **c** Shoreface environment where physical restraints on colonization probably were limited to infrequent storms

**Fig. 5.89** Dense *Ophiomorpha-Thalassinoides* boxwork in hummocky cross-stratified sandstone with upwards-increasing degree of cementation with iron-stained minerals (limonite?). Middle Miocene Tatsukushi Formation (Misaki Group, shoreface), southeast coast of Shikoku Island, southern Japan

the basis of a sequence-stratigraphic analysis in the Upper Cretaceous of Utah, Anderson and Droser (1998) argued that *Ophiomorpha*-related bioturbation is more pervasive in lowstand systems tracts compared to transgressive systems tracts, because of the predominance of marginal- and nearshore-marine, sand-dominated settings as favorable habitats for colonization by *Ophiomorpha* producers.

*Ichnofacies:* Within the shallow-marine realm, *Ophiomorpha* with its significant vertical shafts belongs to the *Skolithos* Ichnofacies, but overlaps with the *Cruziana* Ichnofacies, where the horizontal burrow elements are more pronounced. In the deep sea, the *O. rudis* ichnosubfacies was established by Uchman (2009) as part of the *Nereites* Ichnofacies. There, *O. rudis* characterizes high-energetic proximal and axial turbidite deposits (e.g. Knaust 2009b).

*Age:* Burrowing shrimp appear in the Permian, from which *Ophiomorpha* is known (e.g. Chamberlain and Baer 1973), to the present (e.g. Leaman et al. 2015). Baucon et al. (2014) reported *Ophiomorpha* from Permian fluvial deposits in Italy and suggested that their ghost shrimps producers invaded marine environments during the recovery following the end-Permian mass extinction. From the Late Jurassic onward, *Ophiomorpha* can be also found in deep-marine deposits (Tchoumatchenco and Uchman 2001; Uchman 2009).

*Reservoir Quality:* Given their complex nature and multiple kinds of producers involved, *Ophiomorpha* can tend both improve and diminish the quality of reservoir rock. Because of its preferred suspension-feeding mode of life, shrimp introduce a considerable amount of mud into their sand-hosted dwellings, which can lead to a drastic reduction of porosity and permeability as known from the Vøring case study (Fig. 3.4). In the same area, thin gas-bearing sandstone beds within shaly matrix are vertically connected by long *Ophiomorpha* shafts (Fig. 5.84e). The best documented example of an *Ophiomorpha*- and *Thalassinoides*-dominated reservoir is the Biscayne aquifer in Florida, which relies on the burrowing activity of callianassid shrimp (Cunningham et al. 2009, 2012). In the shallow-marine Cretaceous Ben Nevis Formation off Newfoundland, *Ophiomorpha*-dominated ichnofabrics are documented as important agents of the net-pay intervals (Tonkin et al. 2010). The relatively large caliber of *Ophiomorpha* burrows promotes enhanced fluid flow and thus precipitation of early diagenetic minerals (Fig. 5.89).

## 5.17 *Palaeophycus* Hall, 1847

*Morphology, Fill and Size: Palaeophycus* refers to subhorizontal, essentially cylindrical, straight or slightly curved burrows with a lining and passive fill (Figs. 5.90 and 5.91). The burrows are typically unbranched or have unsystematic branching (Keighley and Pickerill 1995). The size of *Palaeophycus* may range from 1 mm or less to more than 1 cm in diameter.

*Ichnotaxonomy:* The ichnogenus *Palaeophycus* includes about 20 valid ichnospecies of which thin-walled *P. tubularis* and thick-walled *P. heberti* are most relevant in core samples. Other ichnospecies are very thinly lined and differ from each other by varying kinds of striae (Pemberton and Frey 1982).

*Substrate: Palaeophycus* occurs in diverse grain sizes mainly in soft, but also in firm substrate of both, siliciclastic and carbonate deposits.

*Appearance in Core:* The more or less horizontally oriented burrows with their lined wall and the passive fill make

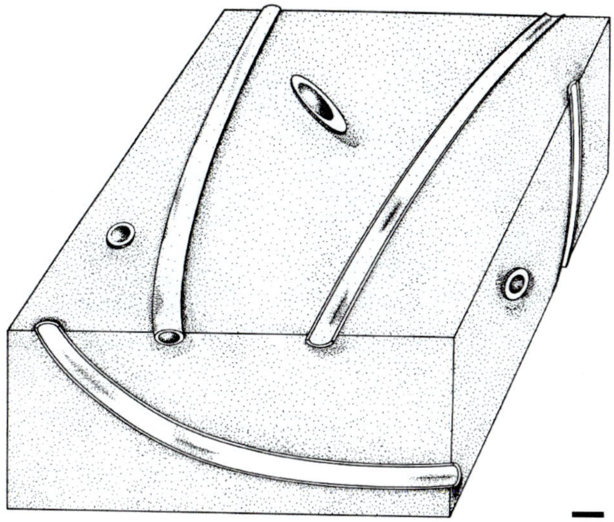

**Fig. 5.90** *Palaeophycus heberti* in a reconstruction by Howard and Frey (1984), republished with permission of Canadian Science Publishing; permission conveyed through Copyright Clearance Center, Inc. *Scale bar* = 1 cm

**Fig. 5.91** *Palaeophycus* in outcrop. *Scale bars* = 1 cm. **a** *P. heberti* in the rim of chert nodules (silicified limestone, vertical section). Middle Permian Kapp Starostin Formation (mixed siliciclastic carbonate ramp), Akseløya, Svalbard. **b** Same as in (**a**), detail view showing crowded burrows. **c** Large, elongate *P.* cf. *heberti* burrows on bedding plane of silty sandstone. Paleocene Grumantbyen Formation (proximal shelf), near Longyearbyen, Svalbard. **d** Same as in (**c**), close-up view. **e** *P.* cf. *alternatus*, burrow with weak annulation. Same locality as in (**c**)

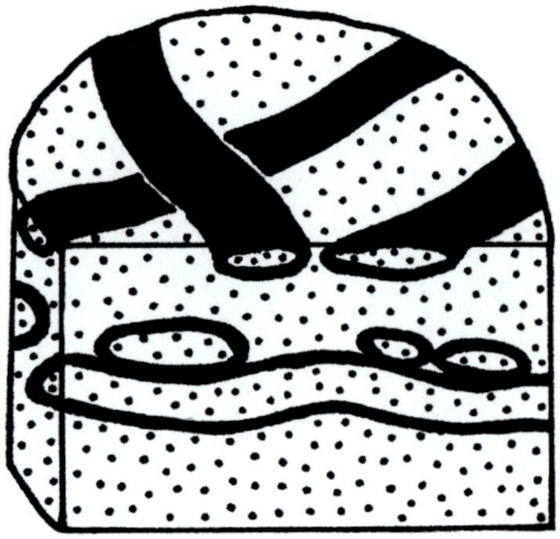

**Fig. 5.92** General morphology of *Palaeophycus* and its appearance in slabbed core

*Palaeophycus* distinctive in core. They occur as circular, elliptical and elongated sections (Figs. 5.92, see also Fig. 5.111b). Their appearance (e.g. wall thickness, size, density, etc.) is related to the encountered ichnospecies, facies and limiting environmental factors (Fig. 5.93). *Palaeophycus* may occur as individual burrows or in high densities.

*Similar Trace Fossils: Palaeophycus* is a relatively simple trace fossil with correspondingly high potential for confusion with similar ichnogenera. *Planolites* and *Macaronichnus* resemble *Palaeophycus* in morphology but differ from it by their active fill. Other lined burrows, such as the horizontal parts of *Ophiomorpha* and *Siphonichnus*, may be similar to *Palaeophycus*, as is the case for parts of more complex trace fossils such as *Hillichnus* (Fig. 5.64b). A detailed study of many specimens commonly reveals the true nature of *Palaeophycus*.

*Producers:* Vermiform animals (e.g. annelids) are the most likely producers of *Palaeophycus*, although other groups of organisms (e.g. arthropods) cannot be ruled out. Modern traces similar to *Palaeophycus* are produced by nereidid polychaetes (Dashtgard and Gingras 2012; Gingras et al. 2012).

*Ethology: Palaeophycus* is commonly interpreted as the dwelling (domichnion) of a predaceous or suspension-feeding animal (Pemberton and Frey 1982).

*Depositional Environment: Palaeophycus* occurs in a wide range of paleoenvironments in marine and continental settings. In the continental realm, *Palaeophycus* is common in (but not restricted to) fluvial and lacustrine deposits (Fig. 5.5c, d). Low-diversity assemblages with diminutive burrows occur in marginal-marine environments with brackish conditions (e.g. estuarine, intertidal and subtidal; Fig. 5.93a, b), while dense *Palaeophycus* assemblages with large burrows are reported from delta deposits (lower delta plain and delta front; Fig. 5.93c). These occurrences indicate that the tracemaker was euryhaline. *Palaeophycus* is also common in shoreface and offshore deposits, where it occurs at much higher ichnodiversities. It has also been reported from continental slopes (Hubbard et al. 2012) and deep-sea fans (Uchman and Wetzel 2012; Fig. 5.93d).

*Ichnofacies:* In the marine realm, *Palaeophycus* belongs to the *Cruziana* Ichnofacies and, subordinately, to the *Skolithos*, *Zoophycos* and *Nereites* ichnofacies, while continental *Palaeophycus* occurs within a wide range of defined ichnofacies (MacEachern et al. 2012; Melchor et al. 2012), of which the *Mermia* and *Scoyenia* ichnofacies are probably most relevant.

*Age:* Aside from dubious reports of *Palaeophycus* from the Upper Proterozoic, reliable occurrences are known throughout the entire Phanerozoic.

*Reservoir Quality:* Under optimal circumstances and if occurring in high densities, passively sand-filled *Palaeophycus* burrows may contribute to an improved reservoir quality.

## 5.18 *Paradictyodora* Olivero et al., 2004

*Morphology, Fill and Size: Paradictyodora* is a complex vertical spreite burrow that widens upwards, displaying a prismatic to conical shape (Figs. 5.94, 5.95 and 5.96). It consists of subvertical folded laminae produced by the lateral migration of a subvertical J-shaped tube (Olivero et al. 2004). *Paradictyodora* is a relatively large burrow reaching several centimeters in vertical and lateral extent.

*Ichnotaxonomy:* Only two ichnospecies of *Paradictyodora* have been described so far: *P. antarctica* and *P. flabelliformis*. *Tursia* D'Alessandro and Fürsich, 2005 was erected approximately at the same time as *Paradictyodora* and is a junior synonym of it (Serpagli et al. 2008).

*Substrate:* All few records of *Paradictyodora* to date are from sandy substrates.

*Appearance in Core:* In horizontal view from the top (bedding-parallel), the laminae of *Paradictyodora* merge into a regularly to irregularly meandering or spiraling band with backfilled meniscate structure. In vertical view, laminae appear as series of regularly imbricate, oblique to subvertical bands (Olivero et al. 2004; Fig. 5.97). The few specimens observed in core show a winding and undulating trace ca. 12–15 cm long in vertical section. Individual spreite burrows are filled with alternating sand and mud, while a band-like trace appears in horizontal section.

*Similar Trace Fossils: Paradictyodora* may be confused with similar vertical spreite burrows such as *Dictyodora*, *Teichichnus*, *Heimdallia*, *Stellavelum*, *Zavitokichnus* and *Euflabella* (Olivero et al. 2004; D'Alessandro and Fürsich

(a)
(b)
(c)
(d)

◀**Fig. 5.93** *Palaeophycus* in sectioned core. *Scale bars* = 1 cm. **a** Sandstone with small specimens in oblique and cross section. Middle Jurassic (Bathonian) Hugin Formation (marginal marine), Gudrun Field, Norwegian North Sea (well 15/3-9T2, ca. 4490.4 m). **b** Argillaceous sandstone with complete bioturbation, consisting of a dense ichnofabric of small *Palaeophycus*. Middle Jurassic (Callovian) Hugin Formation (shallow marine), Sleipner Field, Norwegian North Sea (well 15/9-1, ca. 3532.3 m). **c** Sandstone with dense ichnofabric almost entirely consisting of large *Palaeophycus*. Lower Jurassic (Pliensbachian) Cook Formation (marginal marine, deltaic?), Norwegian North Sea (well 34/2-2, ca. 3670.0 m). **d** Silty sandstone with individual *Palaeophycus*. Upper Cretaceous (Campanian) Nise Formation (deep marine), Aasta Hansteen Field, Norwegian Sea (well 6707/10-1, ca. 3021.25 m)

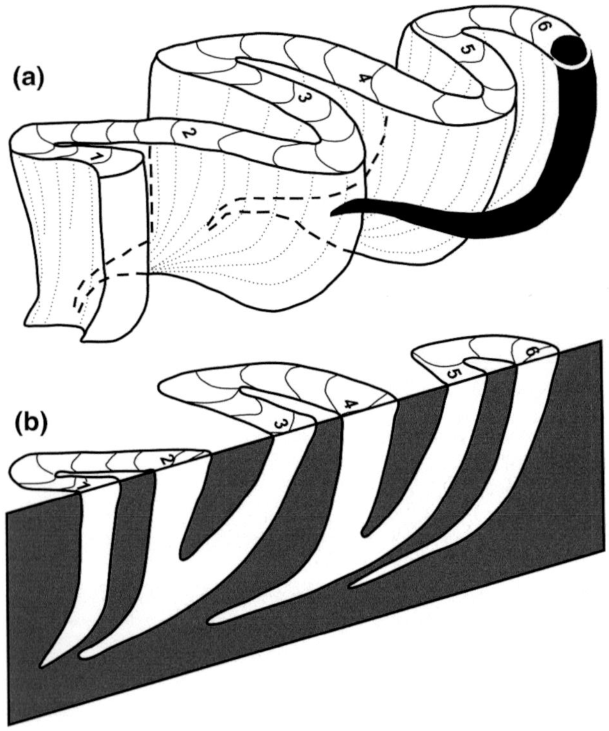


**Fig. 5.94** Reconstruction of *Paradictyodora antarctica*. From Olivero et al. (2004), © Paleontological Society, published by Cambridge University Press, reproduced with permission. **a** Three-dimensional view with the causative burrow in black and successive curved segments along the meandering band (*1–6*). **b** Plan view and vertical section

2005; Michalík and Šimo 2010; Olivero and López Cabrera 2013).

*Producers:* Deposit-feeding activity by the lateral shift of the inhalant siphon of tellinid bivalves and by the lateral displacement of the feeding shaft of the polychaete *Arenicola* were assumed as potential tracemaking processes by D'Alessandro and Fürsich (2005) (Fig. 5.98). Serpagli et al. (2008) argue that the preservation of the inhalant siphon trace (e.g. Fig. 5.97a, b) supports the tellinid bivalve model (Fig. 5.98).

*Ethology:* Deposit-feeding (fodinichnial) behavior can be inferred from the winding spreite burrows (Olivero et al. 2004; D'Alessandro and Fürsich 2005).

*Depositional Environment: P. antarctica* originally is described from fine-grained deposits of deep-marine fan systems (Olivero et al. 2004). *P. flabelliformis* occurs in

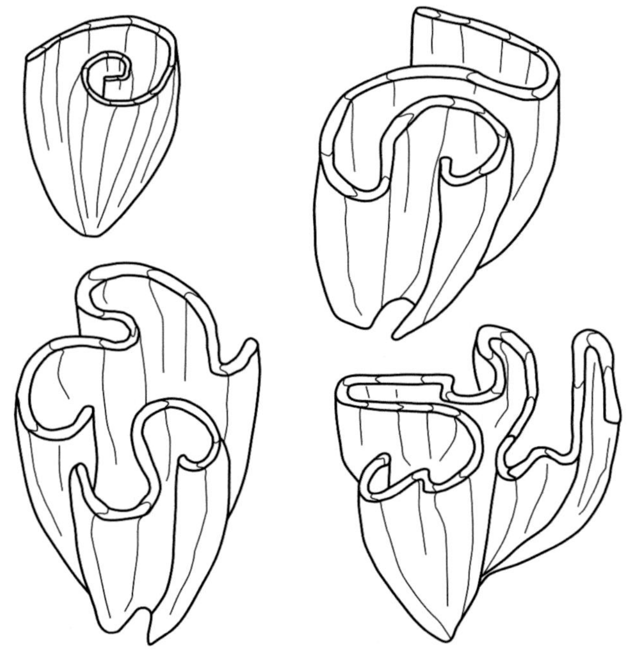

**Fig. 5.95** Variant of *Paradictyodora antarctica* (schematic reconstruction). From Olivero et al. (2004), © Paleontological Society, published by Cambridge University Press, reproduced with permission

inner shelf to lower shoreface settings (Bourgeois 1980; Serpagli et al. 2008) and in protected shoreface deposits (D'Alessandro and Fürsich 2005).

*Ichnofacies: Paradictyodora* seems to be part of the *Cruziana* Ichnofacies.

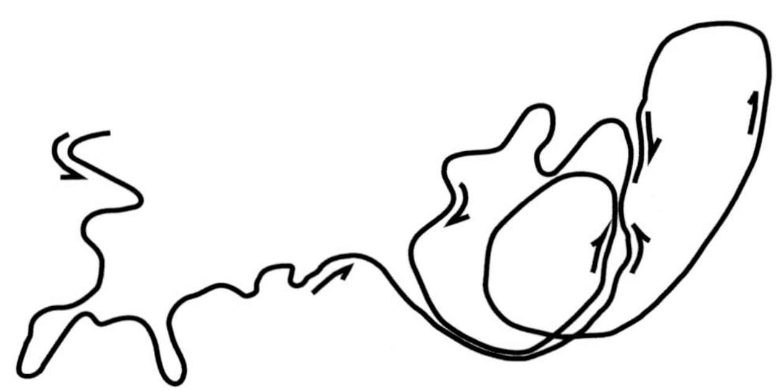

**Fig. 5.96** Plan view of course of lateral displacement of *Paradictyodora* (=*Tursia*) *flabelliformis*. *Arrows* indicate direction of spreite displacement. After D'Alessandro and Fürsich (2005), reprinted by permission of the publisher (Taylor & Francis Ltd., http://www.tandfonline.com)

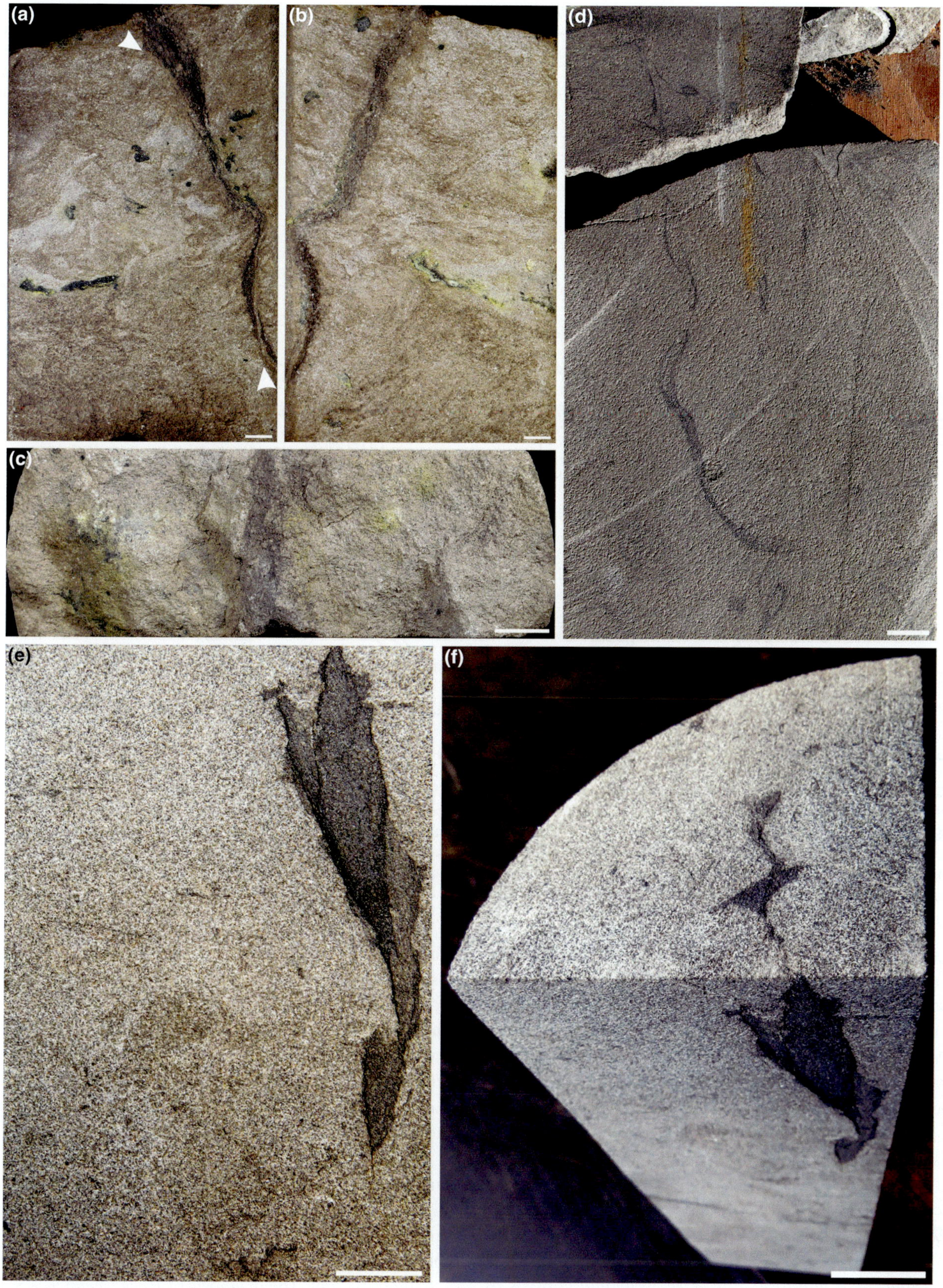
(a)
(b)
(c)
(d)
(e)
(f)

◀**Fig. 5.97** *Paradictyodora* in sectioned core from the Middle Jurassic (Bathonian-Callovian) Hugin Formation (shallow marine, **a–f**) and the Lower Jurassic (Pliensbachian) Amundsen Formation (shallow marine, **g**), Norwegian North Sea. *Scale bars* = 1 cm. **a**, **b** Completely bioturbated sandstone with a vertical section (part and counterpart) of an undulating and vertical spreite burrow, predominantly filled with mud and showing an internal, sand-filled tubular burrow (between arrow heads). **c** Same specimen as in (**a**) and (**b**), horizontal section at the top (well 15/6-4, ca. 10637′). **d** Sandstone with an alternating mud-sand spreite burrow with undulating course in vertical section. Sleipner Vest Field (well 15/9-1, ca. 3544.0 m). **e**, **f** Intensely bioturbated sandstone with a muddy spreite burrow in vertical and horizontal section (well 25/7-2, ca. 4465.3 m)

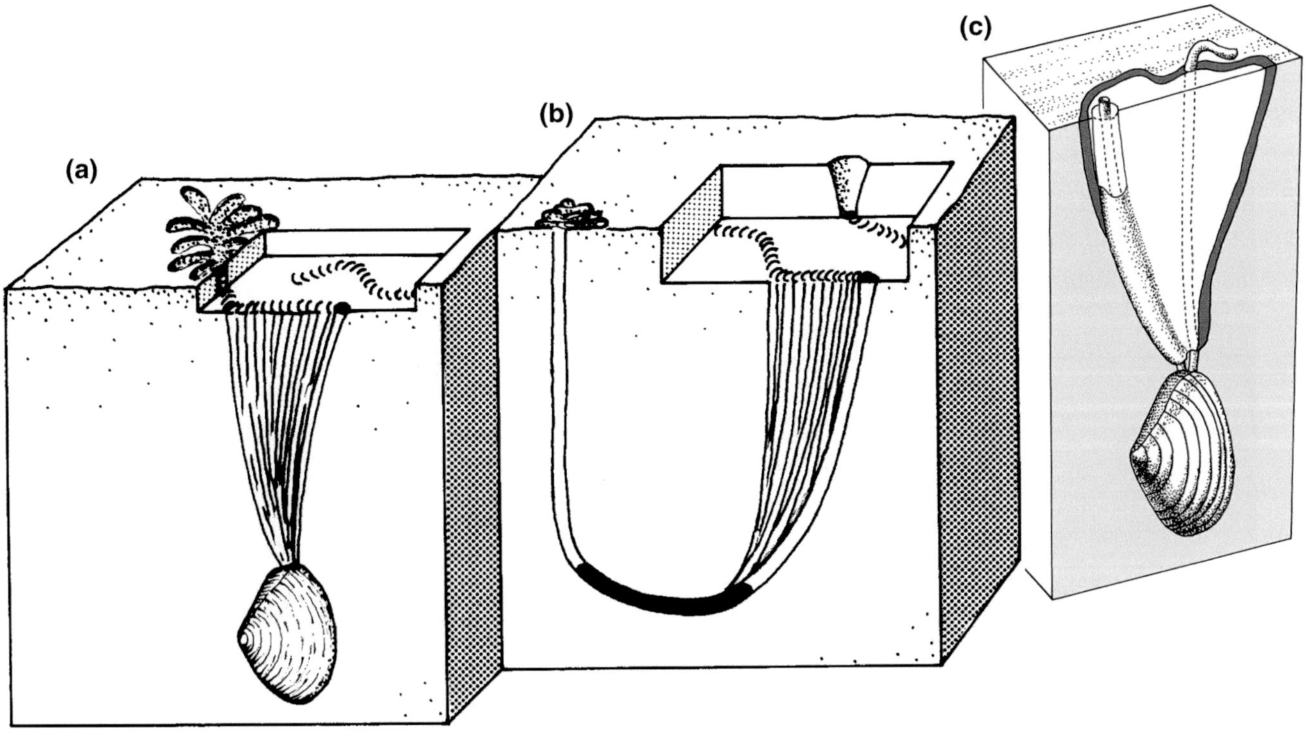


**Fig. 5.98** Alternative ethological models of *Paradictyodora* (=*Tursia*) *flabelliformis*. **a** Interpretation as produced by lateral shift of the siphon of a deep infaunal deposit-feeding bivalve (tellinid model). **b** Interpretation as expression of the lateral shift of one shaft of a vertical U-shaped burrow in connection with the feeding activity of a deposit-feeding worm-like organism (*Arenicola* model). **a** and **b** from D'Alessandro and Fürsich (2005), reprinted by permission of the publisher (Taylor & Francis Ltd., http://www.tandfonline.com). **c** Interpretation as produced by the wandering position of the inhalant siphon of a tellinid bivalve, while the trace of the exhalant siphon (*left*) remains finally preserved. From Serpagli et al. (2008)

*Age: Paradictyodora* has been described from the Late Cretaceous (Bourgeois 1980; Olivero et al. 2004) to the Pleistocene (D'Alessandro and Fürsich 2005). The core material presented herein extends the occurrence of *Paradictyodora* to the Early Jurassic.

*Reservoir Quality:* No evaluation of the reservoir quality influenced by *Paradictyodora* is known so far. Given its mud-rich active fill, reducing reservoir quality can be assumed in places where *Paradictyodora* occurs more densely.

## 5.19 *Parahaentzschelinia* Chamberlain, 1971

*Morphology, Fill and Size: Parahaentzschelinia* consists of numerous small, irregular, mud- and sand-filled tubes radiating vertically and obliquely upward to the sediment surface (Chamberlain 1971), which leads to an overall funnel-shaped burrow. A conical depression may be left on the upper surface of a bed where mud-filled tubes have been extensively developed. The type ichnospecies *P. ardelia* is relatively small (1.5 mm in tunnel diameter, 15–20 mm deep and 15–60 mm in surface diameter) and has an inclined main tunnel at its apex (Fig. 5.99). In contrast, *P. surlyki* Dam, 1990 is larger (4–20 mm in tunnel diameter, up to 50 mm deep and up to 120 mm in surface diameter) and consists of vertically bundled burrows with sand fill and mud lining, radiating vertically or obliquely upward to the sediment surface from a central and vertical main shaft. The tubes may show a distinct meniscate fill (see Głuszek 1998).

*Ichnotaxonomy:* The diagnosis of *P. egesheimense* Schweigert (1998) is only based on size differences and variations in tube density and therefore has to be ascribed to one of the

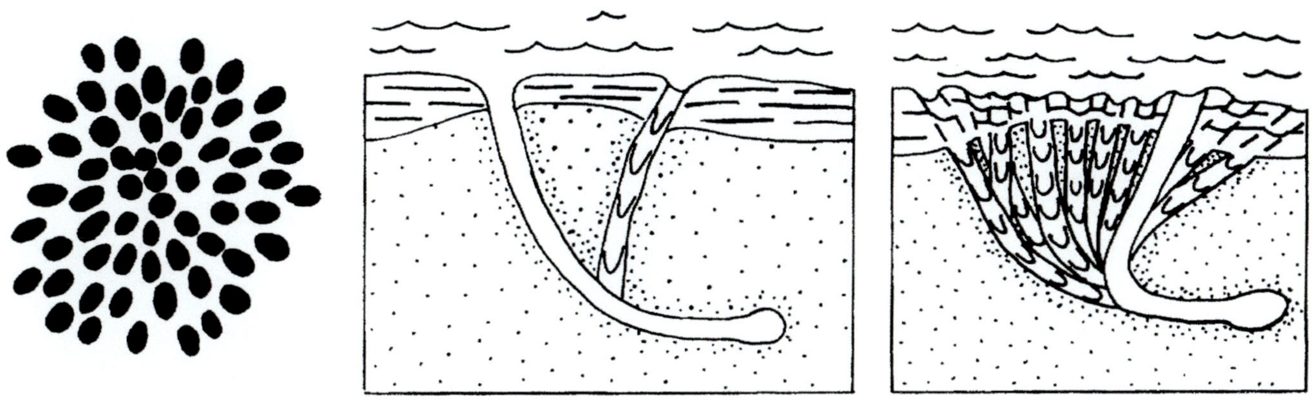

**Fig. 5.99** *Parahaentzschelinia ardelia* in plan view on the bedding surface (*left*), cross section of initial development of burrowing (*middle*), and complete perforation of sediment (*right*). From Chamberlain (1971), © Paleontological Society, published by Cambridge University Press, reproduced with permission

two existing ichnospecies of *Parahaentzschelinia*. Likewise, *Rosselia rotatus* McCarthy, 1979 with its backfilled main tube is similar to *P. ardelia* and may be regarded as junior synonym of it.

*Substrate:* *Parahaentzschelinia* is most common in siliciclastic sandy substrates but also occurs in limestone.

*Appearance in Core:* In vertical core sections, *Parahaentzschelinia* appears as funnel-shaped burrows (if sectioned in an axial position) with the main tube preserved in an apical position (Fig. 5.100). Marginal sections typically miss the basal tube and appear in form of an inverted conical structure. Internally, the burrow is intensely laminated or shows irregular convolution. One (the terminal) or more of the mud-lined tubes are typically preserved and can be either passively or actively filled with sand.

*Similar Trace Fossils:* The overall shape of *Parahaentzschelinia* commonly gives reason for confusion with similar ichnospecies of *Rosselia*, from which it differs by its overall funnel-like shape (instead of bulbous or spindle-like) and a more irregular and laminated internal structure (instead of a concentric geometry). In addition, more than one terminal tube can be preserved and is randomly positioned within the funnel-shaped burrow. Thus, *R. rotatus* McCarthy, 1979 is better accomodated within *Parahaentzschelinia* (Fig. 5.101). *Paradictyodora* is a vertical spreite structure with subvertical folded laminae (Olivero et al. 2004). In vertical sections, compact forms of *Paradictyodora* may be confused with *Parahaentzschelinia* but differ from it by curved and meandering spreite segments.

*Producers:* Originally attributed to the feeding activity of a worm-like animal, *Parahaentzschelinia* is now best explained as the trace of a tellinid bivalve (e.g. Bromley 1996; Fig. 5.102). Tellinid bivalves apply various methods of feeding with the means of their siphons, either in a subsurface position or, most commonly, on surface detritus. In some instances, the casts of the *Parahaentzschelinia*-producing bivalves are preserved below the funnel-shaped siphonal trace (Fig. 5.103). Furthermore, the burrowing behavior of particular holothurians may produce similar traces.

*Ethology:* By analogy with modern counterparts, *Parahaentzschelinia* can be interpreted as the feeding trace (fodinichnion) of a tellinid bivalve. The similar trace fossil *Paradictyodora* probably also results from the feeding activity of a tellinid bivalve (Serpagli et al. 2008).

*Depositional Environment:* *Parahaentzschelinia* was originally described from deep-marine deposits (Chamberlain 1971), where it was found subsequently by other workers (Uchman 1995, 1998; Tunis and Uchman 1996; Monaco 2008; Heard and Pickering 2008; Wetzel 2008; Fig. 5.100c). It is often reported from shallow-marine environments with higher energy conditions (Fig. 5.104), including foreshore (Fürsich et al. 2006), tidal flats (Mángano and Buatois 2004), high-energetic shoal (Knaust 2009a), shoreface (Bann and Fielding 2004), storm-dominated shelf (Dam 1990) and tide-dominated delta (McIlroy 2007). *Parahaentzschelinia* has also been encountered in lithographic limestone deposited in a coastal lagoon with influence of turbidity currents (Schweigert 1998), and from other marginal-marine (estuarine?) settings.

*Ichnofacies:* Shallow-marine occurrences of *Parahaentzschelinia* are typical constituents of the *Skolithos* Ichnofacies, while deep-marine occurrences are associated with the *Ophiomorpha rudis* ichnosubfacies.

*Age:* *Parahaentzschelinia* occurs from the Carboniferous (Chamberlain 1971; Głuszek 1998) to the Holocene (Wetzel 2008). A wide distribution is known from the Jurassic.

*Reservoir Quality:* Due to the incorporation of various amounts of mud into the burrows, *Parahaentzschelinia* may have a slightly negative impact on reservoir quality, which is partly compensated for by the preferred vertical to slightly oblique orientation of the tubes.

(a)
(b)
(c)
(d)
(e)
(f)
(g)
(h)

◄**Fig. 5.100** *Parahaentzschelinia* in sectioned core. *Scale bars* = 1 cm. (**a**), (**b**) and (**f**) from Knaust (2015a), republished with permission of Elsevier; permission conveyed through Copyright Clearance Center, Inc. **a** *P. surlyki*, nicely displaying the central mud-laminated tube and irregular disturbance with mud incorporated into the funnel-like burrow. Middle Jurassic (Bajocian) Ile Formation (marginal marine), Trestakk Discovery, Norwegian Sea (well 6406/3-2, ca. 4075.35 m). **b** A funnel-shaped and sideritic specimen of *P. surlyki*. Middle Jurassic (Bajocian) Ile Formation (marginal marine), Norwegian Sea (well 6406/8-1, ca. 4382.95 m). **c** Small specimen emplaced in a thin turbiditic sand bed, with the bivalve cast preserved below the funnel. Lower Cretaceous (Albian, continental slope), off Tanzania. **d** Large, undulating specimen of *P. surlyki* (outlined with *arrow heads*) with partly pyritized (brown) tubes. Middle Jurassic (Callovian-Oxfordian) Hugin Formation (marginal marine), Johan Sverdrup Field, Norwegian North Sea (well 16/2-7, ca. 1967.6 m). **e** Small specimen in ripple-laminated sandstone. Lower Jurassic (Sinemurian-Pliensbachian) Tilje Formation (marginal marine), Åsgard Field, Norwegian Sea (well 6506/12-K-3H, ca. 4524.5 m). **f** Three funnel-shaped specimens of *P. surlyki* in ripple-laminated sandstone. The burrows are accompanied with and partly cross-cut by *Siphonichnus*. Middle Jurassic (Bajocian) Ness Formation (deltaplain), Valemon Field, Norwegian North Sea (well 34/10-23, ca. 3236.5 m). **g** Small specimen in ripple-laminated sandstone. Lower Jurassic (Pliensbachian) Åre Formation (marginal marine), Skuld Field, Norwegian Sea (well 6608/10-12, ca. 2796.9 m). **h** Dense ichnofabric including *Siphonichnus*, *Teichichnus* and *Schaubcylindrichnus*, overprinted by a relatively small *P. surlyki* (*middle left*). Lower Jurassic (Toarcian) Cook Formation (lower shoreface), Norwegian North Sea (well 34/2-4, ca. 3832 m)

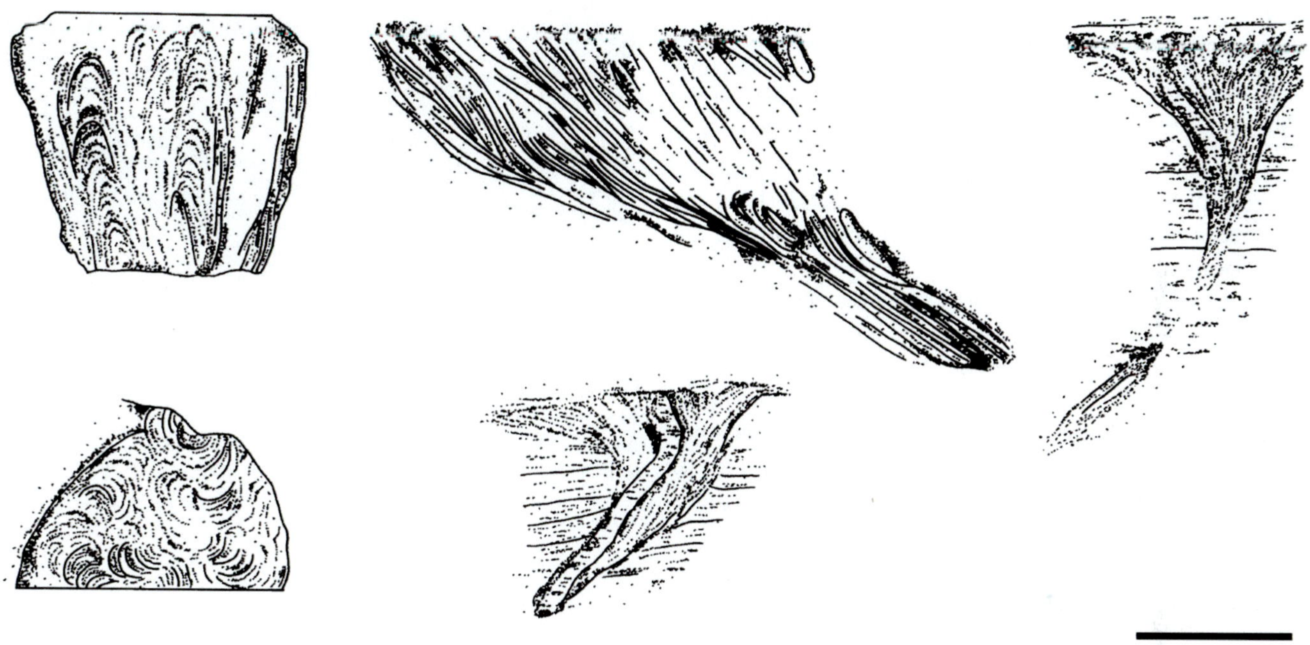

**Fig. 5.101** *Parahaentzschelinia ardelia* (=*Rosselia rotatus*) from Permian shoreface to foreshore deposits from the Sydney Basin, Australia. Holotype of *R. rotatus* in vertical (*upper left*) and horizontal (*lower left*) section, as well as different burrows in vertical section (*middle and right*), showing the overall funnel-like shape, mud-lined tubes, and inclination to bedding. *Scale bar* = 3 cm. Modified after McCarthy (1979), © Paleontological Society, published by Cambridge University Press, reproduced with permission

## 5.20 *Phoebichnus* Bromley and Asgaard, 1972

*Morphology, Fill and Size: Phoebichnus* is a horizontal, star-like burrow system with a relatively thick, vertical central shaft (or boss), from which numerous straight, long burrows radiate (Fig. 5.105). The burrows have a wall with discrete annuli and are actively filled with a meniscate backfill (Bromley and Asgaard 1972; Evans and McIlroy 2015). The central shaft is often subject to diagenetic modification (e.g. cementation). The entire burrow system may reach several decimeters in diameter, while the diameter of individual burrows typically ranges between 1 and 2 cm (Fig. 5.106).

*Ichnotaxonomy:* Beside the type ichnospecies *P. trochoides* from the Jurassic, three other ichnospecies have been added to this ichnogenus. *P. minor* Li et al., 1999 (from the Lower Cambrian of China) and *P. dushanensis* Yang et al., 2004 (from the Carboniferous of China) remain poorly defined and probably belong to other ichnogenera, while the

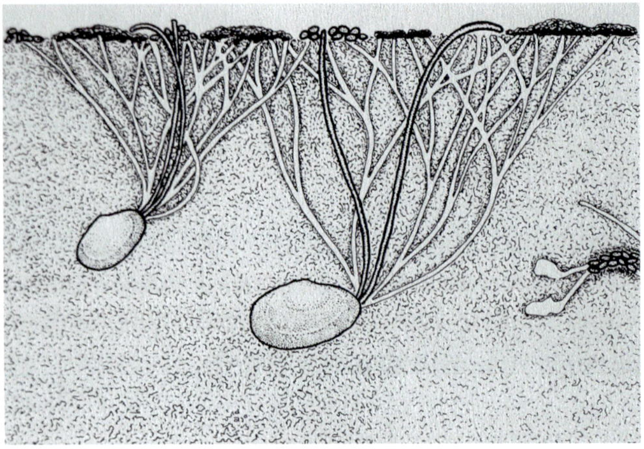

**Fig. 5.102** Feeding activity of deep-sea tellinid bivalves *Abra nitida*, which may result in *Parahaentzschelinia*. Zones of siphonal activity shown as networks of abandoned canals. *A. nitida* feeds on surface detritus and places both pseudofeces and feces on the sea floor. After Wikander (1980), Bromley (1996), reprinted by permission of the publisher (Taylor & Francis Ltd., http://www.tandfonline.com)

radiating burrows in *P. bosoensis* Kotake, 2003 (from the Pleistocene of Japan) lack the diagnostic thick wall and are filled with pellets.

*Substrate: Phoebichnus* is predominantly reported from micaceous sandstone but also from mixed siliciclastic-carbonate sediments (Joseph et al. 2012).

*Appearance in Core:* In core, *Phoebichnus* appears in the form of bundled burrows sectioned in various directions. Most common are cross sections more or less vertical to the burrow axis, which are accompanied by oblique sections

**Fig. 5.103** Core section displaying the cast of a small bivalve (*arrow*) and a series of emerging, funnel-shaped siphon traces above it. Lower Jurassic (Pliensbachian-Toarcian) Cook Formation (shallow marine, well 34/5-1S, ca. 3648.65 m). *Scale bar* = 1 cm

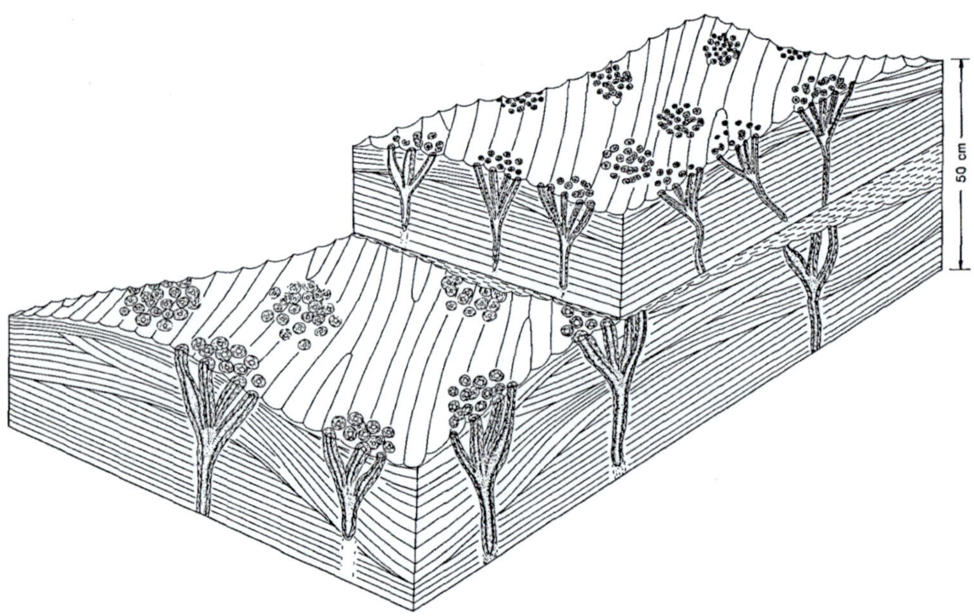


**Fig. 5.104** *Parahaentzschelinia surlyki* associated with hummocky cross-stratified sandstone in the Lower Jurassic Neill Klinter Formation (shallow marine), East Greenland. From Dam (1990), republished with permission of the Geological Society of Denmark

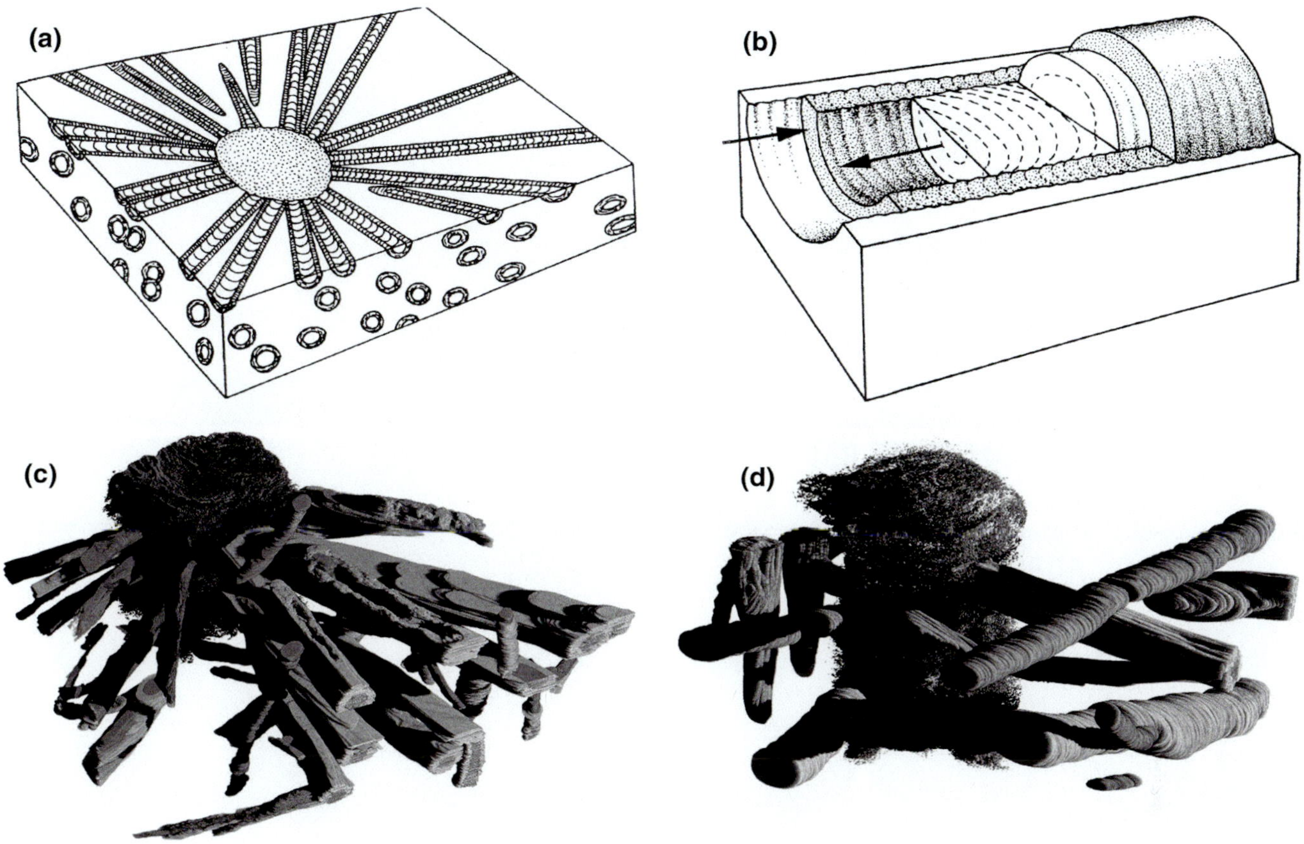


**Fig. 5.105** Bauplan and three-dimensional reconstructions of *Phoebichnus trochoides*. (**a**) and (**b**) from Bromley and Mørk (2000), republished with permission of Schweizerbart (www.schweizerbart.de/series/zgp1). (**c**) and (**d**) based on serial grinding, from Evans and McIlroy (2015), republished with permission of Wiley; permission conveyed through Copyright Clearance Center, Inc. **a** Structure showing vertical axis and the proximal parts of the radial branches. **b** Part of a radial branch, showing the two-zoned backfill. **c** Central boss with radial burrows. **d** Radial burrows with outer wall and laminated fill. Radial burrows are approximately 1 cm in diameter.

with an elliptical outline (Fig. 5.107). The radiating burrows are relatively large (commonly more than 1 cm in diameter) and consist of a thick wall surrounding the actively filled core of the burrow.

*Similar Trace Fossils:* In outcrop, *Phoebichnus* could be confused with other stellate burrows, e.g. *Stelloglyphus*. In core, other large burrows with a thick wall or mantle and an active fill could resemble the radiating burrow segments of *Phoebichnus*, for instance actively filled and thick-walled *Ophiomorpha* as well as *Macaronichnus*. In addition, horizontal sections of large *Siphonichnus* may be confused with individual *Phoebichnus*. Finally, root traces (fossilised root systems) may be in the size range of *Phoebichnus* systems and can produce similar patterns (Gregory and Campbell 2003).

*Producers:* Vermiform organisms (such as Echiura) have been assumed as the producers of *Phoebichnus* (Bromley and Mørk 2000; Kotake 2003), although arthropods (in particular crustaceans) may also be good candidates (Evans and McIlroy 2015).

*Ethology: P. trochoides* is interpreted as the feeding trace (fodinichnion) of a sessile deposit-feeding animal, which is thought to live in the central shaft (domichnion) and producing the radiating feeding burrows (Bromley and Asgaard 1972).

*Depositional Environment: P. trochoides* is a shallow-marine trace fossil with occurrences in shoreface and delta deposits (e.g. Martin and Pollard 1996; McIlroy 2004; Morris et al. 2006; MacEachern and Bann 2008; Pemberton et al. 2012; Joseph et al. 2012) and on the shelf (Heinberg and Birkelund 1984; Pemberton and Frey 1984; Dam 1990; Bromley and Mørk 2000). The dependence of the producer on sediment rich in food particles suggests a low-energy environment (Heinberg and Birkelund 1984), an interpretation supported by the frequent preservation of entire burrow systems.

**Fig. 5.106** *Phoebichnus trochoides* in outcrop. **a** Large burrow system with a cemented central area (vertical shaft), from which numerous burrows radiate. Middle Jurassic (Callovian) sandstone, Scarborough, Yorkshire, UK. **b** Part of a burrow system with the central area and emerging radials displaying the mantle and meniscate fill. Same locality as in (**a**). *Scale bar* = 1 cm. **c** Dense ichnofabric mainly resulting from overlapping burrow systems. Same locality as in (**a**). **d** Large stellate burrow system with cemented vertical shaft. Upper Cretaceous (Turonian) sandstone, Cardium Formation, Seebe Dam, Alberta, Canada. *Scale bar* = 15 cm. See Pemberton and Frey (1984)

*Ichnofacies: Phoebichnus* can be a common part of the *Cruziana* Ichnofacies. The "*Phoebichnus* Ichnofacies" proposed by Heinberg and Birkelund (1984) has not found acceptance among subsequent workers.

*Age: P. trochoides* is a common trace fossil in Mesozoic (Triassic to Cretaceous) deposits (Evans and McIlroy 2015).

*Reservoir Quality:* The complex nature of *Phoebichnus* burrows with their actively created mantle and core indicates increased heterogeneity of the bioturbated substrate, contributing to an overall decrease of potential reservoir properties.

## 5.21 *Phycosiphon* Fischer-Ooster, 1858

*Morphology, Fill and Size: Phycosiphon* is a small spreite burrow consisting of repeated narrow, U-shaped lobes, each enclosing a spreite at a millimetric to centimetric scale, and branching regularly or irregularly from an axial spreite of similar width (Wetzel and Bromley 1994; Fig. 5.108). The lobes typically consist of a mud-dominated marginal string and a silty or sandy spreite (Fig. 5.109). The burrow systems are mainly parallel to bedding, although oblique and even vertical sections may occur too. Three-dimensional reconstructions of "phycosiphoniform" burrows suggest an

**Fig. 5.107** *Phoebichnus trochoides* in sectioned core. *Scale bars* = 1 cm. **a** Heterolithic (silty) sandstone with moderate bioturbation and few burrows in cross section in the upper part (*t*). Middle Jurassic (Bajocian) Ile Formation (shallow marine, prodelta), Norwegian Sea (well 6406/8-1, ca. 4388 m). **b** Highly bioturbated silty sandstone with oblique burrow sections in the middle. Middle Jurassic (Bajocian) Ile Formation (shallow marine, lower shoreface), Norwegian Sea (well 6406/8-1, ca. 4479 m). **c** Heterolithic and ripple-laminated sandstone with a *P. trochoides* ichnofabric, consisting of a cluster of radiating burrows (*center-right*). Middle Jurassic (Bajocian) Ile Formation (shallow marine, prodelta), Njord Field area, Norwegian Sea (well 6407/10-2, ca. 3455.5 m). See also McIlroy (2004). **d** A cluster of *P. trochoides* burrows in cross and oblique sections. Lower to Middle Jurassic (Toarcian-Aalenian) Stø Formation (offshore), Snøhvit Field, Norwegian Barents Sea (well 7120/8-3, ca. 2205.5 m)

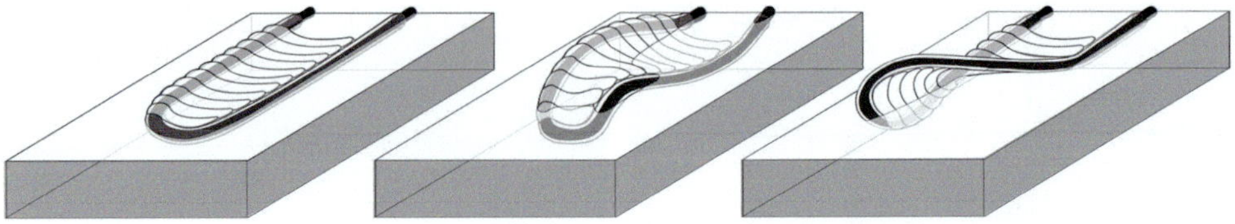

**Fig. 5.108** Conceptual model showing the non-planar orientation of a single *Phycosiphon* burrow lobe with the mantle and spreite shown as being transparent to facilitate viewing of the central mudstone strand. Lobe parallel to the bedding plane (*left*) and possible variations of twisted *Phycosiphon* burrow lobes (*middle* and *right*). From Bednarz and McIlroy (2009), republished with permission of the Palaeontological Association

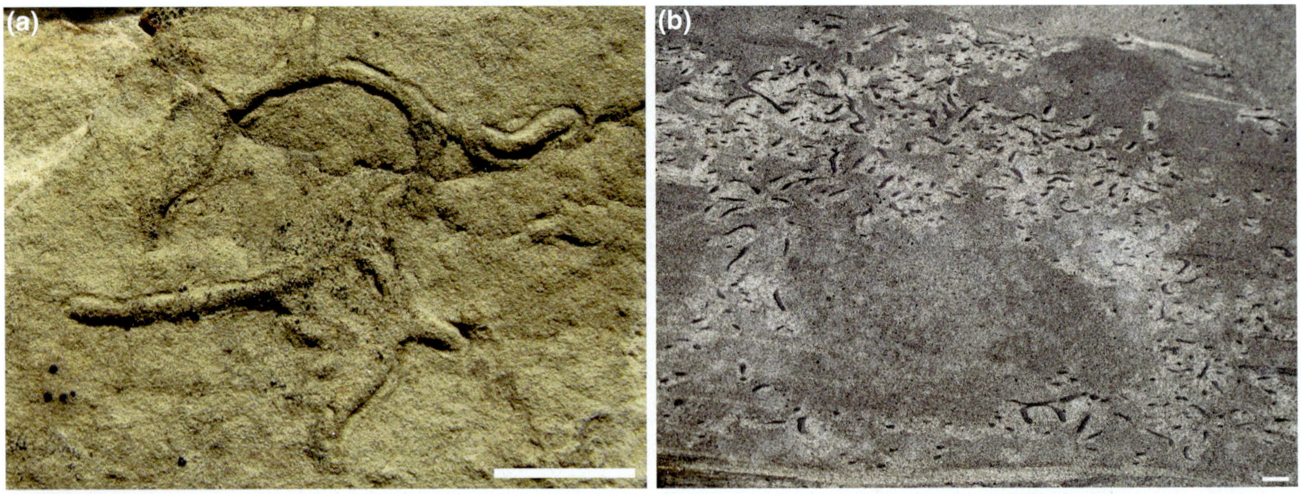


**Fig. 5.109** *Phycosiphon* in outcrop. *Scale bar* = 1 cm. **a** Bedding-plane preservation of a burrow system with empty marginal tunnel (mud fill is weathered out) and enclosed spreite area. Eocene Grès d'Annot Formation (deep marine, turbiditic), southeastern France. From Knaust et al. (2014), republished with permission of Wiley; permission conveyed through Copyright Clearance Center, Inc. **b** Vertical face with various sections of *Phycosiphon*. Miocene Mount Messenger Formation (deep marine, channel-levee system), sea cliff of the Taranaki Peninsula, North Island, New Zealand

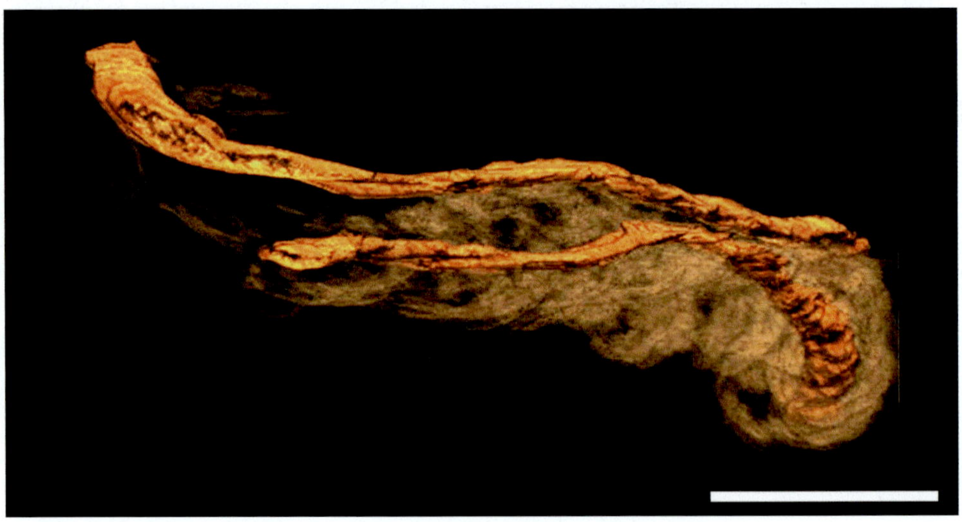

**Fig. 5.110** Three-dimensional reconstruction of a phycosiphoniform burrow (core with halo). *Scale bar* = 1 cm. From Bednarz (2014)

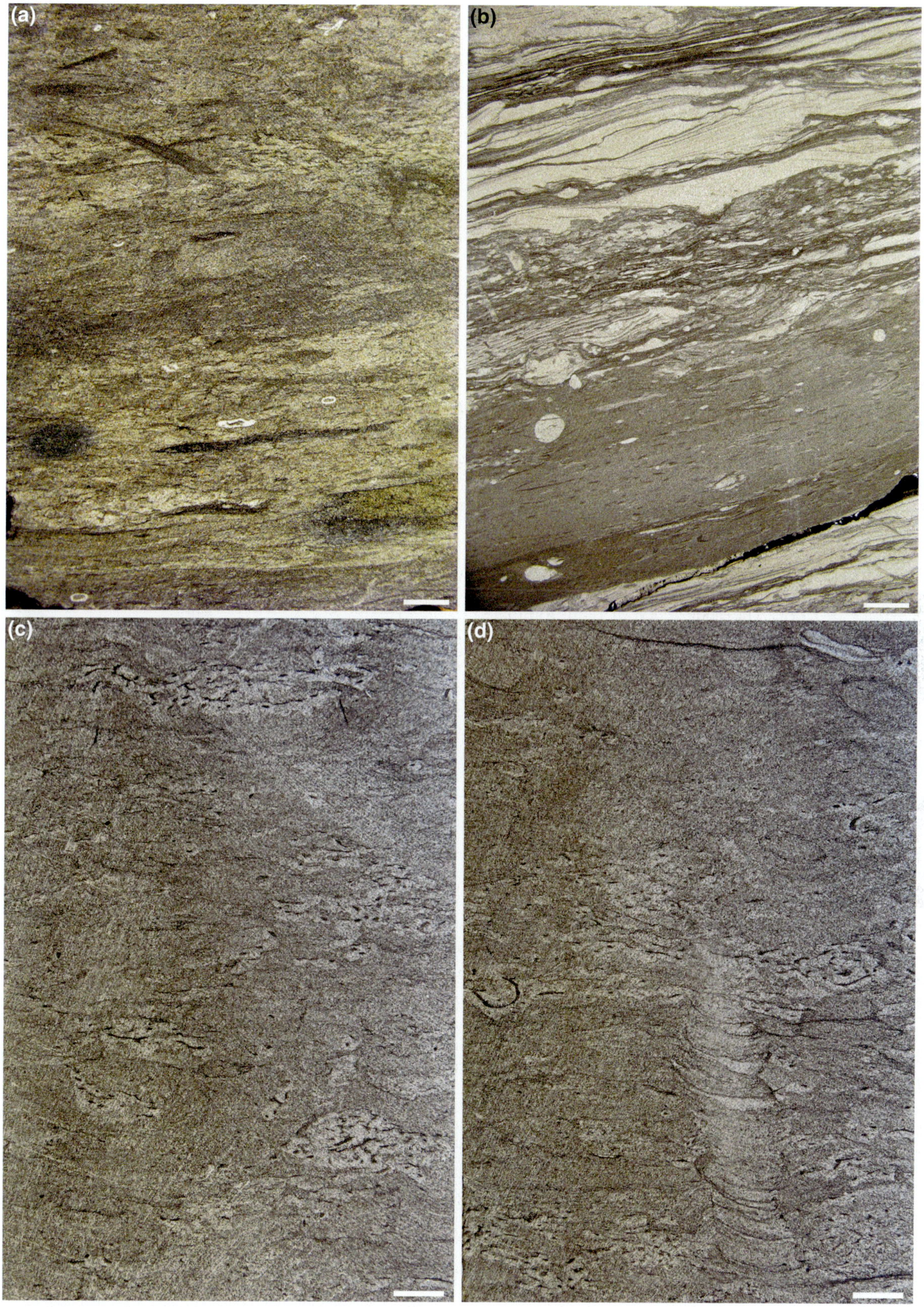
(a)
(b)
(c)
(d)

◀**Fig. 5.111** *Phycosiphon* in sectioned core. *Scale bar* = 1 cm. **a** Heterolithic sandstone with high degree of bioturbation, resulting in an ichnofabric dominated by *Phycosiphon* (dark muddy spots surrounded with light sandy haloes), accompanied by *Teichichnus* and *Schaubcylindrichnus*. Upper Jurassic (Oxfordian) Heather Formation (shelf), Fram Field, Norwegian North Sea (well 35/11-9, ca. 2732.85 m). **b** Heterolithic sandstone with moderate bioturbation. *Phycosiphon* is concentrated within the fine-grained (muddy to silty) layer and is accompanied by *Palaeophycus* and *Teichichnus*. Lower Jurassic (Pliensbachian-Toarcian) Cook Formation (offshore transition), Norwegian North Sea (well 34/5-1S, ca. 3654.5 m). **c, d** Totally bioturbated silty sandstone with *Phycosiphon* concentrated in larger, sand-filled burrows (*Thalassinoides*) and accompanied by *Teichichnus* and *Schaubcylindrichnus*. Lower Jurassic (Pliensbachian to Bajocian) Stø Formation (offshore transition), Iskrystall Discovery, Norwegian Barents Sea (well 7219/8-2, ca. 2986.5 m)

irregularly winding mud tube surrounded by a sandy halo (Naruse and Nifuku 2008; Bednarz and McIlroy 2009; Fig. 5.110).

*Ichnotaxonomy: P. incertum* is the only ichnospecies of *Phycosiphon*. The three-dimensional *Anconichnus horizontalis* is now regarded as a junior synonym of *P. incertum* (Wetzel and Bromley 1994). Bednarz and McIlroy (2009) prepared three-dimensional reconstructions of phycosiphoniform burrows and realized morphological differences from the type material as redescribed by Wetzel and Bromley (1994).

*Substrate: Phycosiphon* typically occurs in silty or fine-grained sandy, mainly siliciclastic deposits with a softground origin. It is more rarely reported from carbonates and chalk.

*Appearance in Core:* The occurrence of mud-dominated "strings", often in high abundance, in combination with reworked patches (spreite) of sand or silt, is a characteristic feature of *Phycosiphon* in core (Fig. 5.111, see also Figs. 5.19c, 5.129a and 5.179a). Because the spreite was often formed slightly deeper in the sediment than the marginal burrow, these reworked patches often occur slightly lowered in comparison with the associated mud-dominated strings. Typical expressions of *Phycosiphon* in vertical sections are "… clusters of closely spaced spots or comma-shaped dots and hooks filled with darker and finer sediments surrounded by narrow (ca. 1 mm) pale mantles. Generally, various orientations of the lobes result in a chaotic arrangement of dark cores in the sections." (Naruse and Nifuku 2008).

*Similar Trace Fossils: Nereites* appears to be similar to *Phycosiphon* (particularly in core sections), but differs from it by the lack of the enclosed spreite and a commonly horizontal course. In general, *Phycosiphon* is smaller than *Nereites*. *Phycosiphon* could also be confused with *Chondrites*, which shows dichotomous branching and lacks the sandy spreite.

*Producers:* It is likely that the tracemakers of *Phycosiphon* are small, vermiform organisms of unknown affinity.

*Ethology: Phycosiphon* burrows are best explained as the deposit-feeding activity of small vermiform organisms that exploit the sediment for organic-rich matter (Wetzel 2010; Izumi 2014; Fig. 5.112).

*Depositional Environment: Phycosiphon* is a characteristic component of offshore (shelf) to lower shoreface deposits, where it commonly occurs in isolation or in higher ichnodiversity within siliciclastic successions (Goldring et al. 1991; Pemberton et al. 2012). It occurs in slope deposits (e.g. Savrda et al. 2001), where it can be related to slump deposits and act as a paleoslope indicator (Naruse and Nifuku 2008). *Phycosiphon* is also common in deep-marine settings such as channel-levee complexes, where it occurs in marginal channel-levee facies (Callow et al. 2013; Fig. 5.136). *Phycosiphon*-producing organisms are among the first colonizers of event deposits which result from storm, turbidity current and bottom current deposition

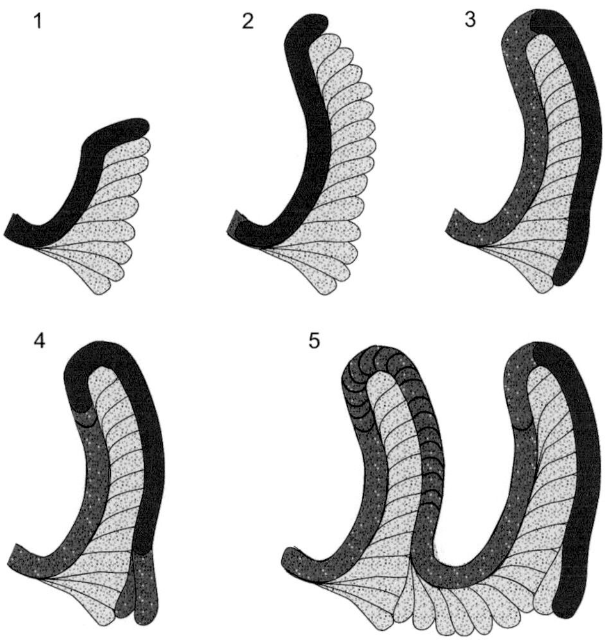


**Fig. 5.112** Reconstruction showing how multiple phases of foraging by an unknown vermiform organism creates phycosiphoniform looped burrows composed of marginal tube and spreite. From Bednarz and McIlroy (2009), republished with permission of the Palaeontological Association. Different shades of gray represent distribution of silt-sized (*light gray*) and mud-sized (*dark gray*) material. *1* Foraging organism creates feeding probes lateral to the marginal tube. *2* Successive probes are made until the organism has produced a marginal tube the length of its body. *3* Outer margin of the loop is produced by the organism moving along previously produced probes. *4* Second loop is stared after the organism body is straight once again. *5* Completion of second loop

(Goldring et al. 1991; Wetzel and Uchman 2001; Wetzel et al. 2008).

*Ichnofacies: Phycosiphon* is characteristic of the *Cruziana, Zoophycos* and *Nereites* ichnofacies.

*Age: Phycosiphon* is a common constituent of Mesozoic and Cenozoic deposits (Goldring et al. 1991) and is formed today (Wetzel 1991, 2008). In their analysis, Callow and McIlroy (2011) observed a dominance of phycosiphoniform ichnotaxa in the Mesozoic and Cenozoic, while *Nereites* prevails in Paleozoic rocks. These authors speculated about a change in an ecological niche from *Nereites*-dominated Paleozoic to *Phycosiphon*-dominated Mesozoic and Cenozoic.

*Reservoir Quality:* The effect of phycosiphoniform burrows on shale hydrocarbon reservoir quality was evaluated by Bednarz and McIlroy (2012), who documented an improved reservoir capacity, permeability and fracturability. In conventional reservoirs, the reworked sandy zone (spreite) in phycosiphoniform burrows may also lead to a slight increase of porosity and permeability.

## 5.22 *Planolites* Nicholson, 1873

*Morphology, Fill and Size: Planolites* is a simple, horizontal to slightly inclined, cylindrical burrow without branching and lining, and with an active (homogeneous) fill (Pemberton and Frey 1982; Keighley and Pickerill 1995; Figs. 5.113 and 5.114). In plan view, *Planolites* appears as straight to tortuous burrows (Fig. 5.178b). The cross section of the burrows is circular to elliptical, which is partly a function of the degree of compaction. An ichnospecies of *Planolites* with both striation and annulation was described by Stanley and Pickerill (1994). Microbially mediated biomineralization along the burrow margin is interpreted as agent in the promoted preservation of Cambrian *Planolites* (Ahn and Babcock 2012). The dimensions of *Planolites* can be quite variable and range from millimetric (e.g. Marenco and Bottjer 2008) to centimetric.

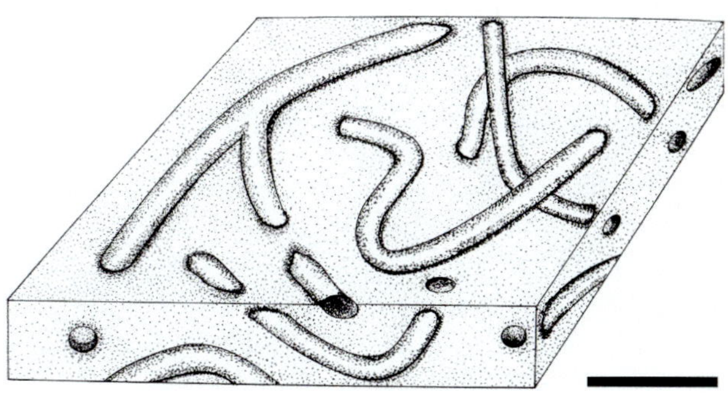

**Fig. 5.113** *Planolites montanus* in a reconstruction by Howard and Frey (1984), republished with permission of Canadian Science Publishing; permission conveyed through Copyright Clearance Center, Inc. Note the false branching due to overlap of two burrows. *Scale bar* = 1 cm

*Ichnotaxonomy:* In their review of the ichnogenus *Planolites*, Pemberton and Frey (1982) recognized only the ichnospecies *P. montanus, P. beverleyensis* and *P. annularis* as valid. Since then, about ten more ichnospecies have been introduced, some of which are poorly known and may turn out to be junior synonyms.

*Substrate: Planolites* is a characteristic component of softgrounds and occurs in siliciclastic and carbonate (including chalk), fine- to medium-grained sediment.

*Appearance in Core:* In core, the elongate horizontal tubes of *Planolites* have circular to elliptical sections that are actively filled but have no lining. Sand-filled *Planolites* often occurs in mud-dominated lithologies and thus is easy to recognize (Fig. 5.115, see also Fig. 4.3). In contrast, recognition of *Planolites* without significant differences in substrate (e.g. sand-filled burrows within sandy host sediment) can be difficult or impossible, and increasing bioturbation may just lead to a diffuse bioturbate texture (Fig. 5.179b, d).

*Similar Trace Fossils:* Their simplicity leads to confusion of *Planolites* burrows with numerous other trace fossils when studied in core. Many other cylindrical burrows are actively filled and have no lining, but are commonly branched, a fact that is often hard to test in core. Confusion with thinly lined *Palaeophycus* may occur, but this ichnogenus differs from *Planolites* by having a passive fill. Similarities do also exist between *Planolites* and *Macaronichnus*, both ichnogenera sharing the same overall geometry. Although *Macaronichnus* also has an active fill, its surrounding mantle differs from the smooth margin in *Planolites*.

*Producers: Planolites* is a relatively simple trace fossil that can be produced by a wide range of organisms belonging to different phyla. Wormlike animals (such as annelids, hemichordates and priapulids) are often interpreted as producers of *Planolites*, but arthropods (e.g. crustaceans) and molluscs (e.g. bivalves) are likewise able to produce such traces. Regarding the long stratigraphic range of *Planolites* it is likely that this trace fossil was constructed by various types of animals over time.

*Ethology: Planolites* is interpreted as the product of deposit-feeders (fodinichnia), which actively process the sediment.

*Depositional Environment: Planolites* has been reported from all aquatic environments in marine as well as non-marine settings. It is a common element of shallow-tier ichnofabrics.

*Ichnofacies: Planolites* is a facies-crossing element and can be found in various ichnofacies. It is, however, a common constituent of the *Cruziana* Ichnofacies and also occurs in the *Nereites* Ichnofacies (e.g. Buatois and Mángano

**Fig. 5.114** *Planolites* from outcrop. *Scale bars* = 1 cm. **a**, **b** Specimen with numerous sand-filled *P. montanus* covered with mud. Upper Carboniferous paralic coal measure, western Germany. Senckenberg coll., Frankfurt Main (holotype; Richter 1937) in bedding-plane (**a**) and cross-section view (**b**). **c** *P. montanus* on bedding plane. Lower Ordovician sandstone-shale heterolithics (deltaic) near Ritland, Rogaland, southwestern Norway. **d** Same as (**c**), vertical section. See Knaust (2004a). **e**, **f** *Planolites* isp. in bioclastic limestone in vertical section (**e**) and on bedding plane (**f**). Lower Cretaceous (Barremian) of Cabo Espichel, Portugal. **g**, **h** *Planolites* isp. on bedding plane of calcareous sandstone. Close association with *Lockeia* isp. and transitional forms with *Protovirgularia* isp. suggest bivalves as producers. Middle Triassic (Anisian) Udelfangen Formation near Trier, western Germany. See Knaust et al. (2016) for geological setting

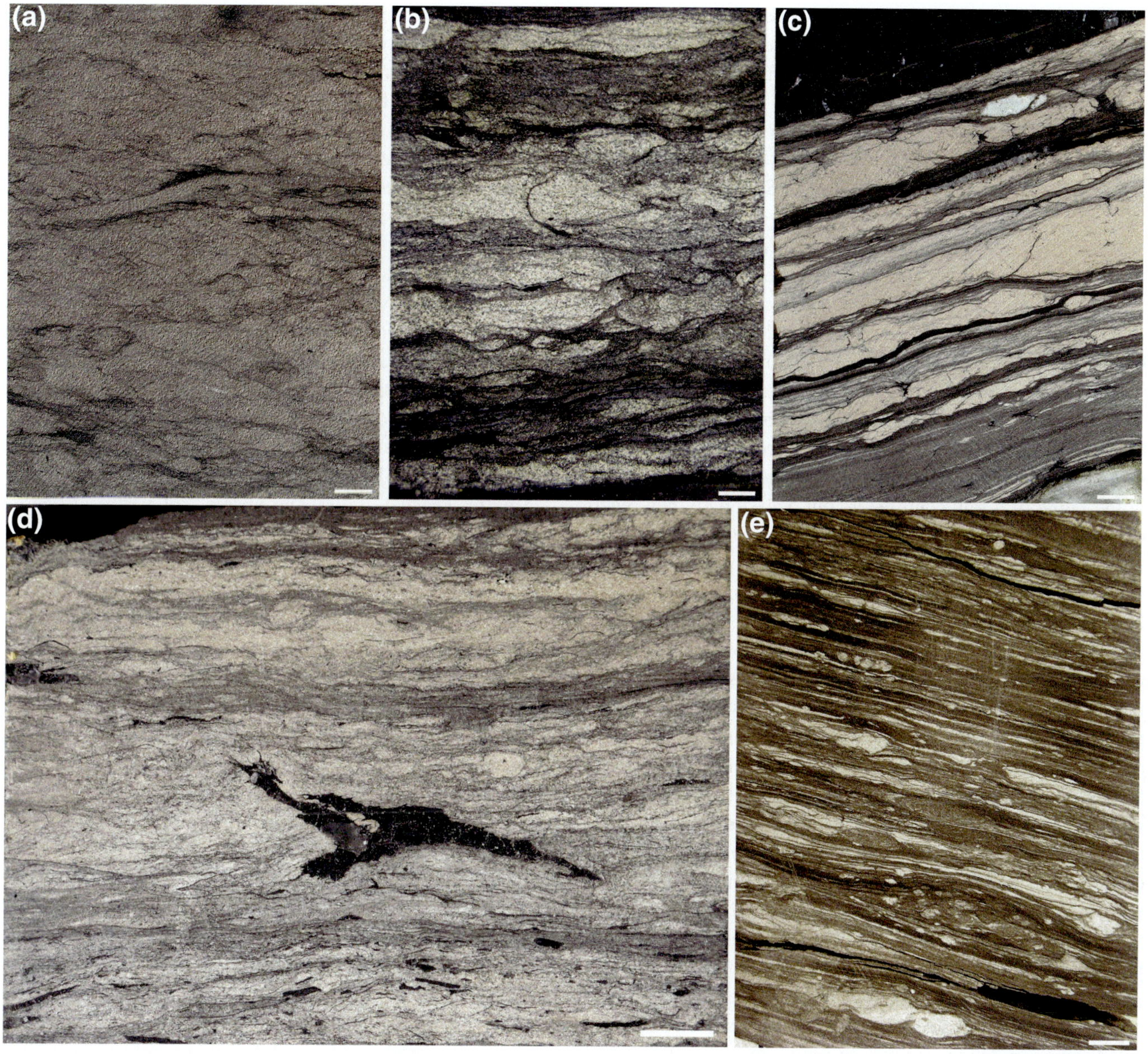


**Fig. 5.115** *Planolites* in sectioned core. *Scale bars* = 1 cm. **a** Dense occurrence of *Planolites* in sandstone. Middle Jurassic (Bajocian-Bathonian) Hugin Formation (shallow marine, shoreface), Sleipner Vest Field, Norwegian North Sea (well 15/9-7, ca. 3565.4 m). **b** *Planolites* ichnofabric with large burrows in heterolithic sandstone. Upper Jurassic (Oxfordian) Heather Formation (offshore, shelf), Fram Field, Norwegian North Sea (well 35/11-11, ca. 2711.5 m). **c** *Planolites* in heterolithic (silty) sandstone with coal seams. Lower Jurassic (Pliensbachian) Åre Formation (marginal marine), Skuld Field, Norwegian Sea (well 6608/10-14S, ca. 2702.5 m). **d** *Planolites* in silty sandstone with coal fragments (debris). Middle Jurassic (Bathonian-Callovian) Hugin Formation (marginal marine), Norwegian Sea (well 15/6-4, ca. 10626.5′). **e** Small *Planolites* in silty mudstone with lamination. Upper Cretaceous (Turonian-Santonian) Lange Formation (deep marine, basin floor), Norwegian Sea (well 6607/5-1, ca. 3408.5 m)

2011). In continental deposits, *Planolites* occurs in the *Scoyenia* and *Mermia* ichnofacies.

*Age: Planolites* is a cosmopolitan trace fossil observed from the Ediacarian (Alpert 1975; McCall 2006) throughout the Phanerozoic.

*Reservoir Quality:* Little is known so far about the impact of *Planolites* on reservoir quality. Dawson (1981) recognized a reduction of reservoir quality when *Planolites* is present. This fact can be explained by the active fill of the burrows, which leads to increased heterogeneity of the sediment. However, sand-filled burrows in muddy matrix may increase connectivity, although preferably in a horizontal direction due to the orientation of *Planolites*.

## 5.23 *Rhizocorallium* Zenker, 1836

*Morphology, Fill and Size: Rhizocorallium* is a U-shaped spreite burrow in which a marginal tube encloses an actively reworked area (spreite; Fürsich 1974c; Basan and Scott 1979). The burrow plane is horizontal to oblique to the bedding (Schlirf 2011). Burrow fill can be both active (e.g. spreite) and passive (e.g. marginal tube), or entirely passive (Knaust 2013). Delicate scratches can be preserved at the burrows surface. Burrow width varies from a few millimeters to several centimeters, while burrow length is typically in the range of a few centimeters but occasionally can reach several decimeters (Fig. 5.116). The ratio of tube diameter to spreite width is comparatively low and approximately 1:2 to 1:5 (Häntzschel 1960; Fürsich 1974c; Worsley and Mørk 2001).

*Ichnotaxonomy:* A high lability in burrow morphology, size and orientation gave reason to the establishment of many ichnospecies of *Rhizocorallium*, which now are regarded as synonyms of only two valid ichnospecies (Knaust 2013; Fig. 2.5). *R. commune* has an active spreite fill and often contains elliptical fecal pellets (*Coprulus* isp.), while *R. jenense* is passively filled and intensively scratched (Fig. 5.117). Uchman and Rattazzi (2016) propose the new combination *R. hamatum* which, however, differs from *Rhizocorallium* by its relatively small marginal tube and widespread branching. Despite the abundant fill with fecal pellets *Coprulus oblongus*, this form is better accommodated in the *Zoophycos* group, with its vertically extended whorls being strongly compacted. The attempt to transfer passively filled and scratched spreite burrows to the ichnogenus *Glossifungites* as done by Belaústegui et al. (2016a) is not supported by the evidence of *R. jenense* as its senior synonym.

*Substrate: Rhizocorallium* is known from siliciclastic and carbonate substrate alike. It occurs in soft and firm substrates, which results in the development of contrasting behaviors that in turn are the ichnotaxobases for differentiating the two ichnospecies (Fig. 2.5). *R. commune* is predominantly created in soft to stiff substrates (Fig. 5.116a), while *R. jenense* is characteristic of firmgrounds (Fig. 5.116b).

*Appearance in Core:* Although *Rhizocorallium* is a common trace fossil of worldwide distribution, it has only rarely been recognized in core, and even some of these reported occurrences remain disputable. However, a clear distinction between the two ichnospecies can be made in core. *R. commune* is best recognized in cross section, where the two marginal tubes with a circular to elliptical cross section and passive fill enclose the flattened spreite with an active fill in between (Figs. 5.116d and 5.118). Additional confidence comes from the association of the burrow with fecal pellets. This is particular helpful in longitudinal sections, where only a thin stripe of reworked sediment can be seen, sometimes with a pronounced bow-shaped spreite (Fig. 5.116c). In contrast, *R. jenense* is inclined to various degrees and has a dumbbell-like or pouch-shaped cross section with a contrasting passive fill. Another aspect is the size of some well-developed *R. commune*, which can reach several centimeters in width and thus make it difficult to be recognized in size-limited core sections.

*Similar Trace Fossils:* The sparse and partly erroneous reports of *Rhizocorallium* from core may be related to difficulties recognizing this trace fossil and potential confusion with similar burrows. Other spreite burrows such as *Diplocraterion* and *Zoophycos* may be mistaken for *Rhizocorallium*, but differ from it by their vertical orientation (*Diplocraterion*) or whorled morphology with lower tube/spreite ratio (*Zoophycos*). *Tisoa* is another U-shaped burrow with passive fill, but differs from *R. jenense* by having two narrow marginal limbs without an intervening spreite; in addition it is relatively deep for its width, and is commonly surrounded by a large concretion. Parts of backfilled horizontal burrows such as *Taenidium* and *Scolicia* may also resemple *R. commune* spreiten but are not U-shaped and do not have a marginal tube.

*Producers:* Three groups of organisms commonly qualify as potential producers of *Rhizocorallium:* decapod crustaceans, annelids and larval mayflies (Knaust 2013). Although *Rhizocorallium* in some earlier studies was interpreted as the product of annelids, most workers have since favored crustaceans as tracemakers because of the presence of scratches (e.g. Seilacher 2007; Rodríguez-Tovar and Pérez-Valera 2008; Neto de Carvalho et al. 2010). However, numerous groups of animals are known to produce scratches. Several lines of evidence suggest annelids (e.g. spionids, eunicids, terebellids) as producers of *Rhizocorallium*, an interpretation that is supported by constructional morphology, accompanying features (e.g. fecal pellets), neoichnological comparison and remains of the fossilized producers preserved in situ (Knaust 2013). An interpretation of marine *Rhizocorallium* as a polychaete burrow is straightforward, while the larvae of emerging mayflies probably account for *R. jenense* produced in fluvial settings (Fürsich and Mayr 1981).

(a)
(b)
(c)
(d)
(e)
(f)
(g)
(h)

◀**Fig. 5.116** *Rhizocorallium* in outcrop and in thin sections. *Scale bars* = 1 cm except in (**g**) = 5 cm. (**b**), (**c**) and (**d**) republished with permission of Elsevier; permission conveyed through Copyright Clearance Center, Inc. **a** *R. commune* in Middle Triassic (Anisian) limestone (type horizon) near Weimar, Thuringia, Germany. **b** *R. jenense* with net-like, crossing and closely spaced scratches. Lower Triassic (Upper Buntsandstein, Pelsonian, *Rhizocorallium* Dolomite, type bed), Jena–Ziegenhain, coll. Mägdefrau, Thüringer Landesanstalt für Umwelt und Geologie, Jena (TLGU 5035-701-202). After Knaust (2013). **c** Thin section of parts of *R. commune* as in (**a**), displaying the actively filled spreite with micritic fecal pellets (*Coprulus oblongus*) and the marginal tube (left) completely filled with pellets. After Knaust (2013). **d** Vertical thin section of a horizontal *R. commune* displaying an actively created spreite between the limbs of the passively filled U-shaped marginal tube. Lower Jurassic, Grimmen, Germany. After Knaust (2012b). **e** *R. commune* in Upper Triassic (Norian) sandstone (storm deposit) of Deltaneset, Templefjorden, Svalbard. **f** Cross section of *R. commune* as in (**e**). **g** Sandstone bedding plane with gregarious *R. commune*. Upper Jurassic (Kimmeridgian, shallow marine), coastal cliffs at Praia do Salgado, western Portugal. **h** *R. commune* in bedding-plane view. Same locality as in (**g**)

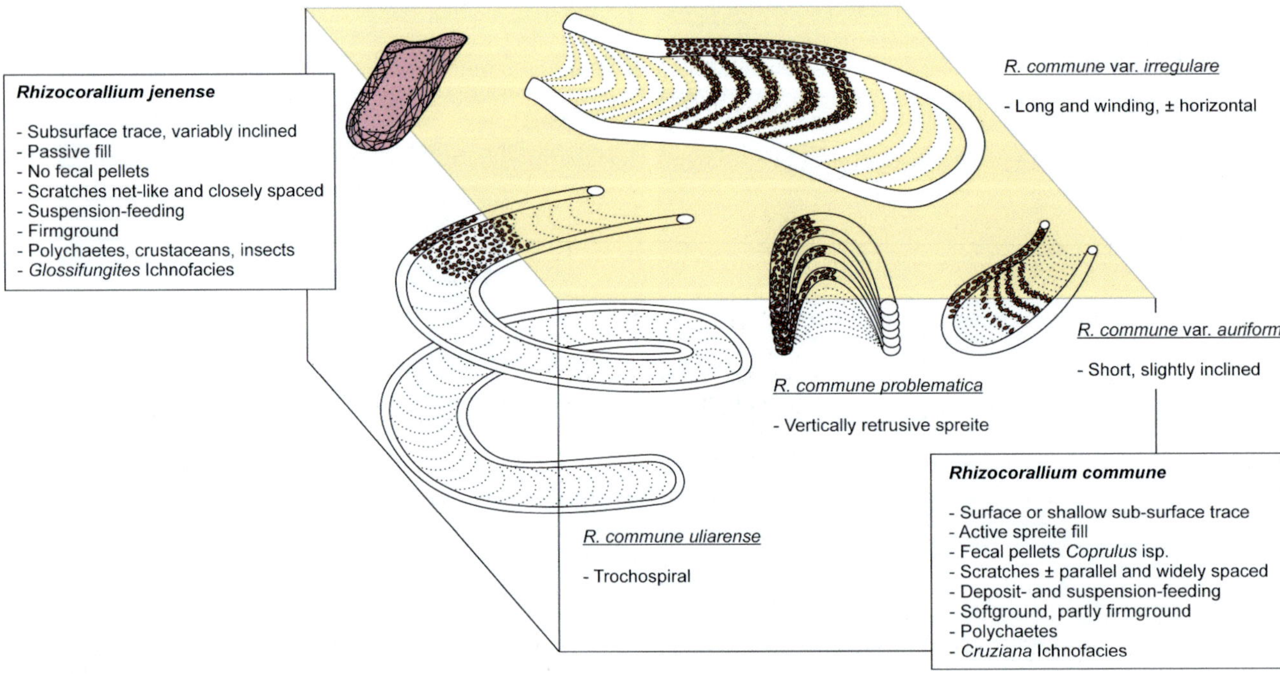


**Fig. 5.117** Classification of *Rhizocorallium* into two ichnospecies (*R. commune* and *R. jenense*). Morphological variability in *R. commune* is captured by two ichnosubspecies, while size differences result in two varieties. From Knaust (2013), republished with permission of Elsevier; permission conveyed through Copyright Clearance Center, Inc.

*Ethology:* Combined suspension- and deposit-feeding behavior is favored by most workers, although tests for gardening and storage behavior (cache) are inferred from oxygenated pyrite framboids found within the *Rhizocorallium* burrow (Zhang et al. 2016).

*Depositional Environment:* The two ichnospecies of *Rhizocorallium* and their varieties may serve as good facies indicators with the potential of supporting sedimentological and paleontological interpretations (Knaust 2013). *R. commune* typically occurs in shelf and nearshore environments (e.g. Farrow 1966; Ager and Wallace 1970; Worsley and Mørk 2001; Rodríguez-Tovar and Pérez-Valera 2008), and since the Triassic also in deep-marine deposits. *R. jenense* is known from shallow- and marginal-marine environments but occurs in fluvial deposits as well (Fürsich and Mayr 1981). Data from the Germanic Basin confirm the preferential occurrence of *R. commune* var. *irregulare* in intertidal and shallow subtidal environments, whereas *R. commune* var. *auriforme* is present in lagoonal environments and in deeper parts of the basin (Fig. 5.119). Furthermore, *R. commune*-associated fecal pellets, *Coprulus bacilliformis*, occur in inter- to supratidal environments and *C. oblongus* is evident in intertidal and deeper environments (Knaust 2013). *R. jenense* typically occurs along omission surfaces in high-energy areas and is often documented along ravinement surfaces exposed during marine transgressions. It is a common constituent of transgressive systems tracts (e.g. Knaust 1998; Knaust et al. 2012; MacEachern et al. 2012).

Several studies have utilized *Rhizocorallium* as a current indicator (Farrow 1966; Schlirf 2000; Worsley and Mørk 2001; Rodríguez-Tovar and Pérez-Valera 2008; Cotillon 2010), with its long axis running parallel or

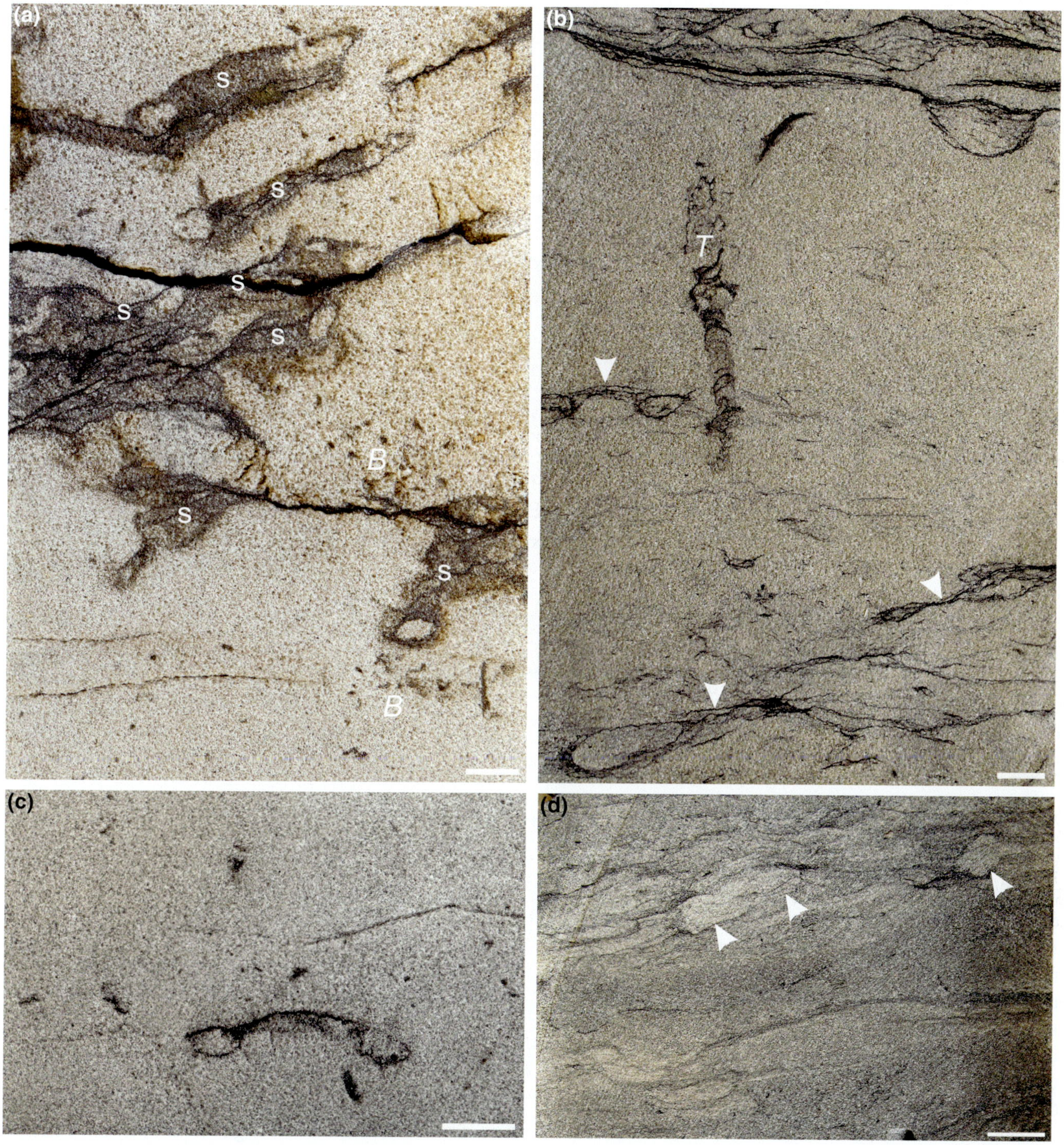


**Fig. 5.118** *Rhizocorallium commune* in sectioned core. *Scale bars* = 1 cm. **a** Cross sections with passively filled (sandy) marginal tubes connected by an actively filled (muddy) spreite (*s*) co-occurring with *Bornichnus tortuosus* (*B*). Lower to Middle Jurassic (Aalenian-Bajocian) Stø Formation (shoreface), Snøhvit Field, Barents Sea (well 7120/6-2 S, ca. 2577.5 m). **b** Several *R. commune* cross sections (*arrow heads*) are accompanied with a vertical spreite burrow corresponding to *Teichichnus zigzag* (*T*). Middle Jurassic (Bathonian-Oxfordian) Hugin Formation (shallow marine), Sleipner Vest Field, Norwegian North Sea (well 15/9-8, ca. 3474.0 m). **c** Cross section with passively filled (sandy) marginal tubes connected by an actively filled (muddy) spreite. Lower Jurassic (Aalenian) Stø Formation (upper shoreface), Snøhvit Field, Barents Sea (well 7120/8-1, ca. 2120.4 m). **d** Weakly defined cross sections (*arrow heads*). Middle Jurassic (Bathonian-Oxfordian) Hugin Formation (shallow marine), Gudrun Field, Norwegian North Sea (well 15/3-9T2, ca. 4497.2 m)

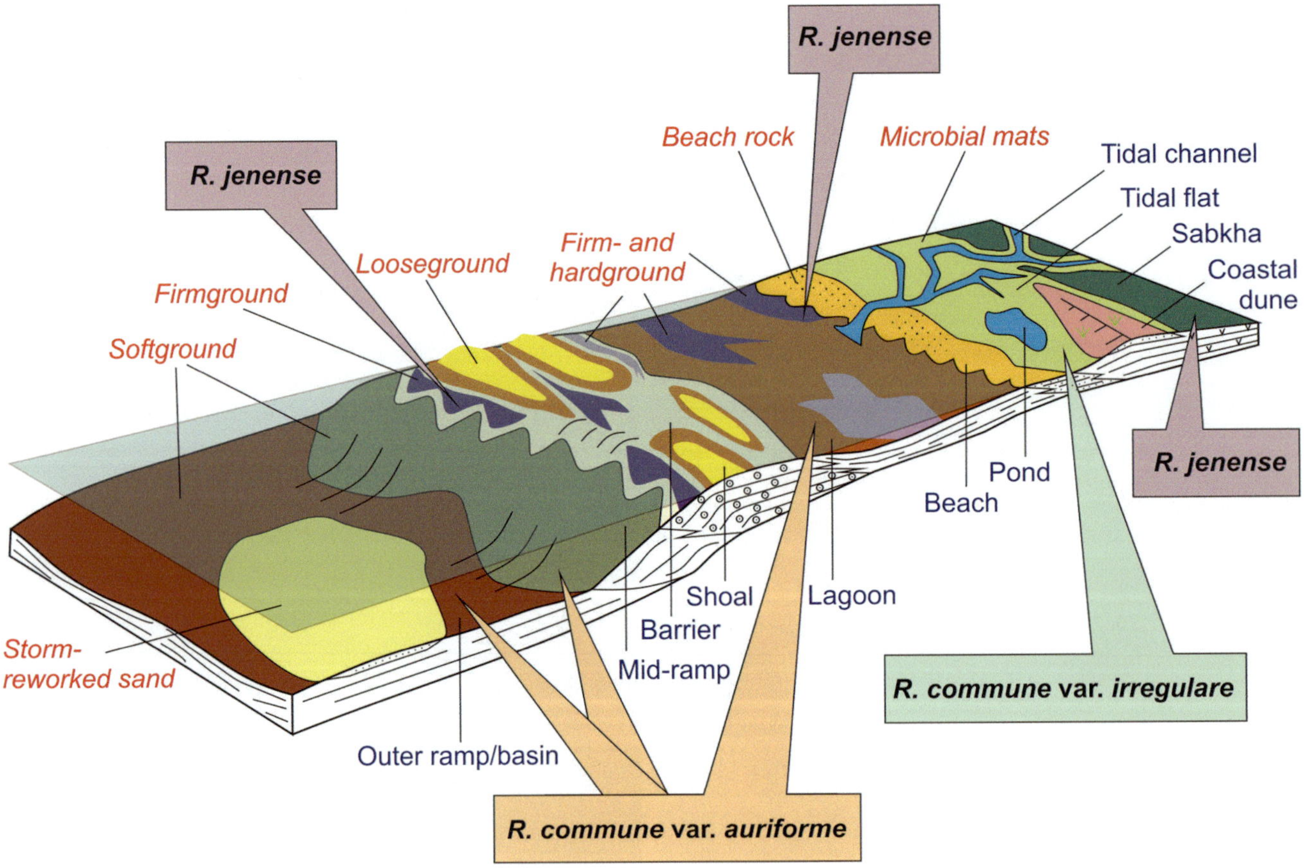


**Fig. 5.119** Distribution of *Rhizocorallium jenense* and *R. commune* (and varieties) and their paleoenvironmental relationships in the Middle Triassic (Lower Muschelkalk) of Thuringia, Germany (type area). From Knaust (2013), republished with permission of Elsevier; permission conveyed through Copyright Clearance Center, Inc.

oblique to the inferred paleocurrent (or the line of the two burrow openings perpendicular to it), and the distal (curved) end directed onshore. This pattern can be obscured by a more random alignment in response to the exploitation of nutrient-rich ripple troughs by the trace-maker. Many marine *Rhizocorallium* are documented as tolerating a wide range of salinity from hypersaline to mesohaline water. *R. commune* occurs in various oxygen-depleted deposits with dysoxic conditions and increasing oxygenation of the sediment (Wignall 1991; Kotlarczyk and Uchman 2012).

*Ichnofacies: Rhizocorallium commune* is a component of the *Cruziana* Ichnofacies, while *R. jenense* is part of the widespread *Glossifungites* Ichnofacies.

*Age:* The ichnospecies *R. commune* is one of the longest-ranging fossils and is known from the Early Cambrian (e.g. Fedonkin 1981; Clausen and Vilhjálmsson 1986; Orłowski 1989) to the Holocene (e.g. Winn 2006), while *R. jenense* first appeared after the end-Permian mass extinction (Knaust 2013).

*Reservoir Quality:* The effect of *Rhizocorallium* on reservoir quality is little studied. However, it can be expected that intense bioturbation with *R. commune* and their active spreite fill and frequent occurrence of muddy fecal pellets reduces the reservoir quality, whereas passively filled *R. jenense* certainly would increase it.

## 5.24 *Rosselia* Dahmer, 1937

*Morphology, Fill and Size: Rosselia* includes subvertical, bulbous- or spindle-shaped burrows with length varying from few centimeters to over 100 cm (Nara 2002). In most cases, *Rosselia* is unbranched, although occasional side-branches occur (Fig. 5.120a, c). The interior of the muddy burrow consists of numerous concentric laminae that are arranged in an onion-like manner (Fig. 5.120b, e–g). The terminal burrow with passive fill and central or marginal position may be preserved in form of a narrow cylindrical shaft (Fig. 5.120c, d). Contrast enhancement by early diagenetic iron mineralization (e.g. goethite, siderite) is common.

*Ichnotaxonomy: R. socialis* is the most common ichnospecies of the ichnogenus *Rosselia* and probably the only valid one. Other ichnospecies such as *R. chonoides* Howard

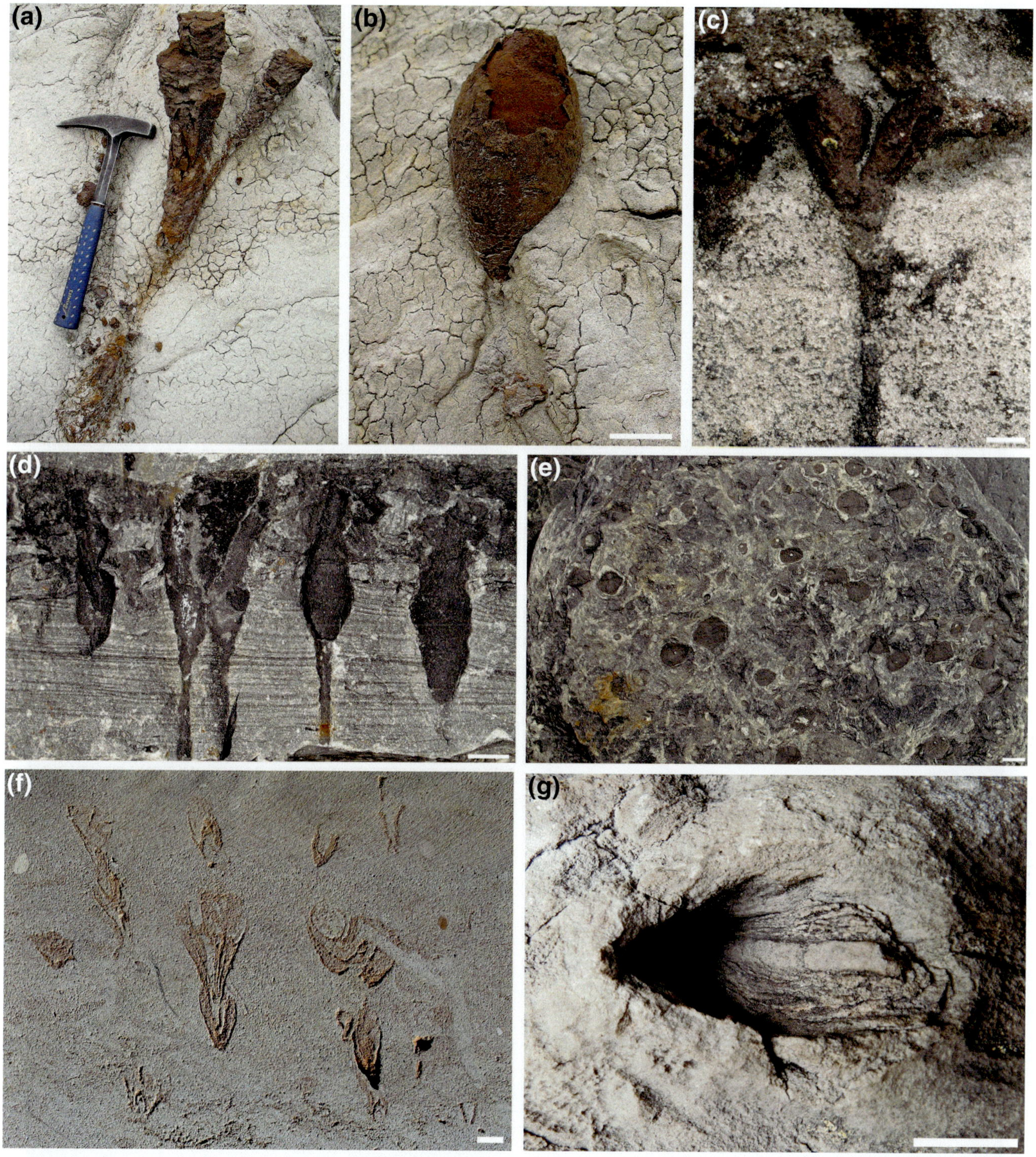


**Fig. 5.120** *Rosselia* in outcrop. *Scale bars* = 1 cm except in (**b**) = 5 cm. **a**, **b** Large *Rosselia* in limonitic preservation and embedded in silty deposits. Flower-like morphology with several off-branches (**a**), and individual bulbous specimen displaying the onion-like laminated interior (**b**). Upper Cretaceous (Campanian) Bearpaw-Horseshoe Canyon Formation (shoreface), near Drumheller, Alberta, Canada. See Zorn et al. (2007). **c** Bulbous *Rosselia* in limonitic preservation within coarse-grained sandstone. Middle Jurassic Scarborough Formation (shoreface), coastal cliff near Scarborough, Yorkshire, UK. **d**, **e** *R. socialis*, crowded in a sandstone bed in vertical (**d**) and horizontal (**e**) section. Lower Ordovician Beach Formation (shoreface), Bell Island, Newfoundland, Canada. See Fillion and Pickerill (1990). **f** Ichnofabric with vertically stacked *R. socialis* in vertical to slightly oblique section. Building stone of unknown age and origin, Ronda, southern Spain. **g** Part of *R. socialis* on a sandstone bedding plane showing the onion-like lamination with the terminal tube. Lower Devonian Taunusquarzit, near Rüdesheim, southern Germany, the type area of *R. socialis*. Senckenberg coll., Frankfurt Main. See Dahmer (1937)

(a)
117
(b)
(c)
(d)

◀**Fig. 5.121** *Rosselia* in sectioned core. *Scale bars* = 1 cm. **a** Slightly off-axial sections of fusiform, mud-lined *Rosselia* cross-cutting a dense ichnofabric mainly consisting of *Teichichnus zigzag* and *Chondrites*. Middle Jurassic (Callovian) Hugin Formation (offshore transition), Gina Krog Field, Norwegian North Sea (well 15/6-9S, ca. 3790.0 m). **b** Slightly bulbous *Rosselia* preserving an elongate basal tube. The truncation of the top of the burrow indicates an erosive bedding boundary that would otherwise not be recognized because of complete bioturbation (*Nereites*)) of the host rock. Upper Jurassic (Oxfordian) Vestland Group (offshore transition in a restricted basin), Johan Sverdrup Field, Norwegian North Sea (well 16/2-13S, ca. 1942.0 m). **c** Ripple-laminated sandstone with *Rosselia* overprinting a bioturbate texture. Lower Jurassic (Pliensbachian) Cook Formation (shallow marine), Norwegian North Sea (well 34/5-1S, ca. 3659.5 m). **d** Spindle-shaped *Rosselia* in carbonates of the Upper Permian Khuff Formation, South Pars Field, Persian Gulf, Iran (well SP9, ca. 3082.5 m). After Knaust (2014a), republished with permission of EAGE

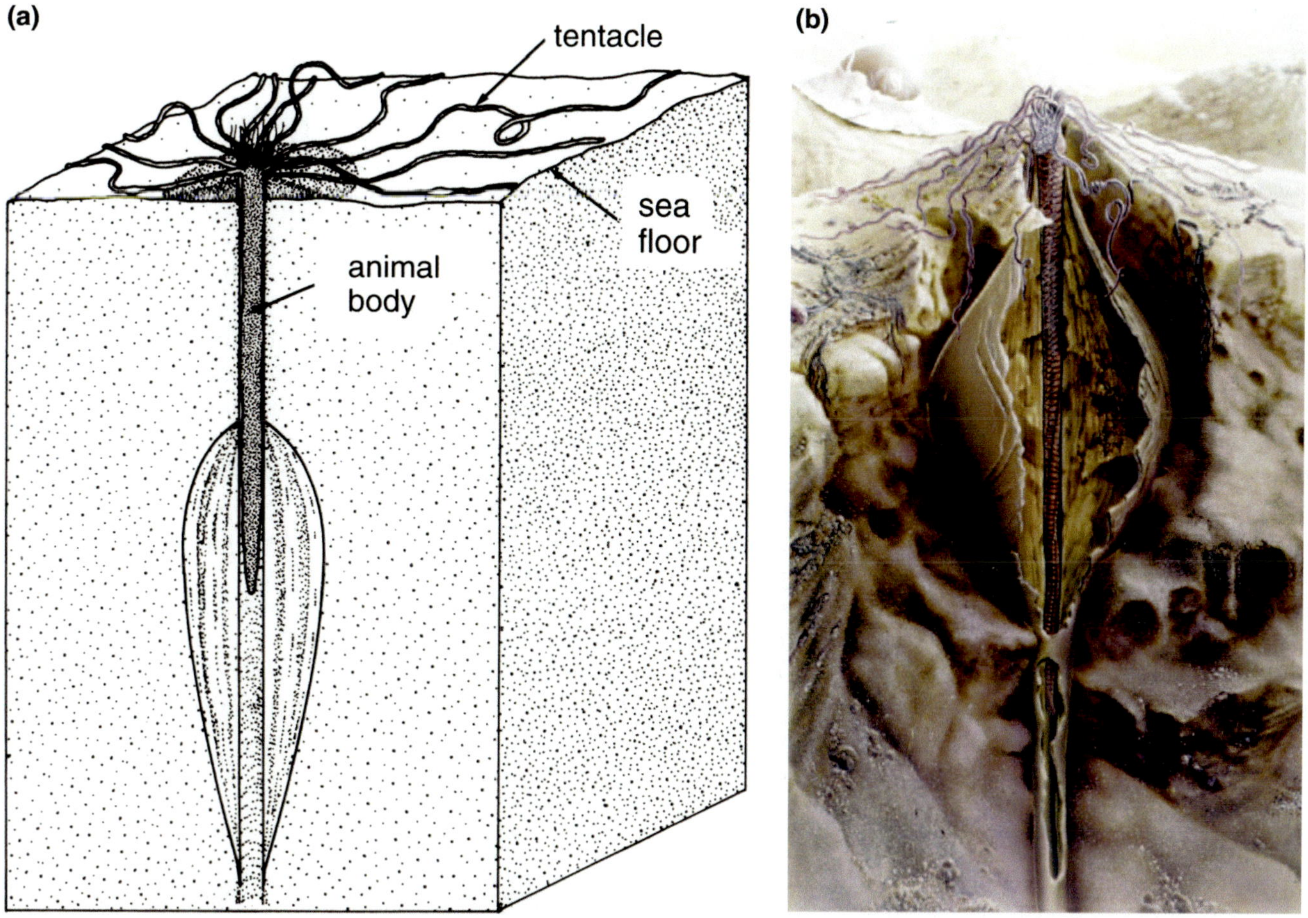


**Fig. 5.122** Reconstruction of *Rosselia* tracemaker as a terebellid polychaete within its burrow. **a** Original interpretation from Nara (1995), republished with permission of Wiley; permission conveyed through Copyright Clearance Center, Inc. **b** Illustration after Pemberton et al. (2001), republished with permission of the Geological Association of Canada

and Frey, 1984 and *R. rotatus* McCarthy, 1979 are morphologically different and may not belong to *Rosselia* (Uchman and Krenmayr 1995; Knaust 2015a).

*Substrate: Rosselia* is a common constituent of siliciclastic softground deposits (Fig. 5.121a–c) but occasionally also occurs in carbonates (Fig. 5.121d).

*Appearance in Core:* The funnel-, bulb- or spindle-like shape of the upper burrow part together with the laminated internal structure makes *Rosselia* easy to recognize in core (Fig. 5.121), although confusion may arise from burrows with similar morphology. *R. socialis* with its bulbous swelling in the upper part of the burrow is the most common ichnospecies of *Rosselia*. Funnel-shaped burrows lacking the internal lamination but exhibiting individual vermiform burrows were attributed to *R. chonoides*, Howard and Frey 1984. It remains unclear if this results from secondarily

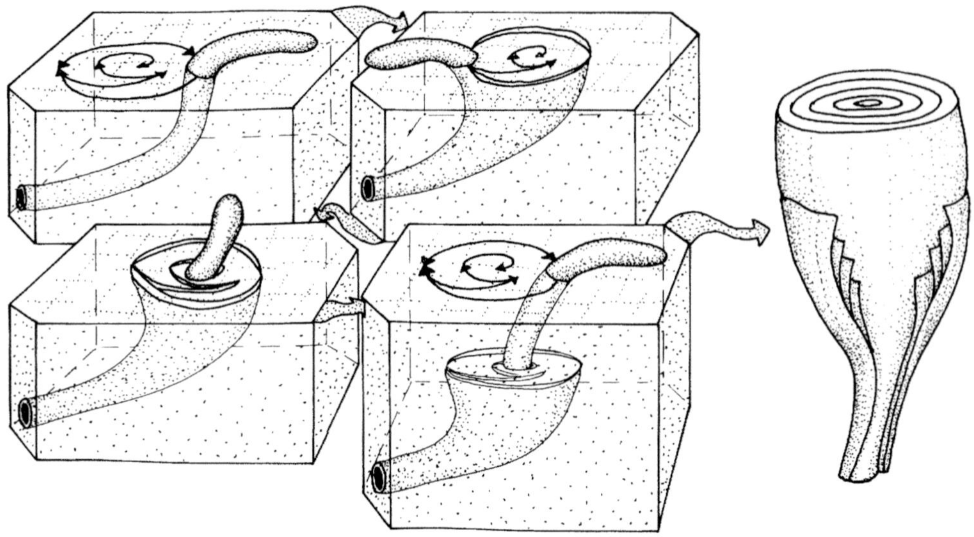

**Fig. 5.123** Burrow construction of *Rosselia socialis* by successive movements of a vermiform animal. After Chamberlain (1971), © Paleontological Society, published by Cambridge University Press, reproduced with permission

reworked burrow fill or if this form belongs to a different ichnogenus (Uchman and Krenmayr 1995).

*Similar Trace Fossils:* The ichnospecies *R. chonoides* Howard and Frey, 1984 and *R. rotatus* McCarthy, 1979 were originally included in *Rosselia* because of their overall similarity with *R. socialis*, but differ from it by having a helicoidal and crescentric fill, respectively. This, in addition to a funnel-shaped instead of bulbous morphology (in *R. rotatus*) makes those forms rather consistent with the ichnogenus *Parahaentzschelinia* than with *Rosselia* (Fig. 5.101). *Asterosoma* is similar to *Rosselia* by its active muddy, concentrically laminated fill. It differs from it by a dominantly subhorizontal orientation and commonly star-shaped morphology with individual spindles branching off from a central area (Bromley and Uchman 2003). Another muddy, concentrically laminated ichnogenus is *Cylindrichnus*, which differs from *Rosselia* by its bow-shaped morphology (Goldring 1996; Goldring et al. 2002). *Lingulichnus* resembles *Rosselia* in having a laminated, funnel-shaped dwelling chamber, which continuous downwards into a pedicle trace (Zonneveld et al. 2007). *Lingulichnus* typically shows an elliptical cross section and perhaps could account for less organized forms. The extensive shaft of *Rosselia* may lead to confusion with *Skolithos* where only partially exposed in core.

*Producers:* Terebellid polychaetes are the most likely producers of *Rosselia* (Nara 1995, 2002), although other polychaete worms (such as Sabellidae) and sea anemones (Actiniaria) have also been considered (Figs. 5.122 and 5.123).

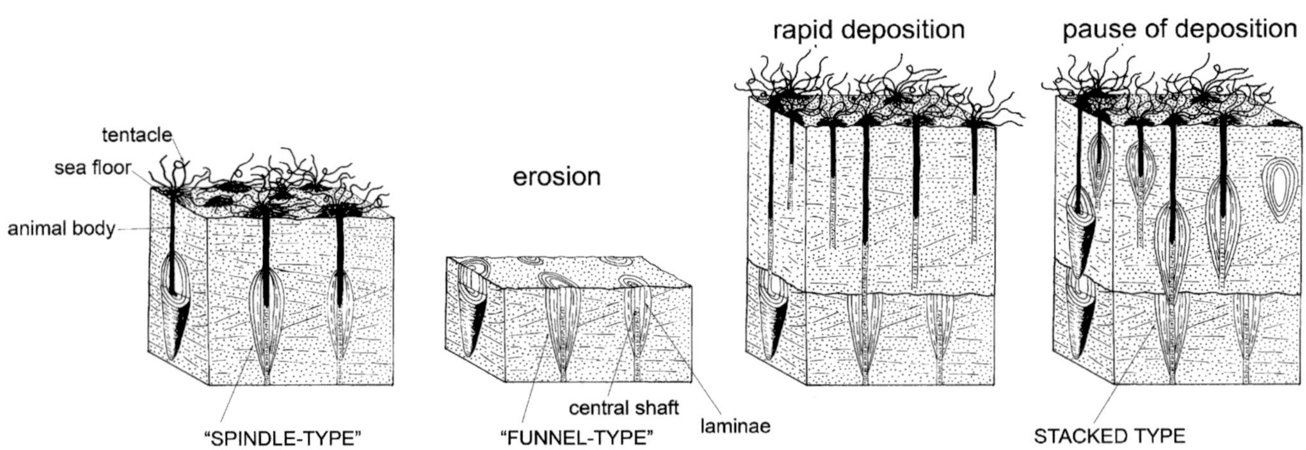


**Fig. 5.124** Idealized model showing the development of stacked *Rosselia socialis* in response to changing deposition and erosion. After Frieling (2007), republished with permission of Springer

*Ethology: Rosselia* is interpreted as the dwelling structure (domichnion) of detritus-feeding polychaetes. Under rapidly changing sedimentation and erosion, the tracemaker tries to adjust such conditions and produces equilibrium traces (equilibrichnia).

*Depositional Environment: R. socialis* is a common trace in inner shelf areas and typically occurs in lower to middle shoreface deposits. It is also reported from marginal-marine environments, including tidal flats, tidal channels, deltas, embayments and lagoons, and flood-tidal deltas (Uchman and Krenmayr 1995; Carmona et al. 2008). Various studies have associated *R. socialis* with high sedimentation rates and repeated erosion (e.g. Nara 1995, 2002; Campbell et al. 2006; Frieling 2007; Netto et al. 2014; Campbell et al. 2016). In such settings, rapid sedimentation leads to vertically exaggerated growth of *R. socialis*, whereas subsequent erosion events cut down into the spindle-like burrow parts and leave behind funnel-shaped structures (Fig. 5.124). This is common during marine transgressions, where *R. socialis* has its highest abundance (Nara 2002), during riverine flood events, when plenty of fine-grained particles are in suspension (Campbell et al. 2006), or during storm events (Netto et al. 2014; Campbell et al. 2016). In marginal-marine environments, size reduction may be prevalent in *R. socialis* communities as a response to lowered salinity (Frieling 2007).

*Ichnofacies: Rosselia* is a common constituent of the *Cruziana* Ichnofacies. Given the preferred ethology of its producer as a suspension-feeder, it may also be associated with elements of the *Skolithos* Ichnofacies to which it can be transitional.

*Age: Rosselia* is reported from the Lower Cambrian (Silva et al. 2014) to the Holocene (Nara and Haga 2007).

*Reservoir Quality:* Because of its suspension-feeding behavior, the *Rosselia* producer incorporates mud-rich sediment into its dwelling and builds a thick lining. This leads to a local accumulation of external mud and therefore *Rosselia* contributes to a reduction of the reservoir volume and quality.

## 5.25 *Schaubcylindrichnus* Frey and Howard, 1981

*Morphology, Fill and Size: Schaubcylindrichnus* is a U-shaped burrow system typically consisting of three parts: an isolated single burrow or a bundle of thickly lined, often interpenetrating tubes that were constructed one after another, a feeding funnel connected to one end of the burrow system, and a fecal mound connected to the other end (Löwemark and Nara 2010; Fig. 5.125). Complete tunnel systems can reach several decimeters in length, while individual burrow diameters are typically in the range of a few millimeters to about 1 cm (Figs. 5.128 and 5.178a). Burrow lining can be around 1–3 mm thick and is typically of white color. Burrow fill is passive.

*Ichnotaxonomy: Schaubcylindrichnus* is now regarded as monoichnospecific; the previously established ichnospecies *S. freyi* and *S. formosus* are morphological variants of the type ichnospecies *S. coronus*, and therefore junior synonyms of it (Löwemark and Nara 2010, 2013; Figs. 5.126, 5.127 and 5.128). The inclusion of *Palaeophycus heberti* in *Schaubcylindrichnus*, as recently proposed by Evans and McIlroy (2016), is not based on the type material of the former and therefore becomes void. In the past, *Schaubcylindrichnus* has sometimes been confused with *Terebellina*, which however, refers to a large agglutinated foraminifer (Miller 1995).

*Substrate: Schaubcylindrichnus* commonly occurs in sandy and argillaceous (silty to muddy) substrates, more rarely also in carbonates.

*Appearance in Core:* Because of its thick lining as diagnostic feature, *Schaubcylindrichnus* is relatively easy to identify in core samples (Fig. 5.129). Typically, clusters of

**Fig. 5.125** *Schaubcylindrichnus coronus* in Miocene offshore-transition deposits outcropping in Japan. Funnel-like and mound-like structures are associated with the bow-shaped specimen. *Scale bar* = 10 cm. After Nara (2006), republished with permission of Elsevier; permission conveyed through Copyright Clearance Center, Inc.

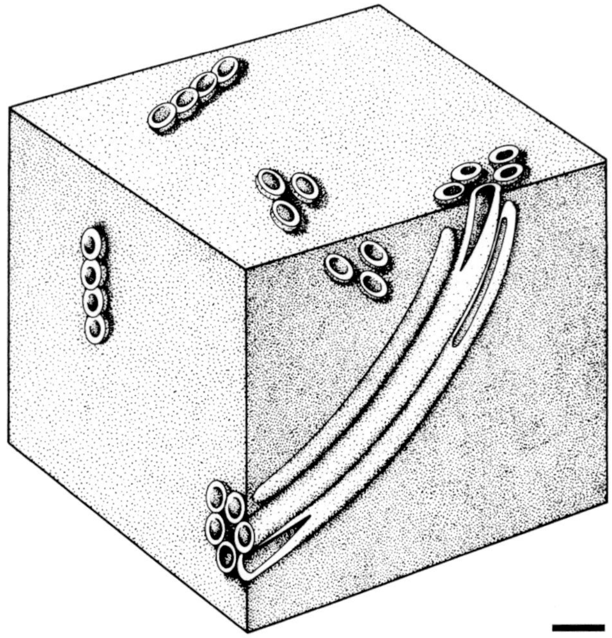

**Fig. 5.126** *Schaubcylindrichnus coronus* in a reconstruction by Howard and Frey (1984), republished with permission of Canadian Science Publishing; permission conveyed through Copyright Clearance Center, Inc. *Scale bar* = 1 cm

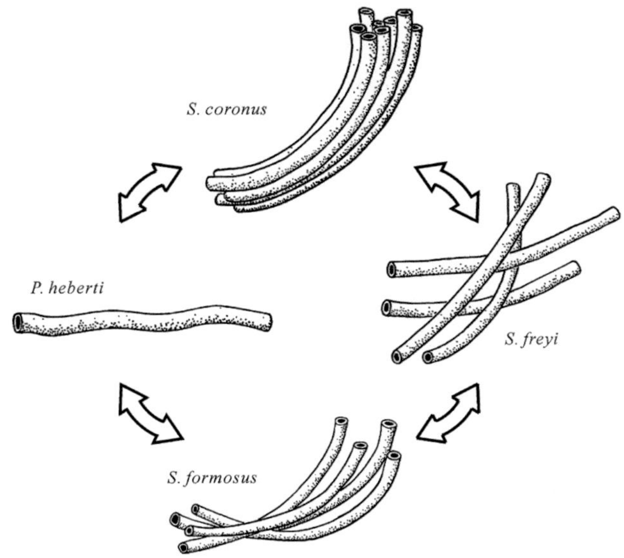


**Fig. 5.127** Morphological variability of *Schaubcylindrichnus*, which in the past gave reason for the erection of different ichnospecies that now are regarded as junior synonyms of *S. coronus*. Individual tunnels are referred to *Palaeophycus heberti*. From Löwemark and Hong (2006), reprinted by permission of the publisher (Taylor & Francis Ltd., http://www.tandfonline.com)

burrows with a thick lining are cut in various directions and thus burrow sections vary from circular to vertically flattened, elliptical and elongate. The characteristically light-colored lining is an eye-catching feature, which hardly can be missed. *Schaubcylindrichnus* also occurs as a late component in complex ichnofabrics and overprints preexisting bioturbation (Figs. 4.3, 5.79d, 5.100h, 5.111a, d and 5.156d).

*Similar Trace Fossils:* The appearance of *Schaubcylindrichnus* is conspicuous even in core, owing to its thick and pronounced burrow lining. Single occurrences may be confused with *Palaeophycus*.

*Producers:* Nara (2006) and Löwemark and Nara (2010) posit an enteropneust worm as the probable producer of *Schaubcylindrichnus*.

*Ethology:* A funnel-feeding behavior of a vermiform animal (e.g. enteropneust, polychaete) can be deduced (Löwemark and Nara 2010; Kikuchi et al. 2016).

*Depositional Environment:* *Schaubcylindrichnus* occurs in a wide range of environments from nearshore to continental slope (Frey and Pemberton 1991a; Löwemark and Nara 2010). It is common in shallow-marine settings, where it occurs especially in lower shoreface to offshore deposits (Frey and Howard 1985, 1990). The intraspecific morphological variability within *S. coronus*, such as the number of tubes within individual specimens, may be indicators for distinguishing distal from proximal position on an offshore-shoreface transect (Löwemark and Nara 2013). An analysis performed by Löwemark and Nara (2013) "… shows a distinct tendency for higher tube numbers in the offshore facies where sediments are characterized by higher silt/mud content, suggesting that *S. coronus* with higher tube numbers were constructed in a calm environment allowing longer dwelling periods. The increased abundance of nested tubes in settings characterized by thin sand layers indicates that the nested tubes are a reparation response to erosional events destroying the feeding funnels at the sediment-water interface."

*Ichnofacies:* *Schaubcylindrichnus* is a common constituent of the *Cruziana* and *Skolithos* ichnofacies.

*Age:* *Schaubcylindrichnus* occurs in strata at least from the Carboniferous to the Pleistocene (Löwemark and Nara 2010).

*Reservoir Quality:* No detailed observations about the influence of *Schaubcylindrichnus* on reservoir quality are available. The composition and architecture of the burrows generally suggest little impact.

**Fig. 5.128** *Schaubcylindrichnus* in outcrop. *Scale bars* = 1 cm. **a**, **b** Completely bioturbated sandstone with scattered *Schaubcylindrichnus* as elite trace fossil. The relatively small size of these burrows seems to be related to oxygen deficiency within the sediment. Paleocene Grumantbyen Formation (shelf) near Longyearbyen, Svalbard. **c**, **d** Sectioned surface of a glauconitic, cross-bedded sand within a fault zone (overturned section) displaying clusters of large *Schaubcylindrichnus*. Lower Cretaceous Arnager Greensand Formation (storm-dominated shoreface) near Rønne, Bornholm, Denmark

## 5.26 *Scolicia* de Quatrefages, 1849

*Morphology, Fill and Size: Scolicia* includes simple (unbranched), winding, meandering to coiling, bilobate or trilobate backfilled burrows with two parallel sediment strings along their lower surface and flattened oval cross section (Uchman 1995, 1998). Burrow diameter is typically in the range of less than 1 cm up to few centimeters. The traces are indefinitely long, capable of reaching at least several centimeters to decimeters in length.

*Ichnotaxonomy:* Several trace fossils are now included as preservational variants in the *Scolicia* group, namely *Taphrhelminthopsis, Laminites, Subphyllochorda* and *Taphrhelminthoida* (Uchman 1995; Fig. 5.130). Three ichnospecies of *Scolicia* can be outlined (Fig. 5.131): *S. prisca, S. plana* and *S. strozzii*.

*Substrate: Scolicia* is common in well-sorted, fine-grained sandstone and siltstone.

*Appearance in Core:* The appearance of *Scolicia* in core is quite diagnostic (Fig. 5.132). Longitudinal and oblique sections of the burrows show a densely meniscate or lamellar backfill, while burrows in oval to reniform cross sections differ from the host rock by their active fill and, in an optimal situation, reveal two sediment strings at their base.

*Similar Trace Fossils: Bichordites* is an echinoid trace fossil similar to *Scolicia* but differs by bearing a central bilobate inner zone (Demírcan and Uchman 2012; Fig. 5.130). In core, *Scolicia* may be confused with

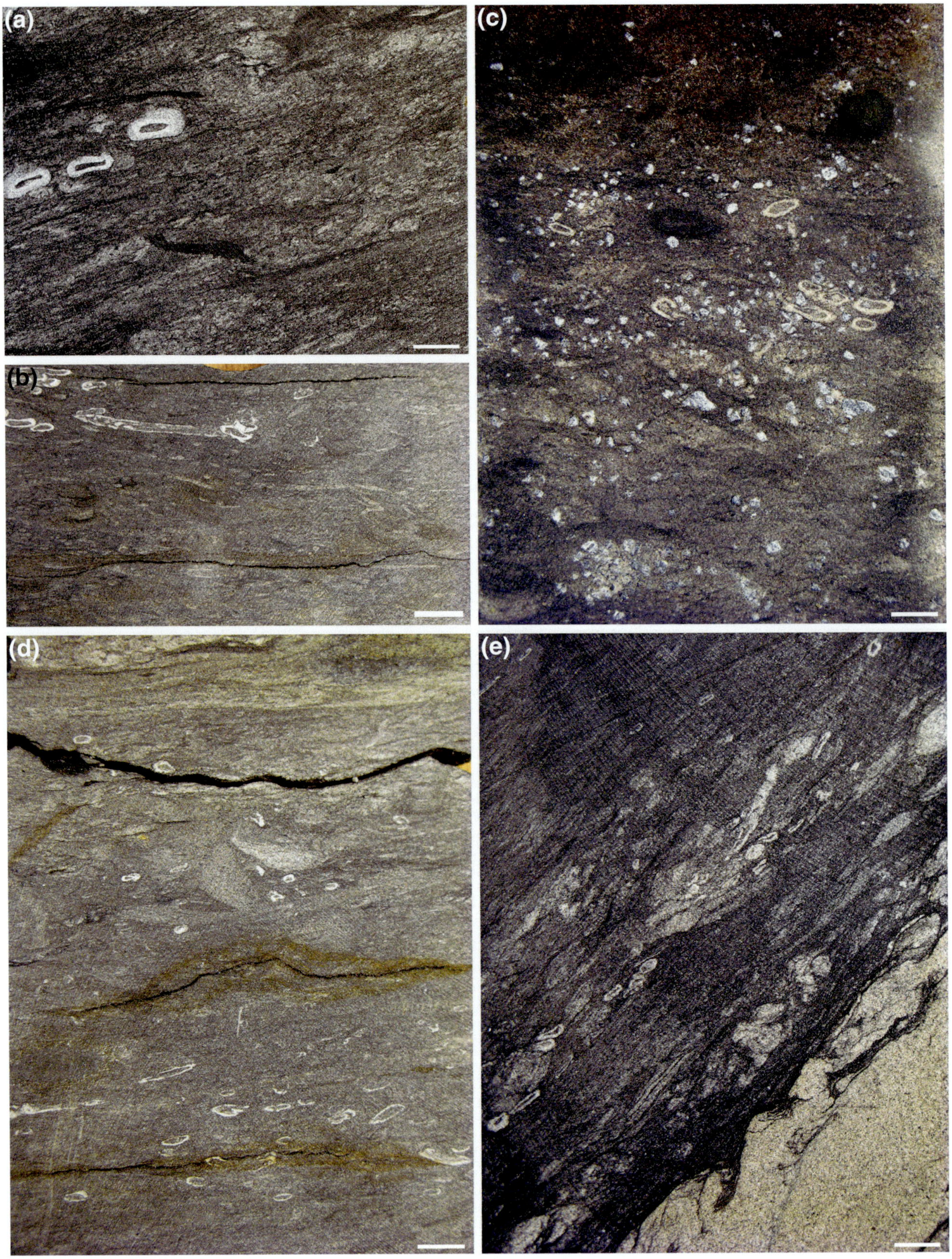
(a)
(b)
(c)
(d)
(e)

◀**Fig. 5.129** *Schaubcylindrichnus* in sectioned core. *Scale bars* = 1 cm. **a** Totally bioturbated sandstone with *Phycosiphon* and *Cylindrichnus* accompanied by thickly lined *Schaubcylindrichnus*, partly compacted. Lower Jurassic (Pliensbachian-Toarcian) Ror Formation (offshore), Åsgard Field, Norwegian Sea (well 6506/12-I-2 H, ca. 4836.7 m). **b** Completely bioturbated siltstone-sandstone with *Teichichnus* and *Nereites*, overprinted with sand-lined *Schaubcylindrichnus* in the lower part and well-defined *Schaubcylindrichnus* in the upper part. Upper Jurassic (Oxfordian) Heather Formation (offshore), Fram Field, Norwegian North Sea (well 35/11-9, ca. 2735.5 m). **c** Completely bioturbated sandstone (partly siderite-cemented at top) with bundles of *Schaubcylindrichnus* in cross section. Upper Jurassic (Oxfordian) Heather Formation (offshore), Fram Field, Norwegian North Sea (well 35/11-11, ca. 2623.9 m). **d** Highly bioturbated sandstone cross-cut by several clusters of *Schaubcylindrichnus*. Upper Jurassic (Oxfordian) Heather Formation (offshore), Fram Field, Norwegian North Sea (well 35/11-9, ca. 2732.9 m). **e** Laminated and weakly bioturbated siltstone (above a sandy turbidite with *Cylindrichnus*) with abundant sand-lined tubes of *Schaubcylindrichnus*. Upper Jurassic (Oxfordian) Heather Formation (offshore), Fram H-Nord Field, Norwegian North Sea (well 35/11-15ST2, ca. 2970.5 m)

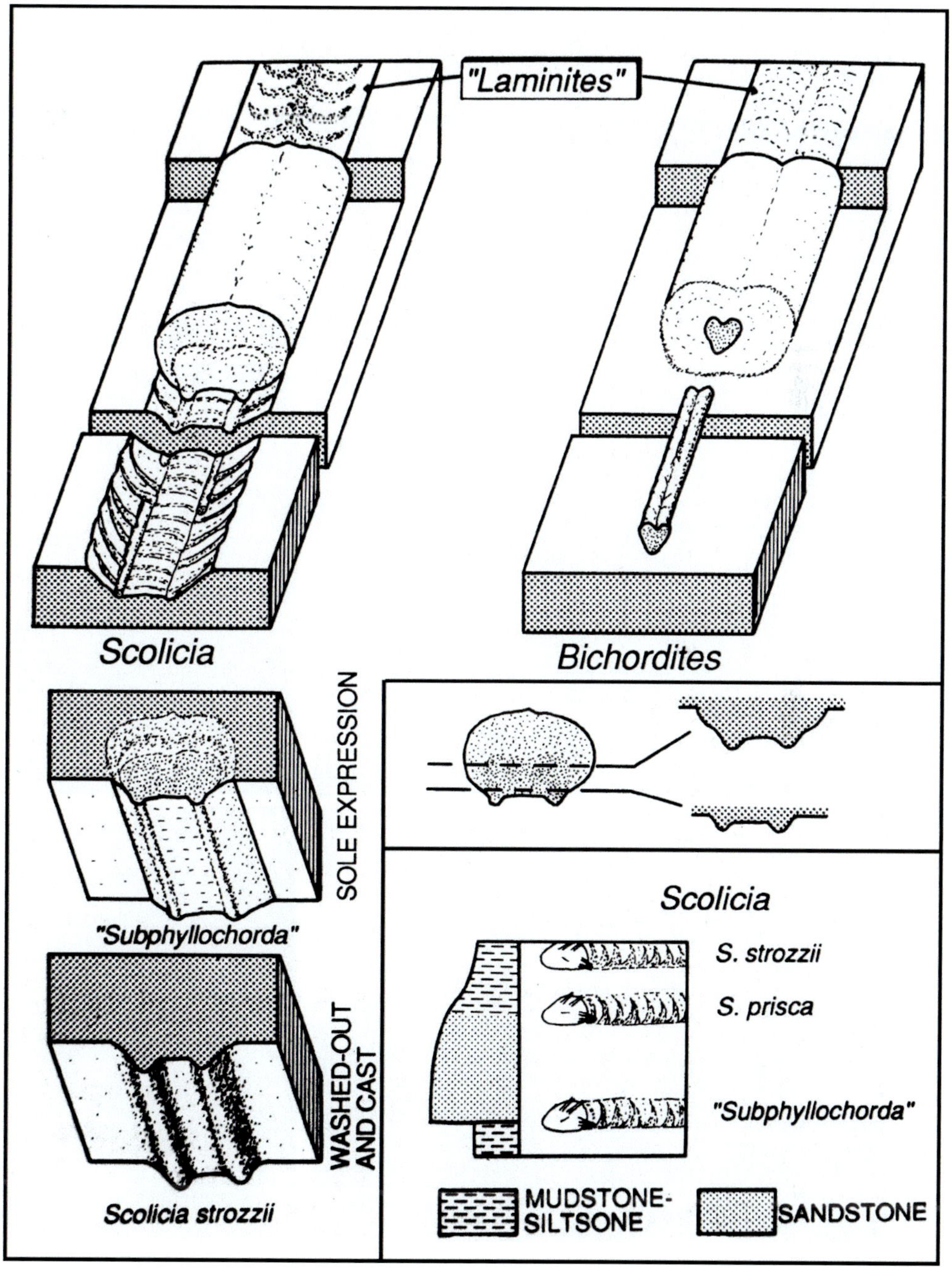


**Fig. 5.130** Characteristic morphology of *Scolicia* and the similar trace fossil *Bichordites*, and preservational variants of *Scolicia*. From Uchman (1995)

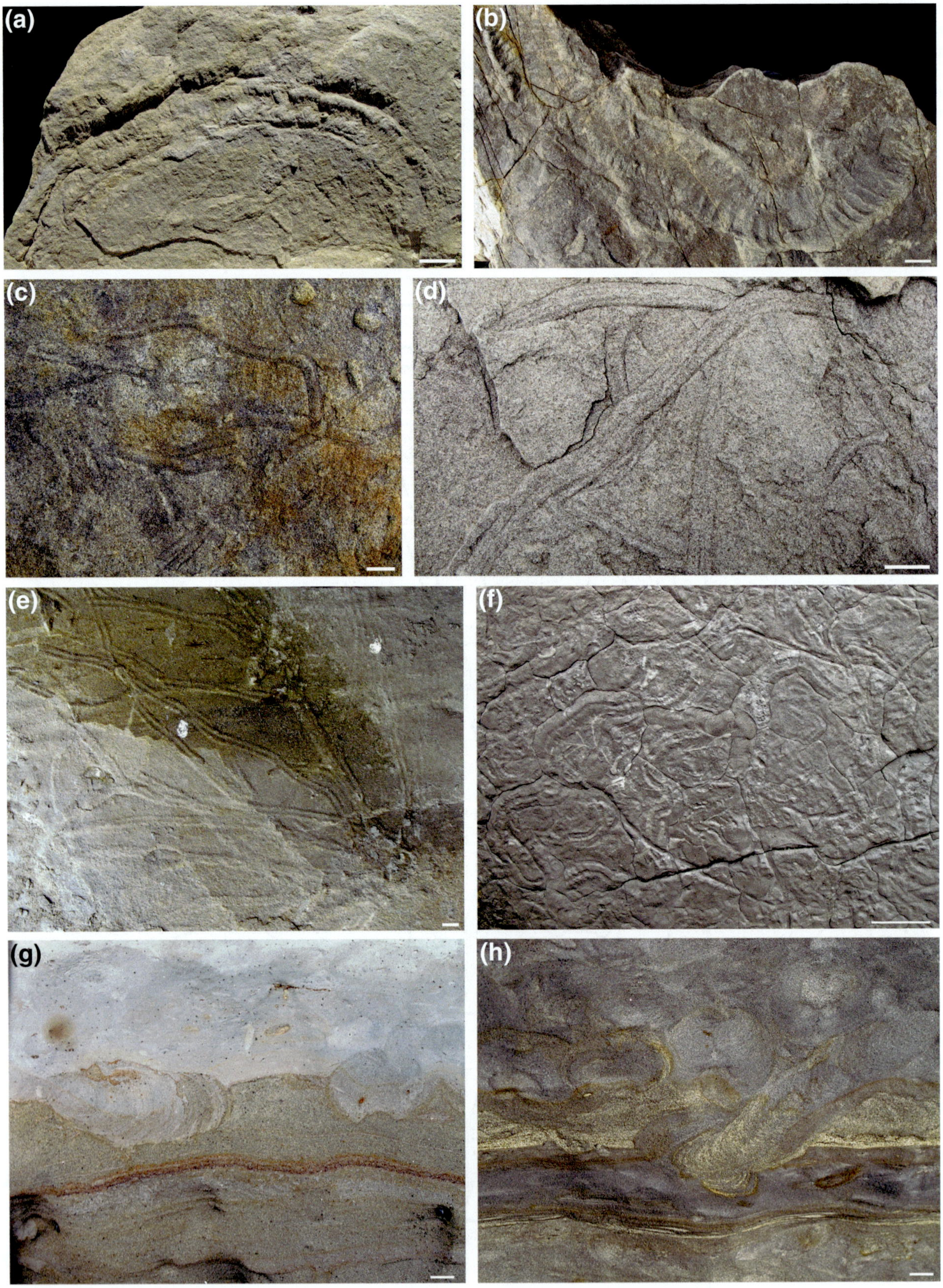


**Fig. 5.131** *Scolicia* from the Eocene Grès d'Annot Formation (deep marine, turbiditic), southeastern France (**a–e**) and in the Miocene Mount Messenger Formation (deep marine, channel-levee complex) in sea cliffs of the Taranaki Peninsula, North Island, New Zealand (**f–h**). *Scale bars* = 1 cm except in (**f**) = 10 cm. (**b**), (**d**) and (**e**) from Knaust et al. (2014), republished with permission of Wiley; permission conveyed through Copyright Clearance Center, Inc. **a**, **b** *S. prisca*. **c**, **d** *S. plana*. **e** *S. strozzii*. **f** Lower bedding-plane with winding *S. plana* along the sand-mud interface. **g** Vertical section with *Scolicia* isp. in longitudinal (left) and cross section (right) atop a laminated sandstone bed. Note the two characteristic sediment strings at the base of the cross section. **h** Thin-bedded, fine-grained sandstone in vertical section with a dense *Scolicia* ichnofabric

(a)
(b)
(c)
(d)
S
N
Z
S
N

◀**Fig. 5.132** *Scolicia* in sectioned core. *Scale bars* = 1 cm. **a** Mudstone-sandstone alternation (inclined) with cross sections of *Scolicia* occurring in the upper part of the thin sandstone layers, where organic sediment is concentrated within ripple troughs. Note the characteristic sediment strings at the base of the burrows (*arrows*). Lower Cretaceous (Albian, deep marine, channel-overbank), off Tanzania. **b** Longitudinal sections of *Scolicia* in heterolithic turbiditic sandstone. Upper Cretaceous (Turonian-Santonian) Kvitnos Formation (deep marine), Aasta Hansteen Field area, Norwegian Sea (inclined well 6707/10-2A, ca. 4462.8 m). **c** Heterolithic intercalation of turbiditic sandstone and hemipelagic mudstone showing several partly deformed specimens of *Scolicia* in longitudinal section. Upper Cretaceous (Campanian) Nise Formation (deep marine, basin floor), Norwegian Sea (well 6607/5-2, ca. 4186.5 m). **d** Mudstone-sandstone alternation with cross-bedding and intense bioturbation. *Nereites* (*N*) in the background is overprinted by large *Scolicia* (*S*), resulting from the exploitation of the sand layers by its producer. All other traces are cross-cut by mud-filled *Zoophycos* (*Z*). Upper Cretaceous (Campanian) Nise Formation (deep marine, fan fringe), Aasta Hansteen Field, Norwegian Sea (well 6707/10-1, 2974.3 m). After Knaust (2009b), republished with permission of Elsevier; permission conveyed through Copyright Clearance Center, Inc.

*Taenidium*, based on the meniscate backfill. However, *Taenidium* is commonly smaller than *Scolicia*, has less regularly packed menisci and lacks the diagnostic sediment strings at the base. Subhorizontal parts of the spreite burrow *Zoophycos* and *Lophoctenium* may also resemble *Scolicia*, but belong to a complex three-dimensional burrow system. Individual *Zoophycos* spreiten are less regularly packed than *Scolicia*, are smaller and lack sediment strings.

*Producers:* On the basis of modern analogs and experimental work, irregular echinoids (e.g. spatangoids, heart urchins) are known to produce *Scolicia*-like traces (Figs. 5.133 and 5.134).

*Ethology: Scolicia* results from the deposit-feeding (fodinichnial) activity of irregular echinoids.

*Depositional Environment: Scolicia* is common in deep-marine deposits belonging to the *Nereites* Ichnofacies and in shallow-marine environments assignable to the *Cruziana* Ichnofacies. Given its preference for sandy to silty, organic-rich sediment, *Scolicia* is a common element of inner levee deposits within deep-marine channel-levee complexes (Callow et al. 2013). Similarly, *Scolicia* is very common in proximal fan and related environments but rare in more distal parts (Heard and Pickering 2008). *Scolicia* occurs in a shallow-tier position and therefore is subject to be destroyed by subsequent burrowers (Figs. 5.9b, 5.135 and 5.136).

*Ichnofacies: Scolicia* is a typical element of the deep-marine *Nereites* and the shelfal *Cruziana* ichnofacies.

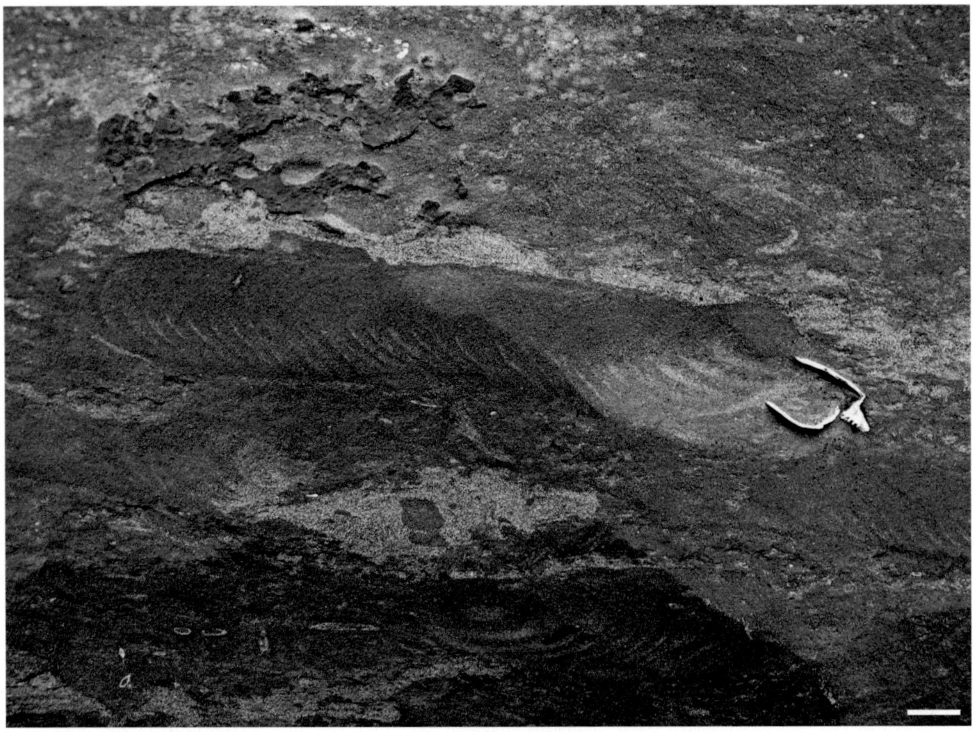

**Fig. 5.133** *Scolicia* in the Miocene Mount Messenger Formation (deep marine, channel-levee complex), one burrow preserving its producer, an irregular echinoid. Sea cliffs of the Taranaki Peninsula, North Island, New Zealand. *Scale bar* = 1 cm

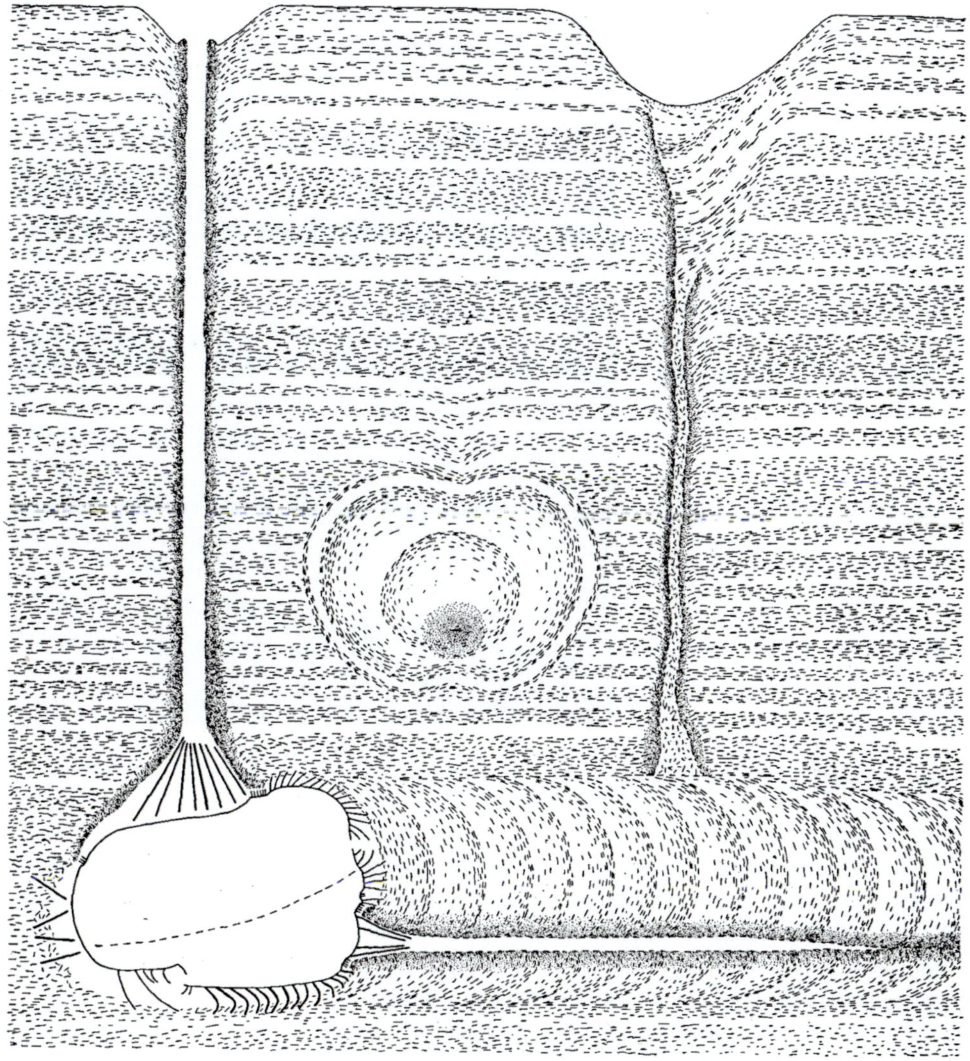

**Fig. 5.134** Ichnological reconstruction of the burrow of *Echinocardium cordatum* in transverse section (center) and longitudinal section. The latter shows two funnels to the surface for deposit-feeding, a functional one and an abandoned, collapsed one. Reorientation of non-spherical sediment grains by the burrowing activity of the animal produces a meniscate backfill. From Bromley and Asgaard (1975), republished with permission of the Geological Society of Denmark

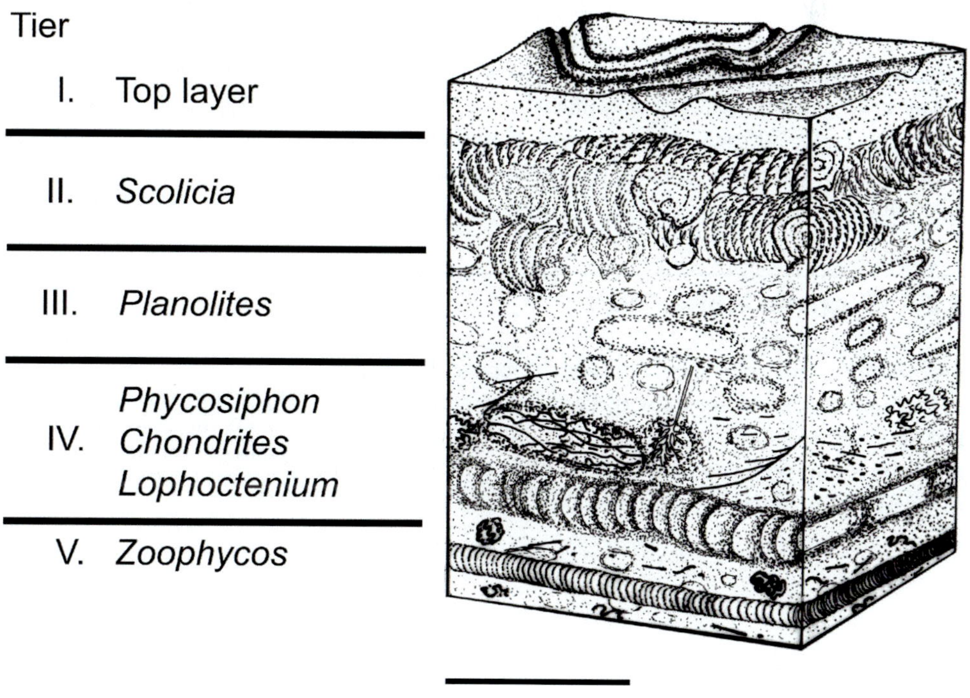


**Fig. 5.135** Tiering diagram with distribution of characteristic biogenic sedimentary structures and traces, based on observations from the modern deep sea. Note the occurrence of *Scolicia* in the upper tier. *Scale bar* = 10 cm. Modified after Wetzel (1981), republished with permission of Schweizerbart (www.schweizerbart.de/9783443190347)

**Fig. 5.136** Laminated heterolithic sandstone with *Phycosiphon*. The upper tier of the ichnofabric is totally bioturbated with *Scolicia*, which leads to a homogenization of the sediment and thus contributes to increased reservoir properties. Parts of the upper tier with *Scolicia* are in turn reburrowed by *Phycosiphon*. Miocene Mount Messenger Formation (deep marine, channel-levee complex), sea cliffs of the Taranaki Peninsula, North Island, New Zealand. *Scale bar* = 1 cm

*Age:* Echinoid-produced *Scolicia* are known since the Jurassic (Seilacher 1986; Fu and Werner 2000), while Paleozoic *Scolicia* (e.g. Benton and Gray 1981; Bjerstedt 1988; Buckman 1992) must have been produced by other organisms (such as slug-like animals). Many such Paleozoic *Scolicia* have relatively simple morphology and composition compared to the complex endichnial *Scolicia* from the Mesozoic onwards (Buatois and Mángano 2004).

*Reservoir Quality:* Sandy to silty, laminated sediment of heterolithic composition is the target of irregular echinoids and becomes homogenized into sand-rich sediment due to the deposit-feeding activity (Fig. 5.136). This happens at a broad scale and this effect is amplified by the presence of relatively large burrows and a dense bioturbate texture (ichnofabric). Therefore, *Scolicia* bioturbation may increase reservoir properties.

## 5.27 *Scoyenia* White, 1929

*Morphology, Fill and Size: Scoyenia* refers to horizontal to inclined burrow elements with faint and irregular lining, longitudinal striation, and a conspicuous meniscate fill, although some portions of a burrow may show transition to passive fill (Bromley and Asgaard 1979; Frey et al. 1984; Retallack 2001; Fig. 5.137). Faint peristaltic annulation of the burrow wall can occur. Burrows are unbranched with a slightly varying diameter, typically between 5 and 20 mm.

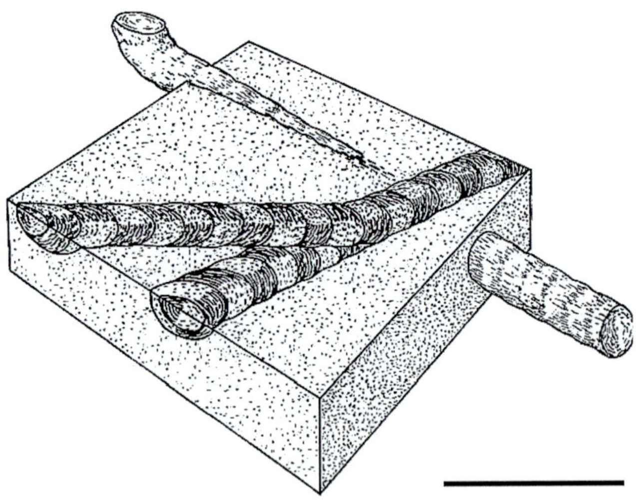

**Fig. 5.137** Structure and morphology of *Scoyenia gracilis*. *Scale bar* = 1 cm. After Bromley and Asgaard (1979), republished with permission of Elsevier; permission conveyed through Copyright Clearance Center, Inc.

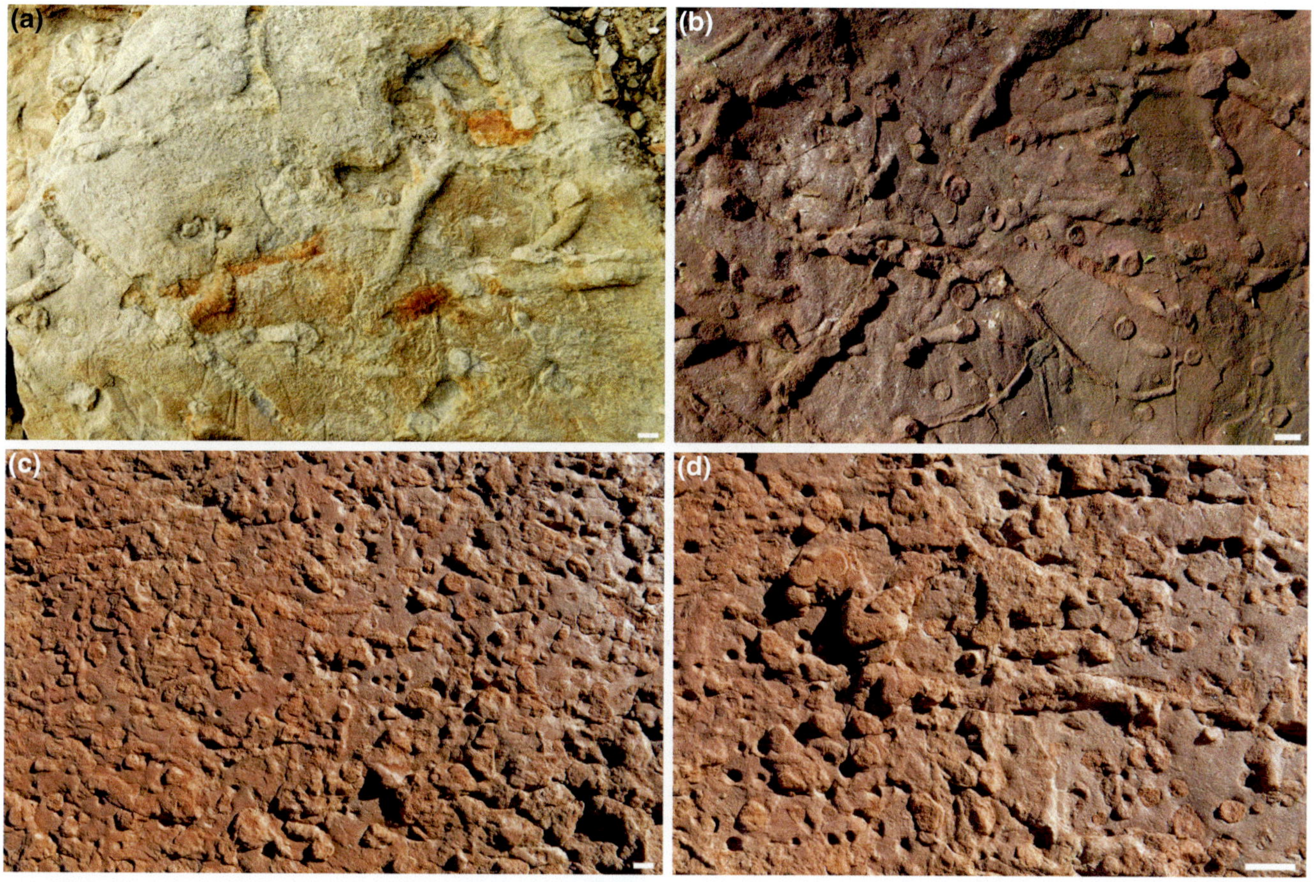


**Fig. 5.138** *Scoyenia gracilis* in outcrop. *Scale bars* = 1 cm. Sandstone bedding planes with tunnels and shafts displaying internal backfill and external scratches. **a** Eocene Aspelintoppen Formation (fluvial), Brongniartfjellet, Svalbard. **b** Lower Triassic Buntsandstein Group (fluvial), near Cracow, southern Poland. **c**, **d** Permian Cutler Formation (alluvial fan), 191 Canyon north of Moab, Utah, USA. Note association with paired burrow apertures likely to be openings of *Arenicolites*

*Ichnotaxonomy:* Only two ichnospecies have been described so far, of which the type ichnospecies *S. gracilis* is the most common (Fig. 5.138), whereas *S. beerboweri* seems to be restricted to Ordovician paleosols (Retallack 2001; Fig. 5.151).

*Substrate: Scoyenia* is often encountered in fine-grained sandstone and siltstone with muddy intervals but may also occur within calcareous successions, particularly those related to paleosols, such as horizons with caliche.

*Appearance in Core:* The relatively large diameter and the appearance of the burrows make *Scoyenia* comparatively easy to recognize in core, although the striated wall is impossible to be proven in sections (Fig. 5.139). Horizontal and shallowly inclined burrows are dominant. The faint lining typically consists of mud but might be poorly developed or preserved. The structured meniscate fill is diagnostic and commonly includes fragments of sediment with contrasting color.

*Similar Trace Fossils: Taenidium* (particularly *T. barretti*) is the most similar trace fossil to *Scoyenia* (e.g. *S. gracilis*) and may be confused with it in core. However, *Scoyenia* has a striated wall (which is difficult to recognize in core) and typically shows a thin mud lining that is rarely present in *Taenidium*. The average diameter of *Scoyenia* is larger than that of *Taenidium*. Most significantly, the meniscate fill of *Scoyenia* appears to be more disorganized into small pieces, while of *Taenidium* is more homogeneous.

**Fig. 5.139** *Scoyenia gracilis* in sectioned core. *Scale bars* = 1 cm. **a** Heterolithic sandstone with oblique and horizontal, actively filled parts of burrows. Eocene Aspelintoppen Formation (fluvial overbank), Brongniartfjellet, Svalbard (Sysselmannbreen well BH 10-2008, ca. 85.5 m). **b** Ripple-laminated sandstone with thinly lined and actively filled parts of burrows. Upper Triassic (Norian-Rhaetian) Lunde Formation (fluvial overbank), Snorre Field, Norwegian North Sea (well 34/7A-9H, ca. 2784.5 m). **c** Iron-stained argillaceous deposit with burrows displaying packets with discontinuous menisci. Upper Triassic (Norian-Rhaetian) Lunde Formation (paleosol), Snorre Field, Norwegian North Sea (well 34/7-1, ca. 2481.7 m)

*Producers:* Arthropods such as insects (e.g. beetles) and millipedes have been considered as producers of *Scoyenia* (e.g. Frey et al. 1984; Retallack 2001; Hasiotis 2010).

*Ethology: Scoyenia* are probably multipurpose feeding, dwelling and breeding burrows (Retallack 2001). More inclined burrows have been inferred to be escape structures (Hubert and Dutcher 2010).

*Depositional Environment: Scoyenia* is a continental trace fossil typical of alluvial, lacustrine and fluvial deposits, such as paleosols, lakes and overbank deposits. It is often related to high soil moisture and inferred to occur in wet seasonal to wet climates (Hasiotis 2010).

*Ichnofacies:* The *Scoyenia* Ichnofacies was originally designed as almost an afterthought of the marine ichnofacies (Seilacher 1967) but now is regarded to be more differentiated (see Bromley 1996; Buatois and Mángano 2011; Melchor et al. 2012).

*Age: Scoyenia* is known from the Ordovician (Retallack 2001) to the Holocene (Hasiotis 2010).

*Reservoir Quality:* The sandy fill of *Scoyenia* burrows makes them a good conductor for fluids and gas, especially if occurring in muddy host sediment. The heterogeneous composition of the meniscate fill may lead to a slight reduction of this property.

## 5.28 *Siphonichnus* Stanistreet et al., 1980

*Morphology, Fill and Size: Siphonichnus* includes vertical, oblique or horizontal cylindrical burrows with a linear, winding or bow-shaped morphology and a circular to oval cross section (Knaust 2015a; Fig. 5.140). It is characterized by a laminated meniscate mantle (active fill), which is penetrated by a homogeneous core (passive fill). The diameter of the complete burrow is typically in the range of a centimeter, although much smaller burrows occur as well. Their length can be several decimeters.

*Ichnotaxonomy: S. ophthalmoides* (Jessen 1950) from the Carboniferous of Germany is the senior synonym of *S. eccaensis* (Knaust 2014b, 2015a). It is currently regarded as the only ichnospecies of *Siphonichnus*, although Zonneveld and Gingras (2013) suggested the inclusion of *Scalichnus phiale* within the ichnogenus *Siphonichnus* and established *S. lepusaures* and *S. sursumdeorsum* as new ichnospecies. *S. lepusaures* and *S. sursumdeorsum* are probably cross sections of *Teichichnus zigzag* and are better accommodated within that ichnogenus. *Siphonichnus* is the eponym of the ichnofamily Siphonichnidae, which encompasses burrows of varying morphology consisting of one or more subvertical tube(s) with passive fill and active lining or mantle (Knaust 2015a). In addition to *Siphonichnus*, it includes the ichnogenera *Laevicyclus*, *Parahaentzschelinia, Scalichnus* and *Hillichnus* (Fig. 5.141).

*Substrate: Siphonichnus* is reported from diverse types of substrate including mudstone, siltstone, sandstone and limestone. It often occurs in variously heterogeneous sandy substrates.

*Appearance in Core:* Although originally described as vertical, *Siphonichnus* is an unbranched trace fossil that can also have oblique and horizontal components. It consists of a wall with tight lamination (meniscate backfill) and a central part, which is occupied by a well-defined tube with passive fill. In core, these cylindrical burrows appear as elongate, elliptical or circular sections (Fig. 5.142; see also Fig. 5.100f, h).

*Similar Trace Fossils:* Varying morphology, orientation and size of *S. ophthalmoides* may give rise to confusion with a number of trace fossils being partly similar. First, the vertical expression of *S. ophthalmoides* may resemble *Skolithos* (Fig. 5.146g) or even *Trypanites*, particularly in cases where the lamination of the mantle is poorly visible. For the same reason, horizontal parts of *S. ophthalmoides*—as observed in core—may appear as *Palaeophycus* and *Macaronichnus*. This situation is further complicated because *S. ophthalmoides* typically co-occurs with *Skolithos* and *Palaeophycus*, all of which may have been produced by a similar tracemaker. The lingulide-produced *Lingulichnus* is probably the trace fossil most similar to *Siphonichnus*. It differs from *Siphonichnus* by its circular instead of elliptical cross section, and the internal lamination is partly penetrated by the passively filled pedicle trace in a less consistent manner. The innermost burrow (core) of *S. ophthalmoides* seems to be wider than the pedicle trace of *Lingulichnus*, and only preserves a relatively small portion of the outer laminated burrow (mantle). *S. ophthalmoides* has also been confused with *Diplocraterion* and, more rarely, the mantle in *S. ophthalmoides* could also be mistaken as the mud lining in *Ophiomorpha*.

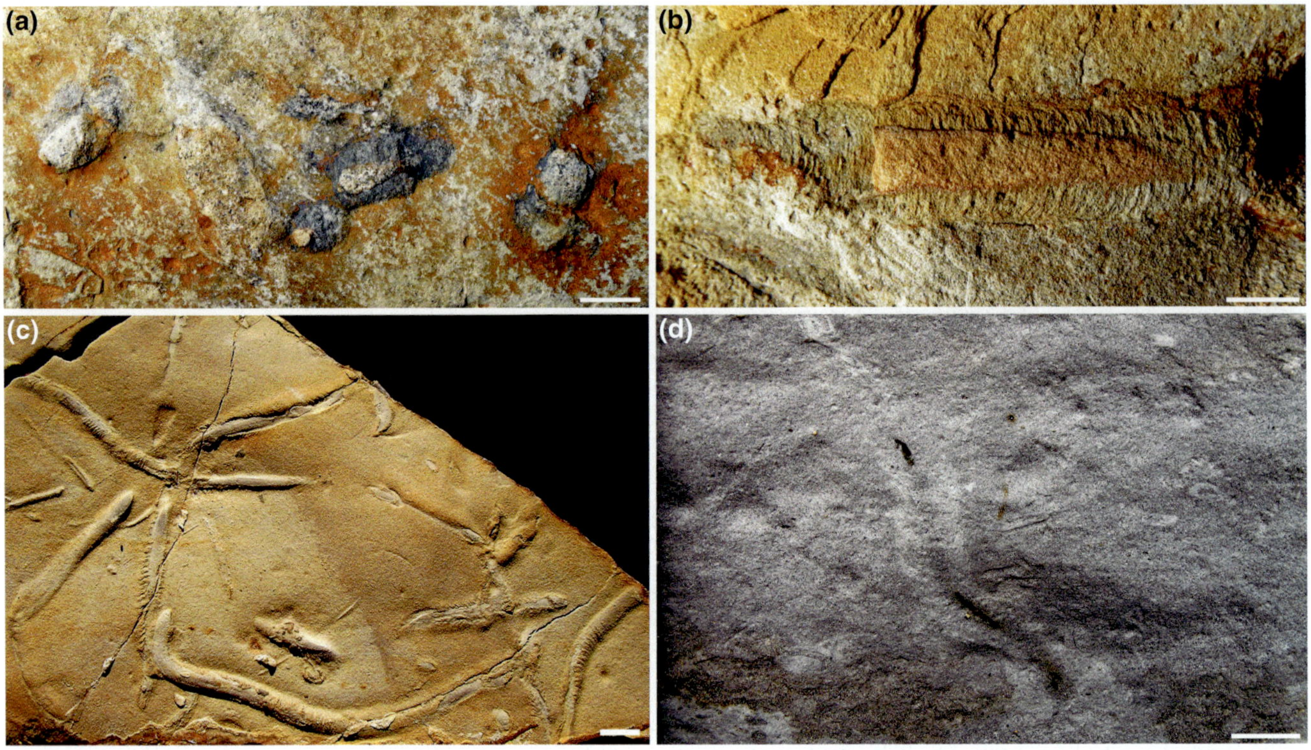


**Fig. 5.140** *Siphonichnus ophthalmoides* in outcrop. *Scale bars* = 1 cm. From Knaust (2015a), republished with permission of Elsevier; permission conveyed through Copyright Clearance Center, Inc. **a** Variably oriented burrows with respect to bedding within micaceous, cross-bedded sandstone of nearshore origin. Lower Jurassic (Hettangian) Höganäs Formation (nearshore), Helsingborg, southern Sweden. **b** Bedding-parallel burrow part. Same locality as in (**a**). **c** Winding burrows revealing core and mantle on a limestone bedding plane. Middle Triassic (Anisian) Meissner Formation (Upper Muschelkalk, carbonate ramp), Troistedt near Weimar, Thuringia, Germany. **d** Single specimen of *S. ophthalmoides* with a pyritized fill within the core. Late Miocene Mount Messenger Formation (deep marine, slope fan), coastal section north of New Plymouth, North Island, New Zealand

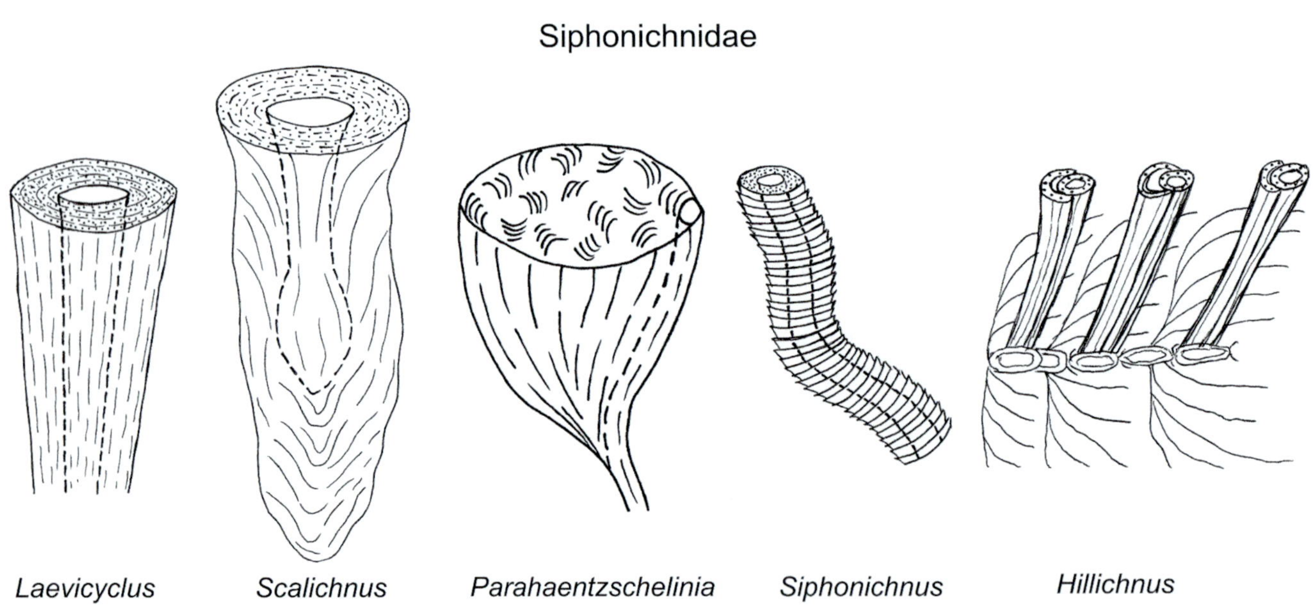


**Fig. 5.141** Line drawings of ichnogenera included in the ichnofamily Siphonichnidae and their morphological characteristics, all of them having one or more subvertical tube(s) with passive fill and active lining or mantle. From Knaust (2015a), republished with permission of Elsevier; permission conveyed through Copyright Clearance Center, Inc.

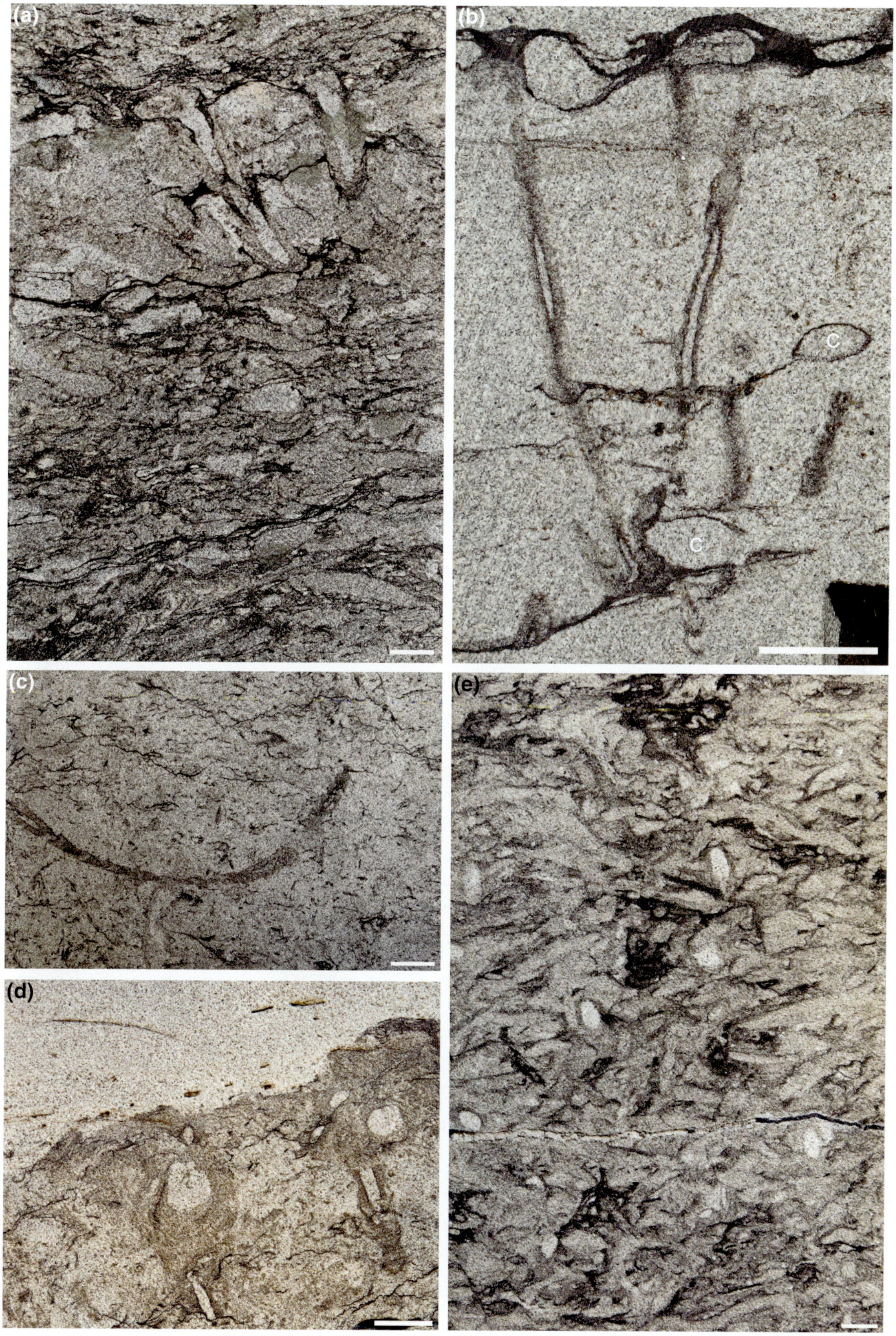
(a)
(b)
C
C
(c)
(d)
(e)

◀**Fig. 5.142** *Siphonichnus ophthalmoides* in sectioned core. **a**, **b** and **e** from Knaust (2015a), republished with permission of Elsevier; permission conveyed through Copyright Clearance Center, Inc. *Scale bars* = 1 cm. **a** Heterolithic sandstone with a crowded *S. ophthalmoides* ichnofabric consisting of subvertical, inclined and horizontal, partly bowed burrows with concave-downwards laminae in the mantle. Lower Jurassic (Sinemurian-Pliensbachian) Tilje Formation (marginal marine), Norwegian Sea (well 6607/12-3, ca. 4217.35 m). **b** Small and elongated *S. ophthalmoides* together with probable casts of the producing bivalves (*c*). Lower Jurassic (Pliensbachian) Tilje Formation (marginal marine), Åsgaard Field, Norwegian Sea (well 6506/12-K-3H, ca. 4459.0 m). **c** Bowed burrow in longitudinal section. Lower Jurassic (Aalenian) Stø Formation, Snøhvit Field, Norwegian Barents Sea (well 7120/8-2, ca. 2097.75 m). **d** Variably oriented burrows displaying a thick mantle. Lower Jurassic (Aalenian) Stø Formation, Snøhvit Field, Norwegian Barents Sea (well 7120/6-2 S, ca. 2583.3 m). **e** Dense *S. ophthalmoides* ichnofabric with total bioturbation and discrete burrow segments trending in all directions. Lower Jurassic (Hettangian-Sinemurian) Nansen Formation, Norwegian North Sea (well 25/10-11T2, ca. 4283 m)

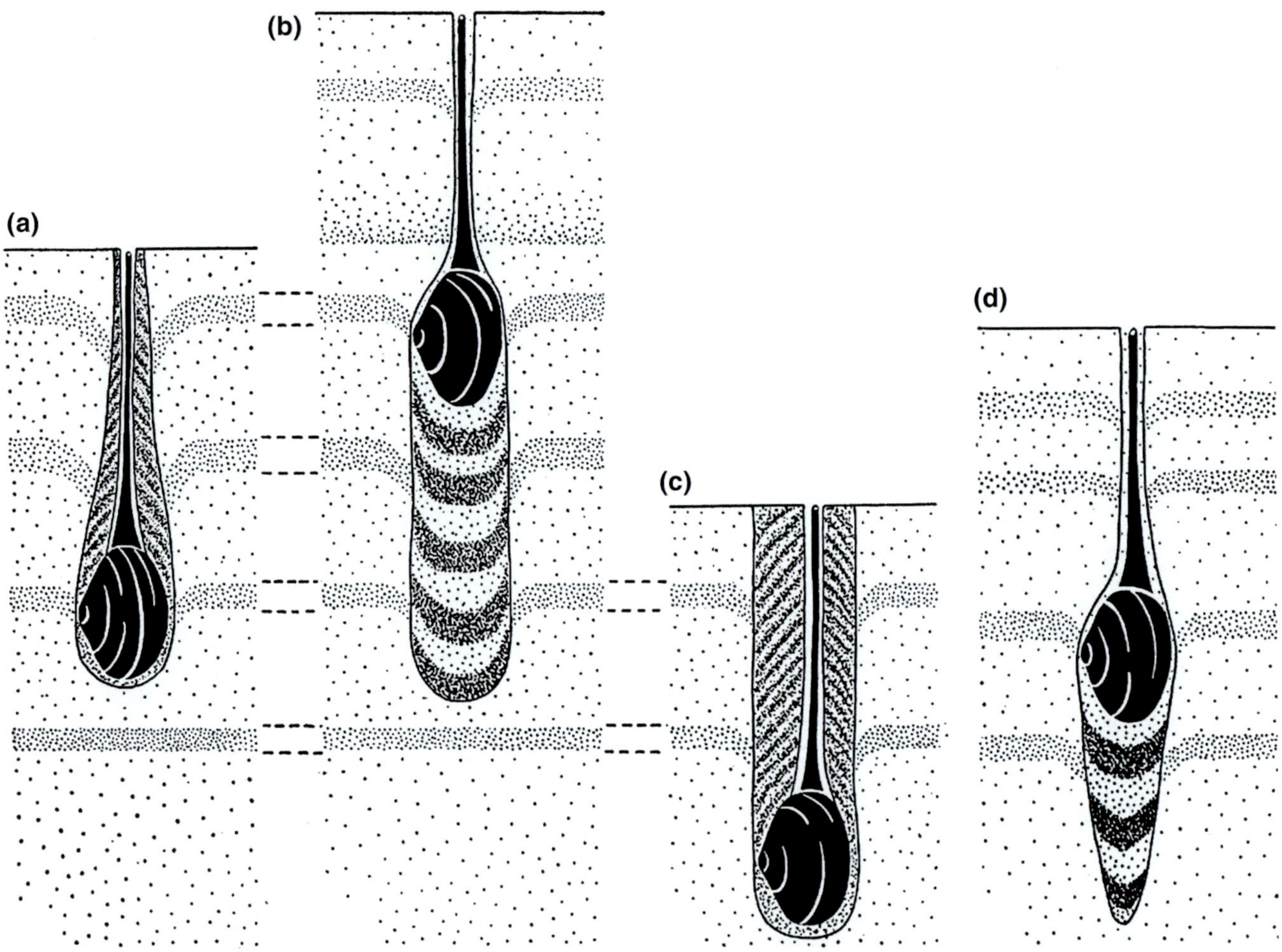


**Fig. 5.143** Modern bivalve burrows in response to deposition. From Reineck (1958), republished with permission of Schweizerbart (www.schweizerbart.de/home/senckenberg). **a** No deposition. **b** Rapid sedimentation. **c** Erosion. **d** Very slow sedimentation

*Producers:* Contemporary *Siphonichnus*-like traces are produced by bivalves adjusting their position within the sediment (Reineck 1958; Fig. 5.143). In some cases, the cast of the bivalve shell or its cavity is preserved at the base of *Siphonichnus* (Dashtgard and Gingras 2012; Gingras et al. 2012a; Fig. 5.142b). The relatively long burrow parts, their wide range of orientations, and the almost perfectly centered core within the mantle are features which make other producers than bivalves possible (e.g. arthropods).

*Ethology: Siphonichnus* is the dwelling trace (domichnion) of a suspension- and deposit-feeder, such as a bivalve. It is able to adjust its position with respect to erosion and deposition (equilibrichnion). The highly variable orientation of *Siphonichnus* burrows apparently results from the activity of a (temporarily?) mobile infaunal tracemaker for the purpose of locomotion and, at least partly, deposit-feeding. Therefore, a combined sedentary lifestyle resulting in more or less vertically oriented burrows, and mobile lifestyle resulting in burrows with variable directions of the producer can be inferred.

*Depositional Environment: Siphonichnus* can be regarded as an indicator of shallow-marine and marginal-marine environments, often related to fluctuating salinity and freshwater influx (Knaust 2015a). Its distribution ranges from proximal offshore and shoreface to deltaic, estuarine and lagoonal environments. *S. ophthalmoides* is a good example of an early application of trace fossils in ichnostratigraphy and the reconstruction of paleoenvironments in connection with industrial coal mining. The assumed producer, an endobenthic marine bivalve with tolerance for freshwater influence and salinity fluctuations, has rapidly colonized the ecological niche along the paleoshoreline with brackish influence. This opportunistic colonization resulted in the abundant occurrence of the "eye"-like burrows (as seen in cross section), while other macrofossils were too sparse for quick characterization of the interval. The importance of *S. ophthalmoides* for the correlation of coal seams in the mines of western Germany on the basis of marine intervals, which are accompanied at their base and top by the "eye"-shale (Augenschiefer), was realized by early coal investigators (e.g. Jessen 1950). Later workers have applied *S. ophthalmoides* as a facies indicator of marginal-marine deposits as well as in association with deposits on the lower delta plain (e.g. interdistributary bay/lagoon) and delta-fed submarine fans. *Siphonichnus* nicely reflects the response of its trace-making bivalve to changing sediment supply. McIlroy (2004) recognized *S. ophthalmoides* in situations with tidal-channel abandonment and delta-lobe abandonment, when a sudden reduction in sediment supply occurs. *S. ophthalmoides* is a common constituent of intertidal and saltmarsh deposits.

*Ichnofacies: Siphonichnus* belongs to the *Skolithos* Ichnofacies and occurs as accompanying constituent in the *Cruziana* Ichnofacies.

*Age: Siphonichnus* is recorded from the Late Devonian (e.g. Angulo and Buatois 2012) to Holocene (e.g. Gingras et al. 2008).

*Reservoir Quality: Siphonichnus* burrows are mainly sand-filled and vertically oriented and therefore can lead to a slight improvement of the vertical connectivity.

## 5.29 *Skolithos* Haldeman, 1840

*Morphology, Fill and Size: Skolithos* is a simple, subvertical cylindrical tube with or without lining and passive fill (Alpert 1974; Figs. 5.144 and 5.145). A funnel-shaped aperture at the top may be developed or preserved (Schlirf 2000). The burrow diameter ranges from millimetric to

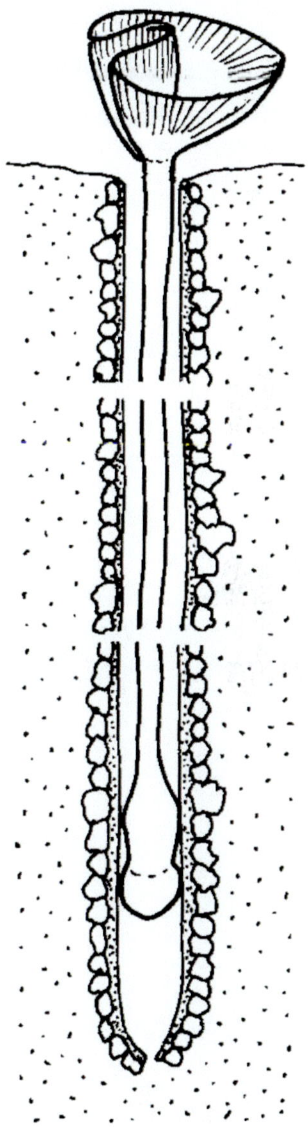

**Fig. 5.144** Sketch of *Skolithos linearis* as produced by a modern Phoronida (in position) in soft sediment. Note the sediment particles adhering to the tube. Not to scale. After Emig (1982), republished with permission of Elsevier; permission conveyed through Copyright Clearance Center, Inc.

**Fig. 5.145** *Skolithos* in outcrop. *Scale bars* = 22 cm (**a**) and 1 cm (**b**)–(**g**). **a**, **b** *S. linearis* from its type locality, Chickies Rock, near Lancaster, Pennsylvania. A thick sandstone bed (tidal-dune set) is penetrated by numerous long burrows (**a**), which show the diagnostic lining (**b**). **c**, **d** Erratic boulder of Lower Cambrian sandstone completely pierced with burrows (piperock) in longitudinal section (**c**) and cross section (**d**). Cliff near Arnager, Bornholm, Denmark. **e** Thick, passively sand-filled *S. verticalis* in modern beach deposits. North Island, New Zealand. **f**, **g** Sandstone with trough cross-bedding and abundant burrows. Lower Ordovician Tosna Formation (high-energy subtidal nearshore environment). Tosna River bank near Nikolskoe village, St. Petersburg region, NW Russia (see Dronov and Mikuláš, 2010)

(a)
(b)
(c)
(d)
(e)
(f)
(g)
(h)
(i)
186

◀**Fig. 5.146** *Skolithos* in sectioned core. *Scale bars* = 1 cm. **a** Laminated micritic limestone with relatively large, deep, sand-filled burrows originating from an erosion surface. Upper Permian carbonates (carbonate platform with tidal flat). South Pars Field, Persian Gulf, Iran (well SP-9, ca. 3082.9 m). From Knaust (2009a), republished with permission of GulfPetroLink. **b** Laminated micritic limestone with sand-filled burrows. Upper Permian carbonates (shoal). South Pars Field, Persian Gulf, Iran (well SP-9, ca. 2909.75 m). After Knaust (2009a, 2014a), republished with permission of GulfPetroLink and EAGE. **c** Bioturbated sandstone with discrete *S. linearis* (lower left). Middle Jurassic (Bathonian) Hugin Formation (mixed tidal flat within an estuarine bayhead delta). Gudrun Field, Norwegian North Sea (well 15/3-9T2, ca. 4493.0 m). **d** Sandstone-siltstone alternation with thin, mud-filled *Skolithos* associated with sparse *Polykladichnus* and *Arenicolites* in the upper part of the siltstone. Lower Jurassic (Pliensbachian) Tilje Formation (tidal flat). Åsgard Field, Norwegian Sea (well 6506/12-K-3H, ca. 4537.85 m). **e** Cross-bedded sandstone with plant debris and large sand-filled burrows penetrating the substrate from an erosion surface. Lower Jurassic (Hettangian) Statfjord Formation (fluvial to marginal marine, mixed tidal flat). Johan Sverdrup Field, Norwegian North Sea (well 16/2-17S, ca. 1925.85 m). **f** Sandstone with a dense cluster (pipe rock) of *Skolithos/Siphonichnus* cut slightly obliquely to their vertical extent. Lower Jurassic (Pliensbachian to Bajocian) Stø Formation (offshore transition), Iskrystall Discovery, Norwegian Barents Sea (well 7219/8-2, ca. 2973.5 m). **g** Intensely bioturbated sandstone with relict bedding preserved (inclined well). The dense *Ophiomorpha* ichnofabric is cross-cut by few *Skolithos* in the center and upper right. Middle Jurassic (Callovian) Hugin Formation (upper shoreface). Gina Krog Field, Norwegian North Sea (well 15/5-7, 3874.2 m). **h** Silty sandstone with numerous long burrows. Middle Jurassic (Bajocian-Bathonian) Tarbert Formation (marginal marine, tidal flat), Valemon Field, Norwegian North Sea (well 34/11-B-13, ca. 4594.5 m). **i** Laminated sandstone with cryptic bioturbation and two mud-filled *Skolithos* (left), accompanied by *Ophiomorpha* (right). Upper Cretaceous (Maastrichtian) Springar Formation (deep marine, lobe). Norwegian Sea (well 6604/10-1, ca. 3647.95 m)

centimetric and its length from a few millimeters to several decimeters. The length/diameter ratio can be used to discriminate *Skolithos* from morphologically similar trace fossils.

*Ichnotaxonomy: Skolithos* is sparse in morphological features. About 20 ichnospecies have been attributed to this ichnogenus based on criteria such as morphology, burrow wall and size, although some of them do not fit the ichnogeneric diagnosis (such as helicoidal and branched forms) and therefore must be excluded. In addition, ichnospecies originally attributed to other ichnogenera such as *Tigillites* and *Sabellarifex* are now regarded as junior synonyms of *Skolithos* (Alpert 1974; Fillion and Pickerill 1990; Schlirf 2000; Schlirf and Uchman 2005). The ichnospecies *S. linearis, S. verticalis* and *S. annulatus* are among the most important forms of *Skolithos*. Despite repeated attempts, the ichnogenus *Skolithos* is in need of ichnotaxonomic revision.

*Substrate: Skolithos* occurs in soft (sandy and muddy) substrates, and subordinately in firm siliciclastic and carbonate sediments.

*Appearance in Core: Skolithos* is relatively easy to recognize in core, where it appears as more or less straight burrows arranged normal to the bedding planes (Fig. 5.144). The fill commonly contrasts with the surrounding rock and a lining of the burrow wall is characteristic.

*Similar Trace Fossils:* Owing to the simplicity of *Skolithos*, it may be confused with similar burrows, parts of other burrows, or even sedimentary and tectonic structures, particularly in core (Fig. 5.189a, f). *Skolithos* resembles *Cylindricum* Linck, 1949 (and the similar *Capayanichnus vinchinensis* Melchor et al., 2010) from fluvial deposits and likely produced by freshwater crustaceans. This ichnotaxon differs from *Skolithos* by a lower length/width ratio (Hasiotis 2008) and thus justifies its own status (Knaust 2012a). In contrast, *Trichichnus* is characterized by a high length/width ratio and occasional branching; moreover, it is commonly pyritized. Very thin *Siphonichnus ophthalmoides* may also be mistaken as *Skolithos* (e.g. Bromley 1996; Gerard and Bromley 2008). Furthermore, burrow parts of other trace fossils such as shafts of *Thalassinoides* and *Ophiomorpha* may easily be confused with *Skolithos* if the three-dimensional context remains unclear (Fig. 5.146).

*Producers:* A wide range of organisms is known to produce *Skolithos*-like traces (Fig. 2.3). Most appropriate are various groups of worms such as priapulids and polychaetes, as well as phoronids, which may account for many Paleozoic and also younger *Skolithos* (e.g. Fenton and Fenton 1934; Emig 1982; Sundberg 1983; Dashtgard and Gingras 2012; Figs. 5.144 and 5.147). Another group comprises crustaceans (e.g. amphipods), which are known to produce *Skolithos*-like traces today (e.g. Dashtgard and Gingras 2012). Anthozoa (sea anemones) may be also capable of producing *Skolithos*-like traces (e.g. Hertweck 1972). In addition, continental *Skolithos* can be produced by plant roots (e.g. Gregory et al. 2006; Fig. 5.183f) and even insects and trap-door spiders (arachnids). In deep-marine environments, holothurians are known to produce *Skolithos* (Dashtgard and Gingras 2012); aplacophorans (molluscs) may also be makers of *Skolithos* in this setting.

*Ethology:* Aquatic *Skolithos* were predominantly built by suspension-feeding organisms for dwelling (domichnia), whereas terrestrial *Skolithos* also include a scavenging behavior.

*Depositional Environment: Skolithos* is a common indicator of relatively high energy, shallow-water, nearshore to marginal-marine environments. In Paleozoic time, the burrows appeared in such high abundance as to build *Skolithos* piperock, with a general decrease in occurrence from

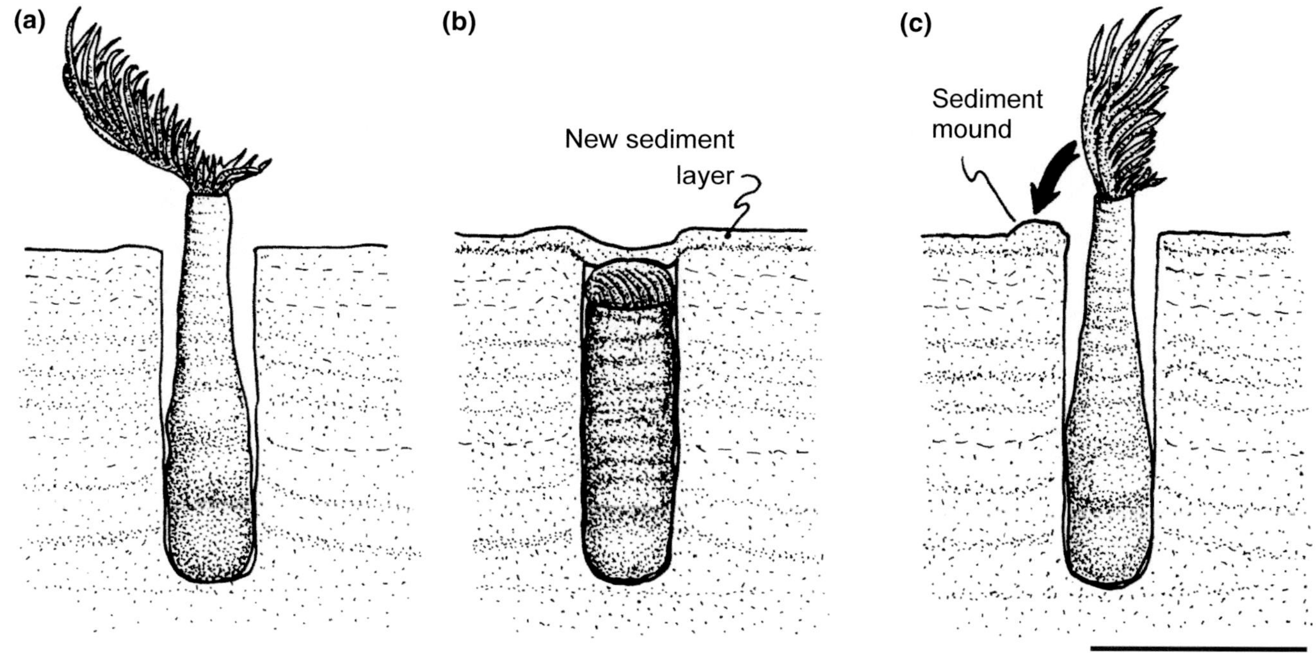


**Fig. 5.147** Reconstructions of the hypothetical organism inhabiting *Skolithos linearis*. Reconstructions based on sedimentary features found in and near *S. linearis* and on life habits of phoronids (Hyman 1959). *Scale bar* = 1 cm. After Sundberg (1983), © Paleontological Society, published by Cambridge University Press, reproduced with permission. **a** Organism during low-turbidity conditions with tentacles fully extended to gather food. **b** Position of organism during time of high-turbidity of sedimentation with tentacles folded inwards. **c** The organism emerging and unfolding its tentacles at the end of a high sedimentation period. The *arrow* indicates the direction of sediment movement of the tentacles to form the sediment mound

Cambrian towards Permian (Droser 1991; Desjardins et al. 2010). *Skolithos* also accompanies other trace fossils in shelfal and deep-marine environments and is a common constituent of fluvial and other continental deposits (e.g. Hasiotis 2010; Melchor et al. 2012).

*Ichnofacies: Skolithos* is the namesake of the *Skolithos* Ichnofacies, but occurs in other marine and continental ichnofacies too.

*Age: Skolithos* is a wide-ranging trace fossil known from late Precambrian time (Alpert 1975; McCall 2006) through the Holocene (Dashtgard and Gingras 2012).

*Reservoir Quality:* Passive burrow fill of *Skolithos* and their tendency of crowded occurrence (so-called piperock) make such horizons as good candidates to influence reservoir quality and vertical connectivity in a positive manner (Knaust 2014a). They have the potential to overcome minor baffles and barriers (at the centimetric-scale) and to connect reservoir layers.

## 5.30 *Taenidium* Heer, 1877

*Morphology, Fill and Size: Taenidium* is a cylindrical meniscate burrow, winding in a predominantly subhorizontal but also subvertical direction (Fig. 5.148). Burrow diameter ranges between 5 and 450 mm (Smith et al. 2008), usually between 5 and 10 mm. Burrows are unlined or very thinly lined and unbranched (D'Alessandro and Bromley 1987). The backfill menisci are often tightly spaced with little contrast in lithology, grain size or color.

*Ichnotaxonomy:* About seven ichnospecies of *Taenidium* are currently distinguished, although this number may change after a thorough review of *Taenidium* and related meniscate trace fossils (Keighley and Pickerill 1994; Rodríguez-Tovar et al. 2016). The distinction of *Taenidium* from the similar trace fossil *Beaconites* is still debated and finally may show that *Beaconites* must be regarded as junior synonym of *Taenidium* (e.g. Goldring and Pollard 1995; Savrda et al. 2000). Likewise, *Naktodemasis* Smith et al., 2008 can probably also be accomodated in this ichnogenus (Krapovickas et al. 2009). Ichnospecies of *Taenidium* are differentiated by the style of meniscate backfill, and *T. serpentinum, T. diesingi* (=*T. satanassi*), *T. cameronensis, T. barretti, T. irregulare* (=*T. crassum*), *T. planicostatum* and *T. bowni* can currently be regarded as valid ichnospecies.

*Substrate: Taenidium* typically occurs in (preferably fine-grained) sandstone and is associated with heterogeneous paleosols (Fig. 5.149). Interactions with incipient limestone nodules (caliche) and many examples with well-preserved and sharp burrow margins indicate cohesive substrate conditions.

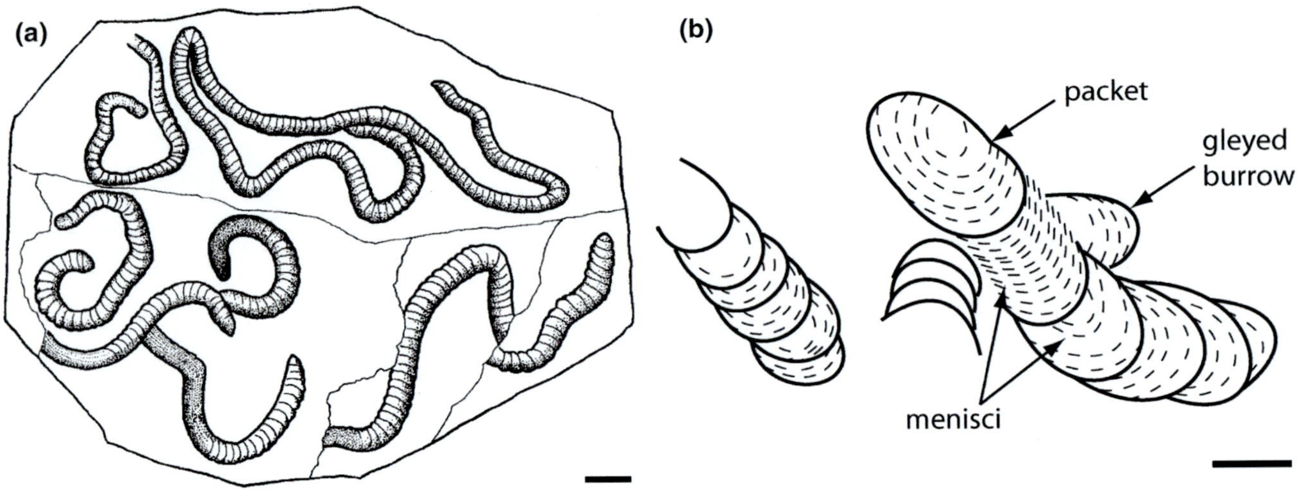


**Fig. 5.148** Morphological features of *Taenidium*. **a** *T. serpentinum* Heer, 1877. *Scale bar* = 1 cm. From D'Alessandro and Bromley (1987), republished with permission of Wiley; permission conveyed through Copyright Clearance Center, Inc. **b** Typical sections of *T. bowni* (Smith, 2007). *Scale bar* = 2 mm

*Appearance in Core:* In core, *Taenidium* appears as circular, elliptical to elongate burrow segments with a winding course and varying orientation (Fig. 5.150). The meniscate backfill of the burrows is conspicuous. Well-developed *Taenidium* ichnofabrics consist of a mixed layer at the top and discrete burrows in the lower tier.

*Similar Trace Fossils: Taenidium* is similar to other meniscate burrows such as *Scoyenia* and *Ancorichnus* (Fig. 5.151; Retallack 2001; Smith et al. 2008). The burrow diameter of *Scoyenia* is commonly larger (10–30 mm) than in *Taenidium* (5–10 mm), and *Scoyenia* is a lined and striated burrow with more irregular menisci. *Ancorichnus* consists of an unlined, backfilled mantle and thick meniscate backfill. In core, *Taenidium* may be confused with *Zoophycos* spreiten, the latter of which typically alternate vertically (where parts of a whorl were cut), or with *Rhizocorallium*, which contains a passively filled marginal tube. Small-scale synsedimentary deformation may be responsible for producing *Taenidium*-like structures (Fig. 5.189c).

*Producers:* Arthropods are probably the main tracemakers of *Taenidium* (Rodríguez-Tovar et al. 2016). By comparison with modern analogs and experimental studies, there is good confidence that continental *Taenidium* is produced by beetle larvae, cicada nymphs, or other insects (Smith et al. 2008; Hembree and Hasiotis 2008; Fig. 5.152), but also earthworms. Marine *Taenidium* may result from the activity of arthropods or worm-like organisms.

*Ethology: Taenidium* is interpreted as burrows produced by a combination of detritus-feeding, locomotion and dwelling behaviors (fodinichnia; Hembree and Hasiotis 2008). It is likely that the tracemakers fed on organic matter and roots within the soil profile (fodinichnion; Fig. 5.152; Smith et al. 2008).

*Depositional Environment:* Insect traces of the kind of *Taenidium* are commonly found in alluvial, fluvial and marginal-lacustrine environments (Savrda et al. 2000; Bedatou et al. 2009; Hasiotis 2010). They are part of many soil deposits, particularly those with a higher amount of moisture but above the water table (A and upper B horizons of soil; Hembree and Hasiotis 2008). *Taenidium* occurs also in shallow- and deep-marine deposits (D'Alessandro and Bromley 1987).

*Ichnofacies: Taenidium* occurs in marine deposits of the *Cruziana* Ichnofacies (e.g. Bromley et al. 1999) and is a characteristic element of the *Scoyenia* Ichnofacies and other continental ichnofacies (Buatois and Mángano 2011; Melchor et al. 2012).

*Age:* Ordovician to Holocene (Keighley and Pickerill 1994).

*Reservoir Quality:* No analysis of the influence of *Taenidium* on reservoir quality is available. The uniformly meniscate fill of the burrows typically is in the same grain-size range as the host rock and therefore little difference may be expected. The homogeneous fill of burrows, which occur in moderate to high density in ripple-laminated sandstone with varying grain-size distribution (e.g. overbank deposits), may favor a better flow behavior than the laminated rock surrounding it (Fig. 5.150c).

**Fig. 5.149** *Taenidium* in outcrop. *Scale bars* = 1 cm. **a**, **b** Completely bioturbated sandstone with *T. barretti* in horizontal (**a**) and vertical section (**b**). Upper Jurassic (Kimmeridgian, fluvial, crevasse splay), La Griega Beach in Colunga, Asturias, northern Spain. **c** Cross-stratified, fine-grained sandstone bed containing brownish *T. barretti* in its deeper tier (ca. 10–30 cm below surface). Upper Triassic Kågeröd Formation (fluvial overbank), Bornholm, Denmark. From Knaust (2015b). **d** Loose slab of sandstone with a reworked caliche nodule and winding *T. barretti* with pronounced backfill. Same locality as in (**c**). From Knaust (2015b). **e**, **f** *T. irregulare* (=*T. crassum*), freely winding and cross-cutting one another along a sandstone bedding plane. Upper Jurassic (Kimmeridgian, shallow marine), coastal cliffs at Praia do Salgado, western Portugal

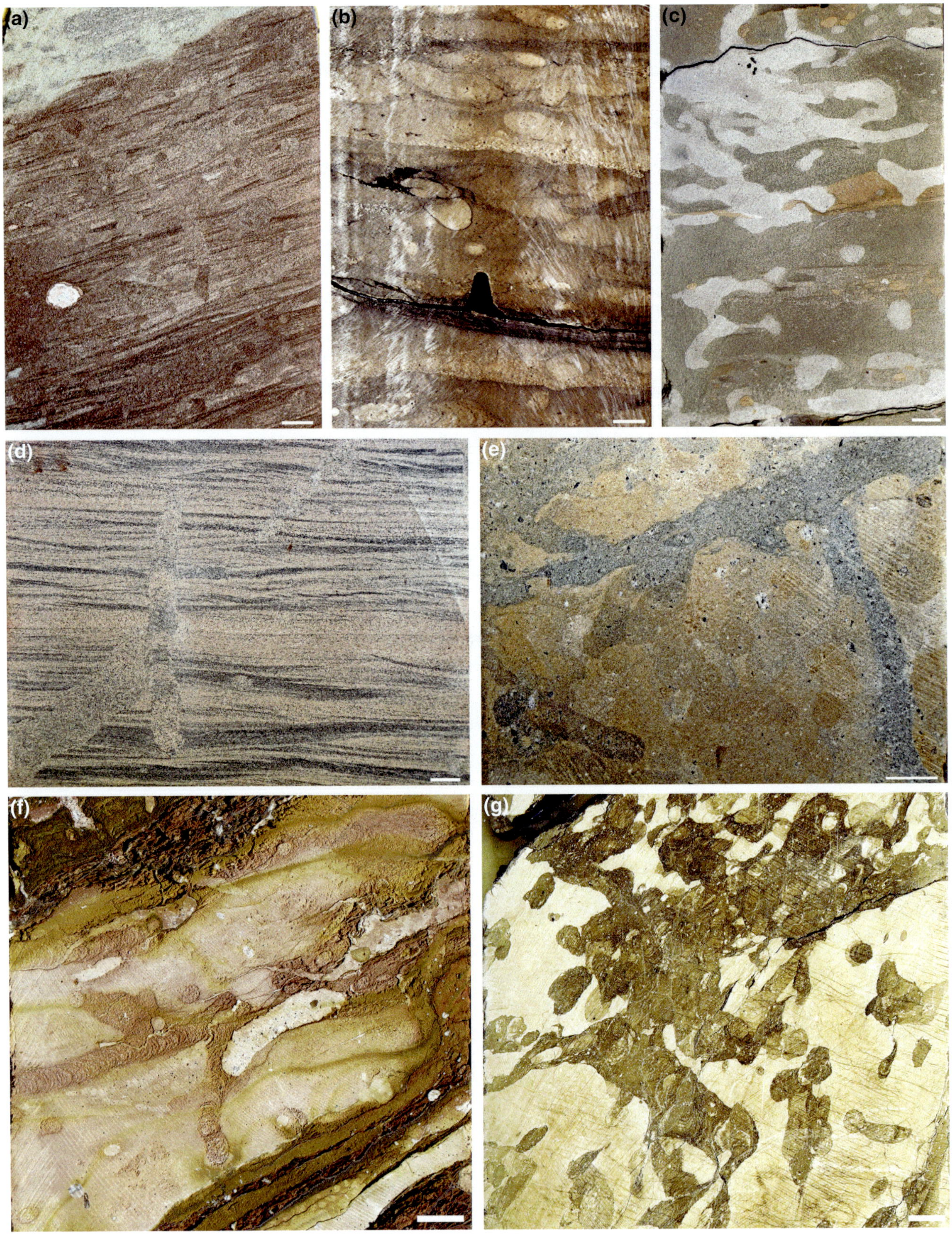
(a)
(b)
(c)
(d)
(e)
(f)
(g)

◀**Fig. 5.150** *Taenidium* in sectioned core. *Scale bars* = 1 cm. **a** *T. barretti* in moderately bioturbated sandstone with a reworked caliche nodule. Upper Triassic (Norian-Rhaetian) Lunde Formation (fluvial, overbank), Snorre Field, Norwegian North Sea (well 34/7-A-9H, ca. 2635.55 m). **b** *Taenidium* ichnofabric in highly bioturbated micritic limestone. Upper Permian Khuff Formation (carbonate platform, restricted lagoon), South Pars Field, Persian Gulf, Iran (well SP-9, 3086.5 m). From Knaust (2009a), republished with permission of GulfPetroLink. **c** Selective carbonate cementation mainly following *Taenidium* burrows as fluid pathways. Triassic Skagerrak Formation (alluvial), Johan Sverdrup Field, Norwegian North Sea (well 16/2-9S, ca. 1954.5 m). **d** Discrete *T. barretti* burrows with meniscate backfill in ripple-laminated sandstone. Upper Triassic (Norian-Rhaetian) Lunde Formation (fluvial, overbank), Snorre Field, Norwegian North Sea (well 34/7-1, ca. 2504.65 m). **e** Pedogenetically modified carbonate rock (calichified dolomitic limestone) preserving *T. barretti* ichnofabric with burrows penetrating the matrix and the cracks. Upper Triassic (Norian) Skagerrak Formation (alluvial), Johan Sverdrup Field, Norwegian North Sea (well 16/2-15, ca. 1975.6 m). **f** Marly limestone with *Taenidium* showing false branching. Lower Cretaceous (Berriasian) Asgaard Formation (shelf), Johan Sverdrup Field, Norwegian North Sea (well 16/2-11A, ca. 2175.5 m). **g** Chalk with a dense *Taenidium* ichnofabric, some of the burrows occupying sediment-filled fractures and/or large burrows. Paleocene (Danian) Shetland Group (shelf), Oseberg Field, Norwegian North Sea (well 30/9-B-44B, ca. 4258.5 m)

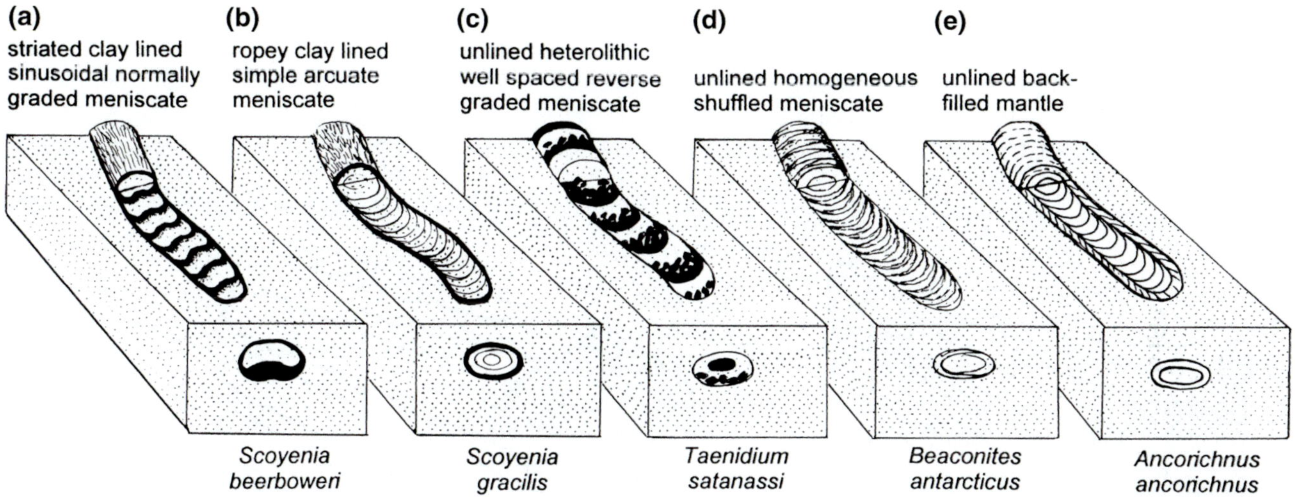


**Fig. 5.151** Differentiation among common backfilled meniscate burrows. After Retallack (2001), republished with permission of Wiley; permission conveyed through Copyright Clearance Center, Inc.

**Fig. 5.152** Insect tracemaker producing meniscate backfill as it moves through the soil. After Smith (2007)

## 5.31 *Teichichnus* Seilacher, 1955

*Morphology, Fill and Size: Teichichnus* is a vertical wall-like spreite burrow with a straight or curved plan view. It commonly includes unbranched (and rarely branched) ichnospecies. Burrow size is quite variable and ichnospecies- as well as age-dependent. It ranges from burrow diameters of a few millimeters to several centimeters, while burrow length can reach more than a meter (e.g. Martinsson 1965).

*Ichnotaxonomy:* More than a dozen ichnospecies were introduced since the erection of *T. rectus* Seilacher, 1955, but many of them are probably morphological variants of a few ichnospecies (Figs. 5.153, 5.154 and 5.155). *T. rectus* (Figs. 5.154a and 5.178d) and *T. zigzag* are the commonest ichnospecies. In addition to burrows with only one opening, U-shaped burrows with two openings occur (Fig. 5.154b).

*Substrate: Teichichnus* preferentially occurs in silty and muddy sand (softground). It is also been described from chalk deposits (e.g. Ekdale et al. 1984; Frey and Bromley 1985; Savrda 2012), limestone successions (e.g. Farrow 1966; Frey 1970b), and is rarely seen in shale (e.g. Jordan 1985).

*Appearance in Core: Teichichnus* is a deep-tier trace fossil and thus has high preservation potential. Because of its diagnostic features it is relatively easy to recognize in core (Fig. 5.156; see also Figs. 4.3, 5.100h, 5.111a, b, d, 5.118b, 5.121a, 5.129b and 5.179a). In vertical sections, *Teichichnus* commonly appears as patches of vertical to steeply inclined spreiten, which may be stacked on top of each other due to different sedimentation and colonization events (Fig. 2.6c). The spreiten are densely laminated and contrast in color with the host rock. Their width is highly variable and depends on the position of the section with respect to the elongate burrow. Most spreiten are concave-up, although concave-down spreiten may also occur, even within the same burrow (e.g. in *Teichichnus* with a sigmoidal shape, Fig. 5.154a). In elongate sections, only horizontal lamination may be

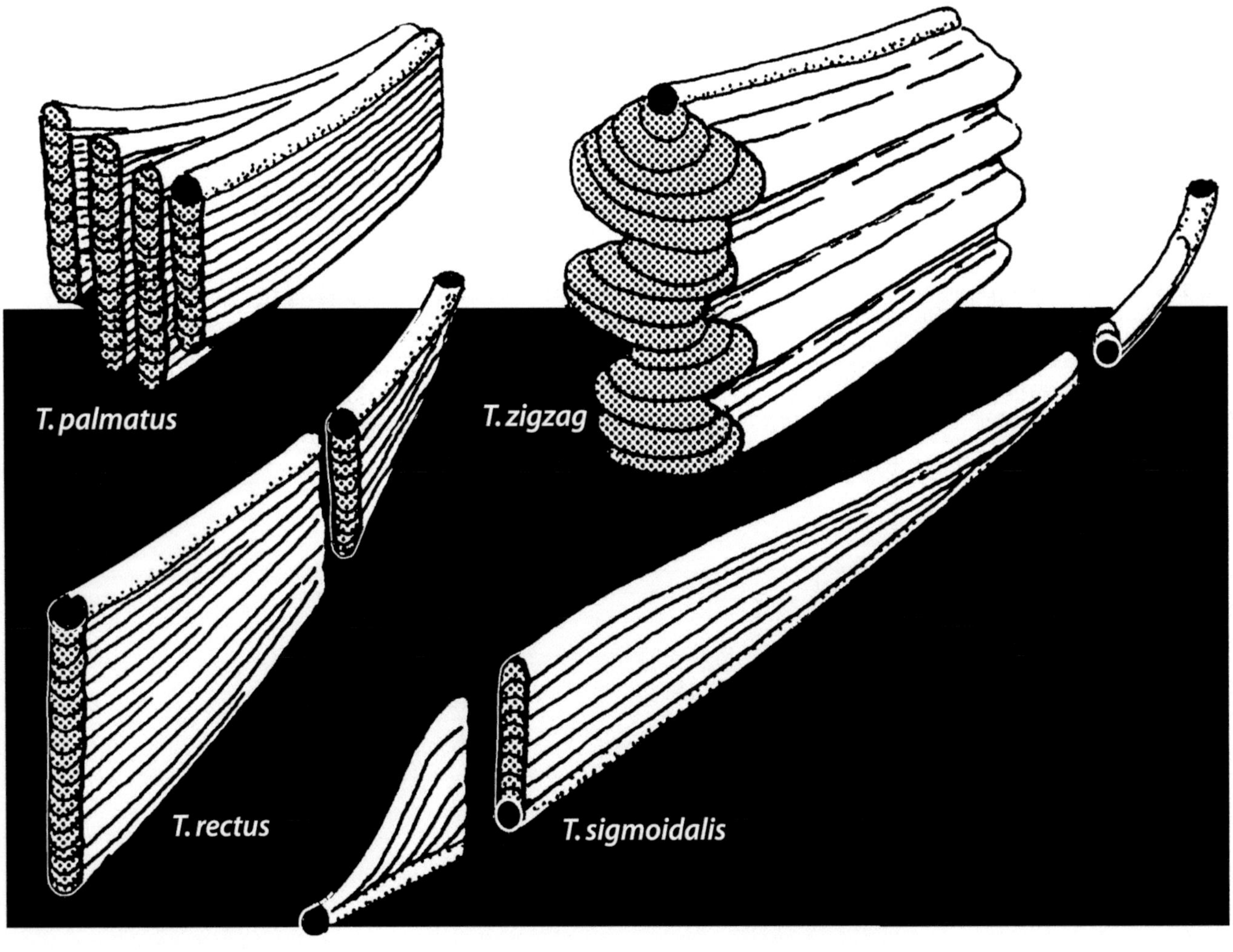


**Fig. 5.153** Idealized reconstruction of most common ichnospecies of *Teichichnus*. From Seilacher (2007), republished with permission of Springer

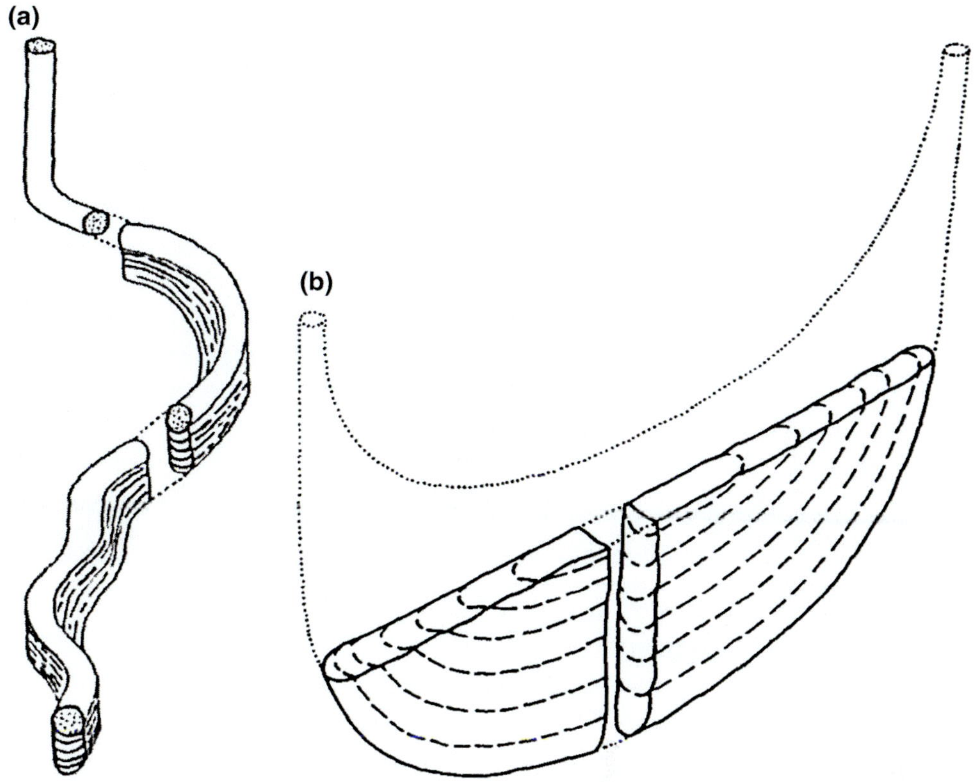


**Fig. 5.154** Examples of two different reconstructions of *Teichichnus*. Reprinted by permission of the publisher (Taylor & Francis Ltd., http://www.tandfonline.com). **a** *T. rectus* from the Carboniferous of Ireland. From Buckman (1996). **b** U-shaped *Teichichnus* from the Holocene of Norway. From Corner and Fjalstad (1993)

evident. Retrusive spreite burrows are common, although protrusive ones may occur too. In some cases, the terminal burrow is preserved as a passively filled lumen and thus contrasts with the actively filled spreite.

*Similar Trace Fossils:* Particularly in core, *Teichichnus* can resemble other spreite traces such as *Diplocraterion* and *Rhizocorallium* but differs from them by lacking a wide marginal tube. U- to L-shaped *Trichophycus* are also similar to *Teichichnus* but commonly have loosely packed segments with poorly developed spreite (Jensen 1997). *Phycodes* is another trace fossil with off-branching, U-shaped spreite burrows, and sections of *P. palmatum* and *P. parallelum* may resemble *Teichichnus*. Some sections of the bulbous, downward tapering to bow-shaped, laminated *Cylindrichnus* may be also confused with *Teichichnus*. The J-shaped cylindrical burrow *Artichnus* displays a vertically stacked spreite structure in its basal part, which differs from *Teichichnus* in wrapping around a central lumen and continuing above it (Zhang et al. 2008; Ayranci and Dashtgard 2013). Nevertheless, distinction of these ichnogenera in core might be difficult and *Artichnus* may be overlooked and misidentified as *Teichichnus*. Sections of many observed specimens suggest an overall broad, U-shaped morphology as described in *Catenichnus* McCarthy, 1979, which closely resembles *Teichichnus* and might be synonymous with it (cf. Corner and Fjalstad 1993; Fig. 5.58). Rather oblique *Teichichnus* spreiten may be confused with the vertical to oblique spreite burrow *Paradictyodora*, which consists of an irregularly folded, fanlike to subconical sheet (Olivero et al. 2004; D'Alessandro and Fürsich 2005). The entrance burrow of some *T. rectus* is a simple, passively filled shaft similar to *Skolithos* (Buckman 1996). Bann et al. (2004) have documented *Rosselia* with *Teichichnus*-like basal extensions, while *Teichichnus* can be part of various trace fossils (e.g. *Ophiomorpha* and *Thalassinoides*; Bertling et al. 2006). Many two-dimensional core sections appear with figures described as *Siphonichnus sursumdeorsum* and *S. lepusaures* (Zonneveld and Gingras 2013; Knaust 2015a; Fig. 5.156e, f, i), both which may turn out simply to be cross sections of broad, bow-shaped burrows (i.e. *T. zigzag*) instead of columnar structures.

*Producers:* Worm-like animals (e.g. annelids) and arthropods (e.g. trilobites, crustaceans) are commonly regarded as potential producers of *Teichichnus*. The modern polychaete *Nereis* sp. is known to produce spreite burrows similar to *Teichichnus* (Seilacher 1957; Dashtgard and Gingras 2012) and many Paleozoic forms probably result from the activity of trilobites. However, X-ray radiography of cores from Holocene sediments of the Baltic Sea has demonstrated that some bivalves (e.g. *Arctica islandica*) may leave *Teichichnus*-like traces by their crawling-plowing mode of locomotion (Werner 2002; Fig. 5.157), while others (e.g. *Solecurtus strigilatus*) can produce spreite-like structures by the periodic shifting of

**Fig. 5.155** *Teichichnus* in outcrop (**a–e**) vertical sections, (**f**) lower bedding plane. *Scale bars* = 1 cm. **a** Individual *T. rectus*. Lower Ordovician Beach Formation (shoreface), Bell Island, Newfoundland, Canada. See Fillion and Pickerill (1990). **b** *T. palmatus* cross sections in silicified limestone. Middle Permian Kapp Starostin Formation (mixed siliciclastic carbonate ramp), Akseløya, Svalbard. **c** Glauconitic sandstone and siltstone with a dense ichnofabric of *T.* cf. *zigzag*. Lower Cambrian Norretorp Member (Læså Formation, shallow marine), Julegård cliff section, Bornholm, Denmark. See Clemmensen et al. (2011). **d** *Teichichnus* ichnofabric in vertical section in sandy substrate. Paleocene (shallow marine), south Shikoku, southern Japan. **e** Retrusive spreite burrows together with the cast of a large bivalve (**b**). Same locality as in (**d**). **f** Straigth specimens of *T. duplex*. Upper Triassic (Middle Keuper, fluvial to lagoonal) of Eppingen near Heilbronn, southern Germany. Image courtesy of Thomas Schulz (Heilbronn)

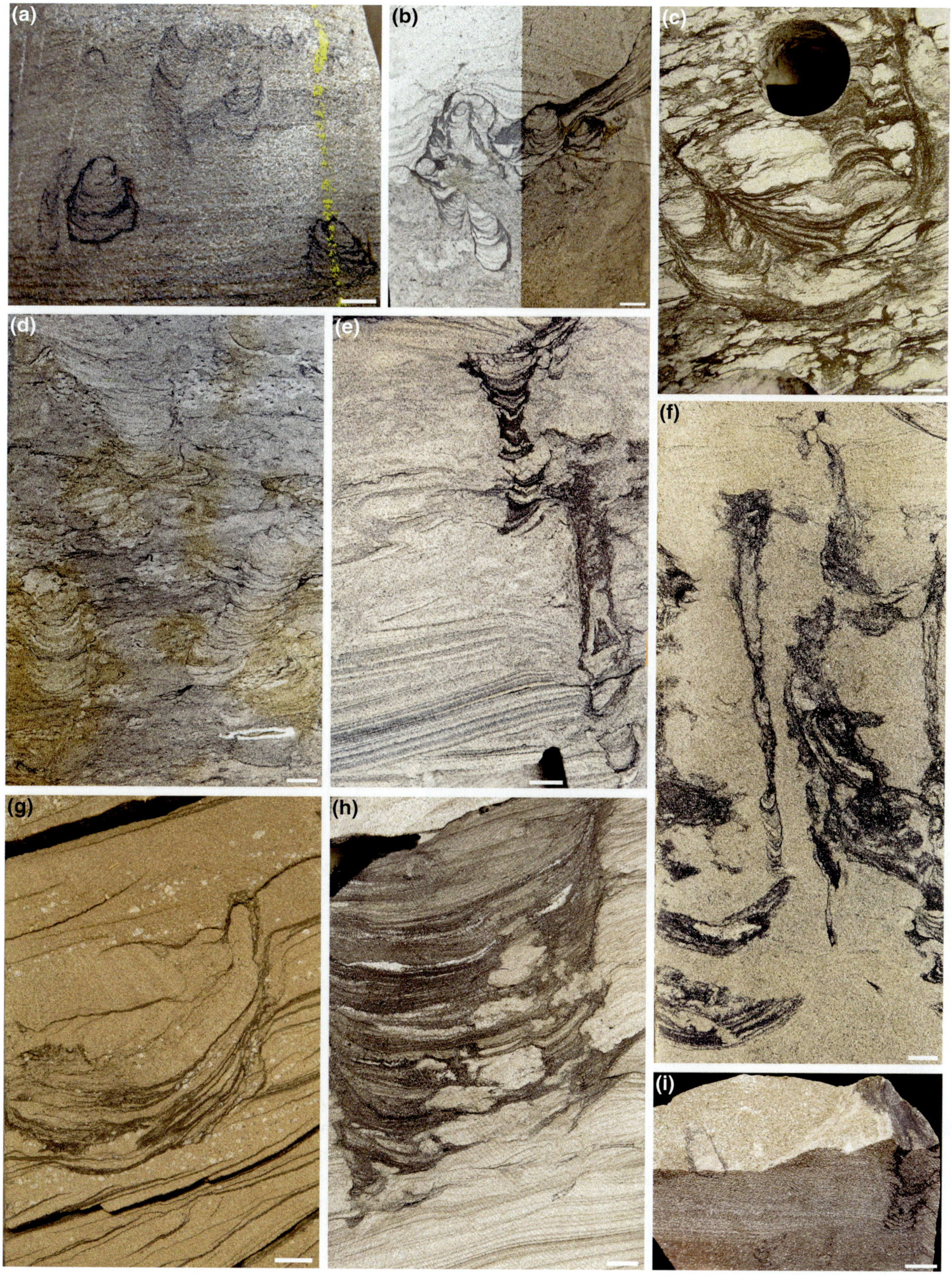
(a)
(b)
(c)
(d)
(e)
(f)
(g)
(h)
(i)

◀**Fig. 5.156** *Teichichnus* in full (**a**) and sectioned core (**b–i**). *Scale bars* = 1 cm. **a**, **b** Various sections of *T. zigzag* on the outside of a core (**a**) and on two vertical core faces trending perpendicular to each other (**b**). Middle Jurassic (Bathonian-Oxfordian) Hugin Formation (marginal marine), Norwegian North Sea (well 25/7-2, ca. 4475.15 m). **c** U-shaped *Teichichnus* isp. Lower Jurassic (Pliensbachian-Toarcian) Cook Formation (marginal marine), Visund Field, Norwegian North Sea (well 34/8-1, ca. 3653.4 m). **d** Completely bioturbated ichnofabric with *T. zigzag*, partly reburrowed with *Nereites* and accompanied by *Schaubcylindrichnus*. Lower Jurassic (Toarcian) Stø Formation (offshore transition), Norwegian Barents Sea (well 7120/8-2, ca. 2153.6 m). **e** Two *T. zigzag* in cross section. Middle Jurassic (Bajocian) Tarbert Formation (sandy tidal flat), Oseberg Sør Field, Norwegian North Sea (well 30/9-10, ca. 2813.15 m). **f** *Teichichnus* ichnofabric with longitudinal, oblique and cross sections of several burrows. Middle Jurassic (Bajocian) Tarbert Formation (sandy tidal flat), Oseberg Sør Field, Norwegian North Sea (well 30/9-F-26, ca. 4462.0 m). **g** *T. zigzag* in longitudinal section. Note the retrusive spreite in the lower burrow part and the sand-filled cast of the (holothurian?) producer in the upper part. Middle Jurassic (Bajocian) Tarbert Formation (sandy tidal flat), Oseberg Sør Field, Norwegian North Sea (well 30/9-4S, ca. 3346.6 m). **h** *T. zigzag* in longitudinal section. Middle Jurassic (Bajocian) Tarbert Formation (sandy tidal flat), Askja Øst Discovery, Norwegian North Sea (well 30/11-9A, ca. 3846.85 m). **i** *T. zigzag* in cross section (*lower part*) and along bedding plane (*upper part*). Middle Jurassic (Bajocian) Tarbert Formation (sandy tidal flat), Oseberg Sør Field, Norwegian North Sea (well 30/9-13S, ca. 3135.5 m)

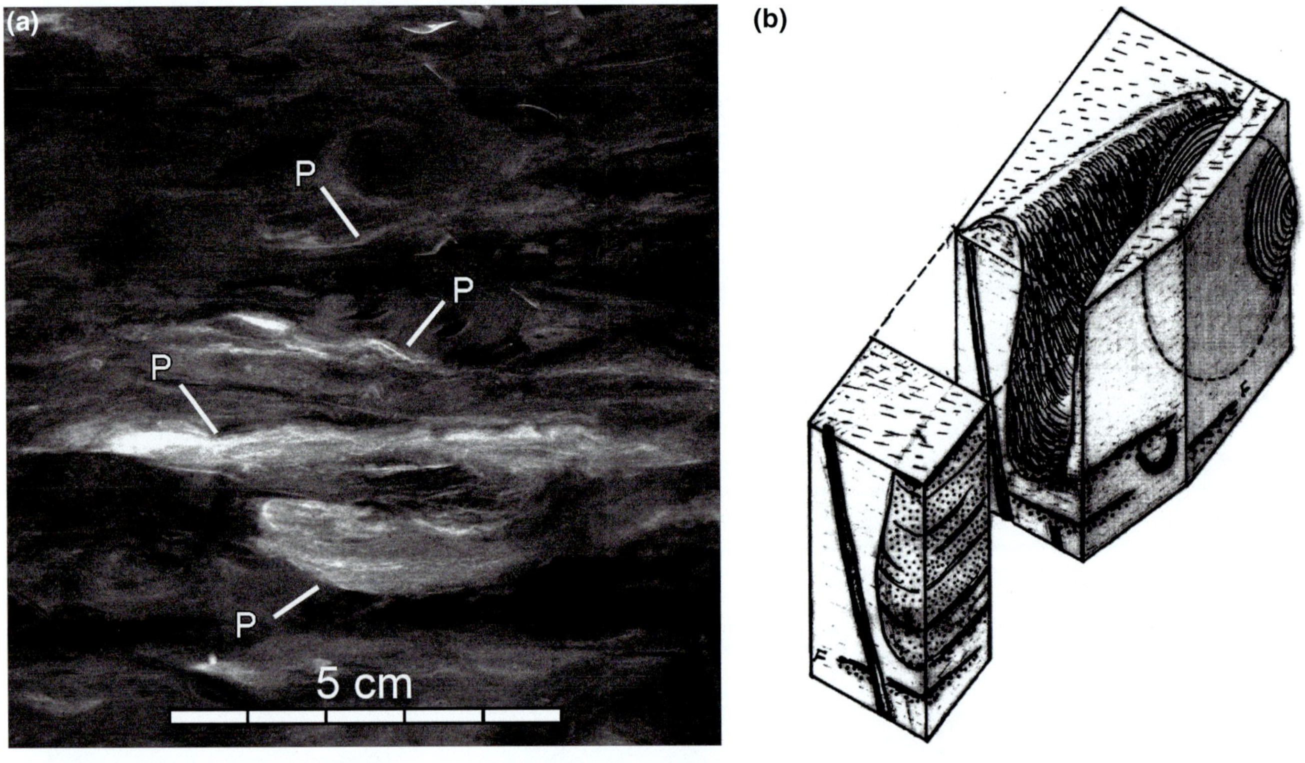


**Fig. 5.157** Comparable modern analogs of *Teichichnus* that were probably produced by bivalves. **a** X-ray radiograph (negative) of Holocene core from the Kiel Bay in the western Baltic Sea, showing spreite-like traces (plow-sole traces, *P*) similar to *Teichichnus*, produced by the bivalve *Arctica islandica*. Ca. 27 m water depth and 0.75–1.0 m core depth. Courtesy of the Radiography Database, University of Kiel, http://www.ifg.uni-kiel.de/Radiographien/radiographien.phtml. **b** Reconstruction of the crawling-plowing mode of locomotion of *A. islandica*, which leads to the origin of a *Teichichnus*-like trace ("fossil" plow-sole trace *F*). After Werner (2002)

the burrow (Bromley 1996; Fig. 5.158). *Catenichnus*- and *Artichnus*-like *Teichichnus* are consistent with features produced by holothurians and might be a result of such.

*Ethology: Teichichnus* mainly results from the deposit-feeding (fodinichnial) activity of its producer and includes dwelling of a non-vagile tracemaker (MacEachern et al. 2012). By adjusting its vertical position, *Teichichnus* partly served as an equilibrichnion.

*Depositional Environment:* The ichnogenus *Teichichnus* seems to be a typical element of siliciclastic systems, where it occurs frequently in association with delta deposits (e.g. Tonkin 2012). Its producer(s) can be regarded as euryhaline organisms that are able to adapt to a wide range of salinity. For this reason, *Teichichnus* has often been recognized in environments associated with lowered salinity. Such conditions are common in marginal-marine settings

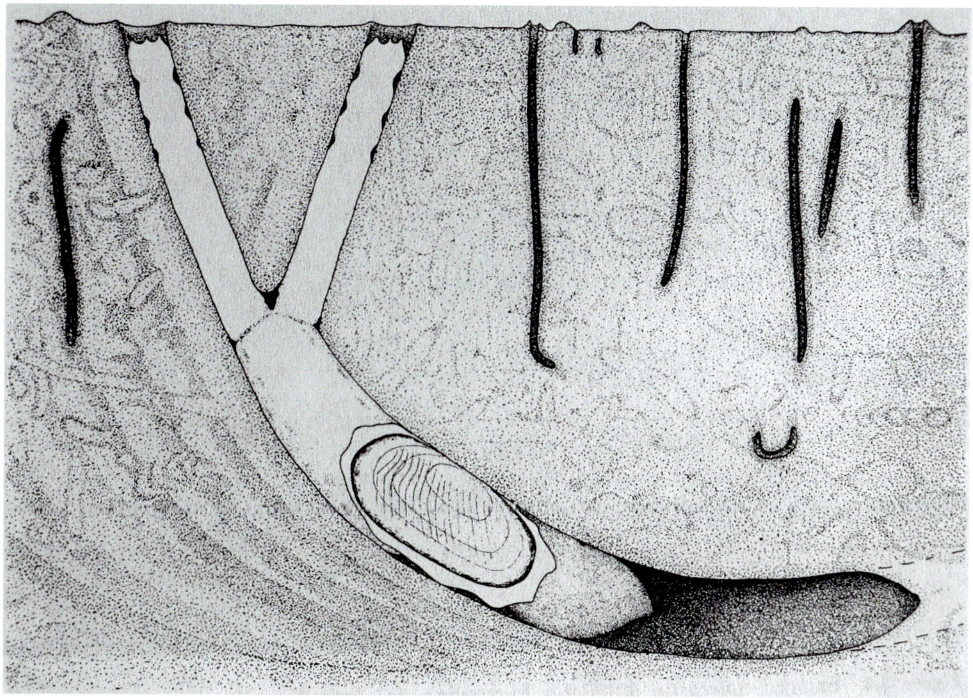

**Fig. 5.158** *Solecurtus strigilatus* in feeding position. Periodic shifting of the burrow leaves a spreite-like structure that may resemble *Teichichnus* (although more oblique). The *upper parts* of this structure are rapidly obliterated by shallower bioturbation. Depth of burrow 30 cm. Broken lines at lower right indicate the path taken during escape reactions. From Bromley (1996), reprinted by permission of the publisher (Taylor & Francis Ltd., http://www.tandfonline.com)

(e.g. embayments, estuaries, lagoons, ponds etc.), where *Teichichnus* is widespread (Fig. 5.156a–c). Although ichnodiversity in such assemblages is low, simple deposit-feeding traces like *Planolites* may occur along with it (Pemberton and Wightman 1992; Buatois et al. 2005; MacEachern and Gingras 2007; Gingras et al. 2012c). Furthermore, *Teichichnus* is often present in tidal deposits (including dunes and bars in brackish settings; Desjardins et al. 2012; Fig. 5.156e–i) and hyperpycnal flow deposits (Buatois et al. 2011).

In contrast to such marginal-marine occurrences with low ichnodiversity, *Teichichnus* is a characteristic element of lower shoreface to offshore (shelf) deposits (Pemberton et al. 2012). There, however, it occurs in highly diverse trace-fossil associations (Fig. 5.156d). In this setting, low- to moderate-energy conditions are prevailing, typical for fair-weather rather than storm deposits (Pemberton et al. 1992). Likewise, *Teichichnus* has been reported from chalky shelf deposits (Frey and Bromley 1985). It also occurs as accompanying trace fossil in assemblages of slope deposits (Hubbard et al. 2012) and deep-sea deposits (Wetzel and Uchman 2012). Given this latter (lower shoreface, shelf and deep marine) distribution, *Teichichnus* has been used in sequence-stratigraphical analysis as indicator of flooding events (Pemberton et al. 1992; Taylor et al. 2003) and is characteristic of transgresssive systems tracts.

*Ichnofacies: Teichichnus* is a typical constituent of the *Cruziana* Ichnofacies but also occurs in the *Zoophycos* Ichnofacies.

*Age: Teichichnus* is known from the Early Cambrian (e.g. Loughlin and Hillier 2010) to the Holocene (e.g. Wetzel 1981; Corner and Fjalstad 1993).

*Reservoir Quality:* The producer of *Teichichnus* creates an actively filled spreite by incorporating and packing fine-grained material (e.g. mud, silt and organic-rich matter) from the host sediment into its burrow. This process results in locally reduced permeability and thus leads to deteriorated reservoir quality (Tonkin et al. 2010).

## 5.32 *Thalassinoides* Ehrenberg, 1944

*Morphology, Fill and Size:* Similar to the genetically related *Ophiomorpha*, *Thalassinoides* consist of boxworks constituting horizontal maze with vertical shafts (Fig. 5.159), although many related burrow architectures can be produced (e.g. Kennedy 1967; Bromley 1967, 1996; Fürsich 1974a). The burrows are circular to elliptical in cross section. Branching is Y- and T-shaped, typically with bulbous enlargement of the junctions. Passive fill is common, although transitional burrow elements can be actively filled (Fig. 5.160). Burrows are unlined and contain a sandy, more rarely also a muddy fill. *Thalassinoides* is a relatively large trace fossil as the boxwork can cover a volume of over 1 $m^3$, although its burrow diameter can range from few millimeters to several centimeters (mean about 1 cm). Complete burrow systems can penetrate substrate to several meters.

*Ichnotaxonomy:* From the about 15 erected ichnospecies, the following are most representative (cf. Myrow 1995):

- *T. suevicus*—predominantly horizontal form that may contain enlargements at Y-shaped bifurcations (Figs. 5.159a and 5.160a, b)
- *T. paradoxicus*—boxwork burrows, highly irregular in size and geometry (Fig. 5.159b and 5.160c–f)
- *T. saxonicus*—large form with tunnels 5–20 cm in diameter.

*Substrate: Thalassinoides* occurs in a wide range of soft to firm substrates and is reported from mudstone, siltstone, sandstone, conglomerate, limestone and dolomite. It can also be associated with lithified (hard) substrates.

*Appearance in Core:* The three-dimensional geometry of *Thalassinoides* is hard to resolve in cored material and therefore an unequivocal attribution is seldom possible with certainty. The relatively large burrow size differs from that of associated ichnotaxa and the burrows appear in circular and elliptical cross sections together with (rarely) vertical shafts (Figs. 5.161a, b and 5.168). The passive fill usually creates a contrast with the host sediment, although in other cases the fill is the same as the host sediment. It can be subject of reburrowing by other tracemakers, for instance resulting in the occurrence of *Chondrites* (Fig. 5.35c and 5.161f). Depending on the substrate, transitions to related ichnogenera (e.g. *Ophiomorpha and Gyrolithes*) are common (Fig. 5.161c, d, see also Fig. 2.6b). Because the producer is capable of bioeroding, *Thalassinoides* may also occur in hard substrates such as cemented layers (Fig. 5.161e).

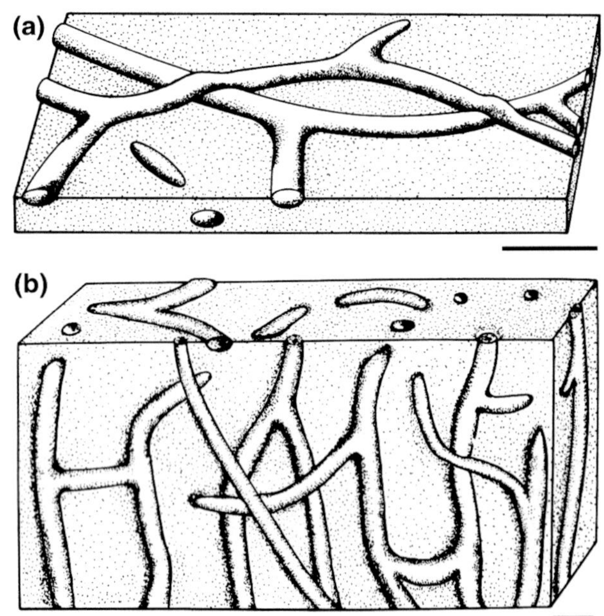


**Fig. 5.159** Morphological variation of *Thalassinoides* ichnospecies. **a** *T. suevicus* (maze). **b** *T. paradoxicus* (boxwork). *Scale bars* = 5 cm. After Howard and Frey (1984), republished with permission of Canadian Science Publishing; permission conveyed through Copyright Clearance Center, Inc.

*Similar Trace Fossils: Thalassinoides* in core may be confused with other large and passively filled burrows such as *Spongeliomorpha* (scratched), *Psilonichnus* and *Parmaichnus* (vertically Y-shaped), *Camborygma* (chambered, dominantly vertical), and others. Sections of actively filled burrow elements, however, could resemble *Planolites*, *Asterosoma* or even *Artichnus*. Processes of methane and hydrocarbon seepage may result in branched pipes similar to *Thalassinoides* (Fig. 5.189g).

*Producers:* Similar to *Ophiomorpha*, comparison with modern analogs shows that the producers of *Thalassinoides* belong to thalassinidean shrimp, particularly callianassids (Figs. 5.25, 5.85–5.87). This is principally true for Permian to modern *Thalassinoides*, whereas other arthropods (e.g. trilobites), sea anemones and worms (e.g. enteropneusts) have been interpreted as tracemakers of Paleozoic *Thalassinoides* (e.g. Ekdale and Bromley 2003; Cherns et al. 2006). Some of these early *Thalassinoides* probably belong to other ichnogenera such as *Balanoglossites* (Knaust and Dronov 2013).

*Ethology: Thalassinoides*-producing shrimp are primarily suspension-feeders, and their extensive burrow systems were excavated as dwellings (domichnia). Combined suspension- and deposit-feeding behavior may apply for many *Thalassinoides*. Other representatives actively collect seagrass and other organic matter and store it in burrow chambers as cache (Griffis and Suchanek 1991).

*Depositional Environment:* Shrimp-produced *Thalassinoides* is most common in shallow-marine environments such as shoreface, deltas and others (e.g. Nickell and Atkinson 1995). Given the ability of the producer to tolerate fluctuations in salinity, *Thalassinoides* can be found in brackish environments, e.g. in estuarine settings and fan deltas (e.g. Swinbanks and Luternauer 1987). *Thalassinoides* is often associated with firm substrates (*Glossifungites* Ichnofacies), where it occurs in a wide range of environments from marginal marine to the deep marine (Monaco et al. 2009). It is a common constituent of carbonate and chalk environments (Bromley 1967; Ekdale and Bromley 1991).

*Ichnofacies: Thalassinoides* is a common constituent of the *Cruziana* Ichnofacies, where it occurs in relatively cohesive substrates. *Thalassinoides* is also a component of the substrate-controlled, firmground trace-fossil suite (the so-called *Glossifungites* Ichnofacies).

*Age: Thalassinoides* is frequently reported from the Ordovician (e.g. Myrow 1995; Ekdale and Bromley 2003; Cherns et al. 2006) to the Holocene (e.g. Swinbanks and Luternauer 1987; Nickell and Atkinson 1995), although tracemakers of different phyla must be invoked.

(a)
(b)
(c)
(d)
(e)
(f)

◀**Fig. 5.160** *Thalassinoides* in outcrop and building stone. *Scale bars* = 5 cm. **a** *T. suevicus* preserved on lower bedding plane. Upper Jurassic sandstone (Lower Coralline Oolite, shallow marine, shoreface) coastal cliff near Scarborough, Yorkshire, UK. **b** *T. suevicus* preserved on lower bedding plane. Upper Jurassic (Upper Oxfordian, lagoonal platform carbonate), Krzemionki Opatowskie Neolithic flint mine, southern Poland. **c** Bedding plane with extensive system of *T. paradoxicus*. Cretaceous (platform carbonate), coastal cliff near Taghazout, western Morocco. **d** *T. paradoxicus* in vertical section. Same locality as in (**c**). **e** Bedded and bioclastic limestone with stylolites and *T. paradoxicus* ichnofabric in vertical section. Cretaceous (shallow marine), Lisbon, Portugal, building stone. **f** *T. paradoxicus* burrow system in limestone. Cretaceous (shallow marine), Lisbon, Portugal, building stone

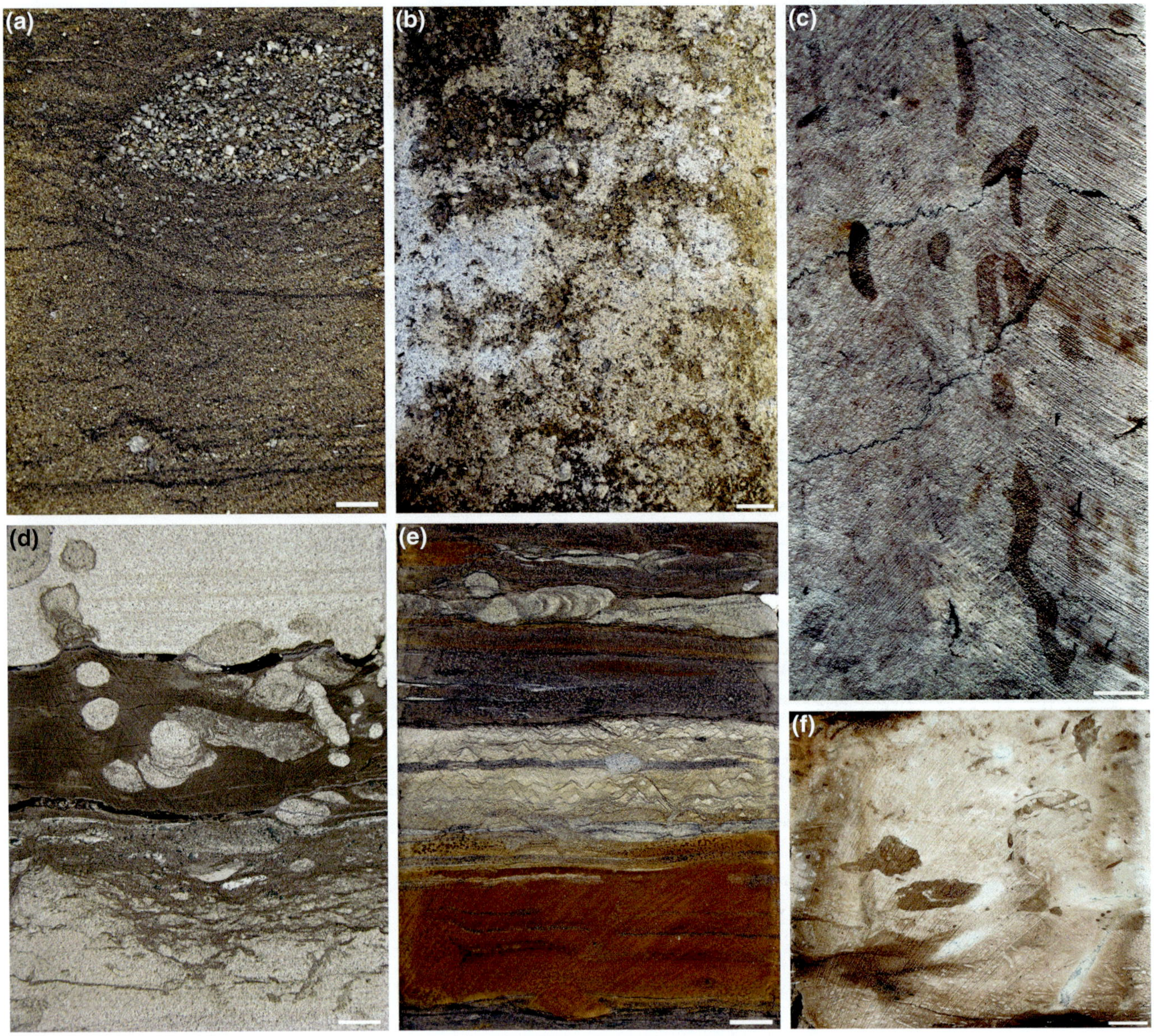


**Fig. 5.161** *Thalassinoides* in sectioned core. *Scale bars* = 1 cm. **a** Argillaceous sandstone exposing a large-diameter burrow in cross section, which is passively filled with coarse-grained sand. Upper Jurassic (Tithonian) Draupne Formation (fan delta, slope), Johan Sverdrup Field, Norwegian North Sea (well 16/2-15, ca. 1943.5 m). **b** Completely bioturbated coarse-grained sandstone with *Thalassinoides* boxwork preserved due to partial cementation of the host rock. This stratigraphical interval represents an omission surface within the region. Upper Jurassic (Tithonian) Draupne Formation (fan delta, slope), Johan Sverdrup Field, Norwegian North Sea (well 16/5-2S, ca. 1958.5 m). **c** Chalky limestone with mainly vertical burrow segments (shafts) with a slightly helical appearance as known from the related ichnogenus *Gyrolithes*. Lower Cretaceous (Berriasian) Åsgard Formation (shelf), Norwegian North Sea (well 16/4-6S, ca. 1945.2 m). **d** Mudstone interlayer between turbidite sandstone, containing large passively filled burrows. Note the occurrence of spreite due to retrusive burrow adjustment and the transition to actively filled and lined *Ophiomorpha* burrows. Upper Cretaceous (Maastrichtian) Springar Formation (deep marine, fan system), Norwegian Sea (well 6604/10-1, ca. 3647.5 m). **e** Sandstone with early diagenetic carbonate cementation. The originally hard palisade calcite band in the middle is partly eroded with a sand-filled tunnel coming from the mudstone layer in between. Upper Cretaceous (Campanian) Nise Formation (deep marine, fan system), Aasta Hansteen Field, Norwegian Sea (well 6707/10-1, ca. 3133.5 m). **f** Chalky limestone with horizontal burrow parts (tunnels), which are mainly mud-filled and intensively reburrowed with *Chondrites*. Lower Cretaceous (Berriasian) Åsgard Formation (shelf), Johan Sverdrup Field, Norwegian North Sea (well 16/2-15, ca. 1899.75 m)

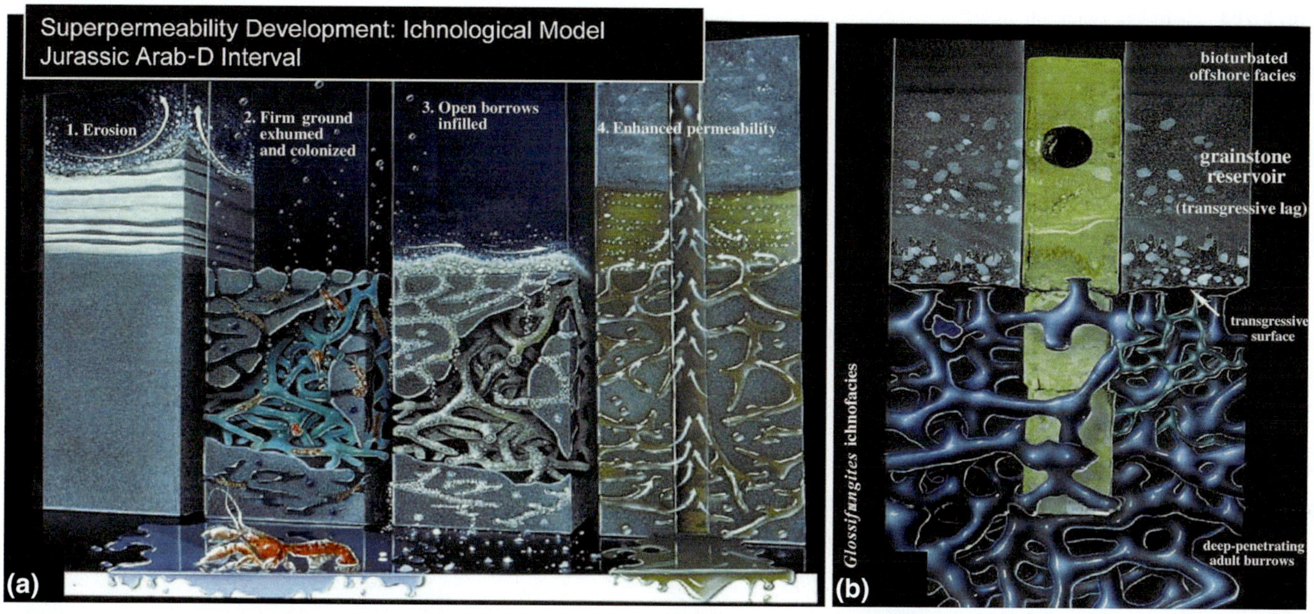


**Fig. 5.162** Burrow-enhanced permeability created by *Thalassinoides* boxworks. From Pemberton and Gingras (2005), republished with permission of AAPG; permission conveyed through Copyright Clearance Center, Inc. **a** Development of super-K in the Jurassic Arab-D, Ghawar Field, Saudi Arabia, is a function of the interaction of a ravinement bed developed on a regional transgressive surface of erosion. The *Glossifungites* Ichnofacies is represented by firmground *Thalassinoides* systems 1–2 cm in diameter, penetrating up to 7 ft (2.1 m) below the transgressive surface of erosion. In some cases, the burrow fill consists of sucrosic dolomite, and the matrix consists of mosaic dolomite, resulting in a large difference in permeability. This leads to a charging effect that may become a problem wherein oil in the matrix is bypassed and water is drawn from the aquifer. **b** Super-permeability forms where the assemblage attributed to the *Glossifungites* Ichnofacies is emplaced at the surface, with a ravinement bed above capped by an offshore bioturbated facies. When present, the flowmeters indicate that 70% of the production comes from this single unit

*Reservoir Quality:* Its wide distribution, large diameter, ramification pattern and passive fill make *Thalassinoides* susceptible for enhanced fluid flow, which may result in concretional cementation (e.g. Bromley 1967; Fürsich 1973; Fig. 5.89), but also in a good friend of the reservoir geologist. The burrows can considerably increase the reservoir quality and commonly interconnect otherwise isolated parts of the reservoir. The reservoir-improving character of *Thalassinoides* has been outlined by various studies. For instance, Pemberton and Gingras (2005) reported on *Thalassinoides*-perforated firmgrounds that act as "super-permeable" horizons in the Arab-D of the world's largest oil field, the Ghawar Field in Saudi Arabia. Sandstone-filled *Thalassinoides* in mudstone of the Ben Nevis Formation off Newfoundland (Canada) creates vertical and horizontal macropore networks with the potential to act as flow conduits (Tonkin et al. 2010; Fig. 5.162). Other examples with a similar size range documenting *Thalassinoides*-dominated ichnofabrics having a critical impact on the fluid-flow properties of hydrocarbon reservoirs and groundwater aquifers (Cunningham and Sukop 2012; Cunningham et al. 2012).

## 5.33 *Tisoa* de Serres, 1840

*Morphology, Fill and Size: Tisoa* refers to very long, elongate U-shaped burrows with subvertical, oblique and subhorizontal orientation. The two limbs of the burrow are positioned very close to each other (Fig. 5.163), which results in cross sections with a dumbbell or figure-of-eight shape (e.g. Gottis 1954). *Tisoa* burrows are passively filled, commonly with sulfide mineralization, and are often encased in carbonate concretions. Branching has been recorded but is not common. Burrow diameter commonly ranges between 3 and 20 mm: the length of the burrow is often hard to determine because of incompleteness, but can reach several decimeters.

*Ichnotaxonomy: T. siphonalis* is the most common ichnospecies (Fig. 5.164), although a few other ichnospecies were erected subsequently based on morphological variation.

*Substrate: T. siphonalis* is often reported from argillaceous substrates, such as black shale. It also occurs in redeposited sediments, for instance debrites and mass-transport deposits, and coal. In outcrop, *Tisoa* is often recognized based on its encasement within a tubular carbonate concretion.

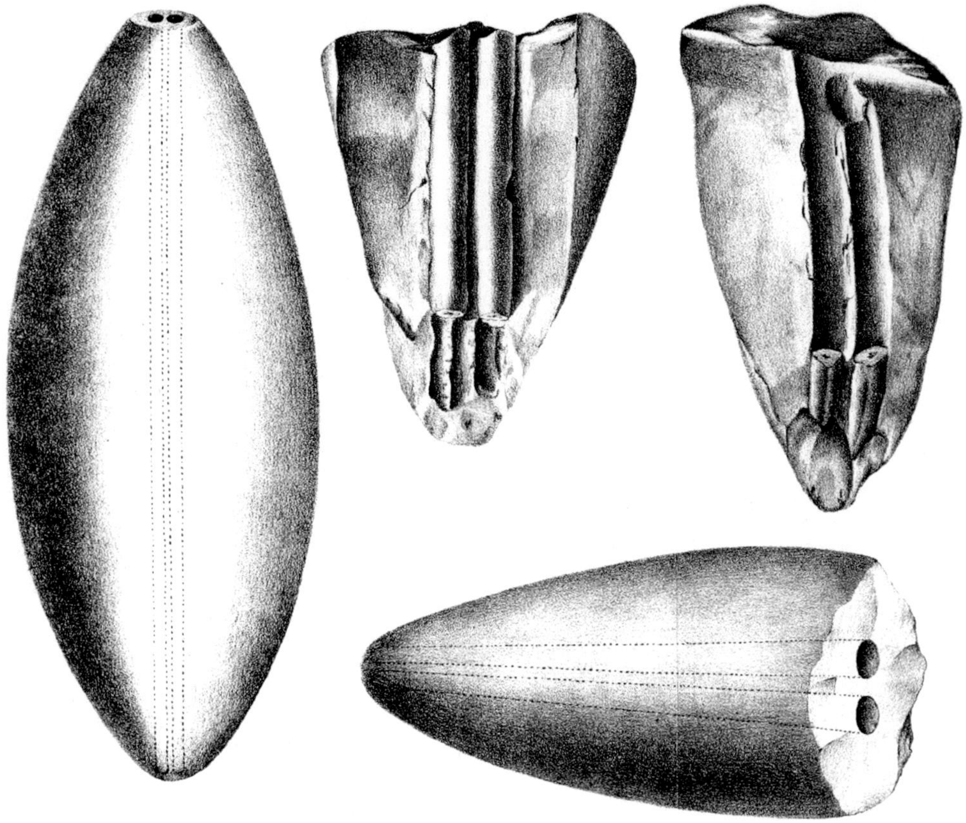

**Fig. 5.163** Historical figure of the type ichnospecies of *Tisoa*, *T. siphonalis* encased in concretionary limestone, in its original description by de Serres (1840)

*Appearance in Core:* A clear indication of *Tisoa* in core is the occurrence of coupled and passively sand-filled burrows (Fig. 5.165). Beside that, elongate oblique and circular cross sections of burrows occur in association. *Tisoa* ichnofabric may reach a relatively high density and may be traced to considerable depth (several decimeters).

*Similar Trace Fossils:* Given the extreme length of the burrows, their U-turn is rarely preserved, which may give the impression that vertically oriented specimens are not trace fossils but conduits for escaping seep gases (van de Schootbrugge et al. 2010). So far, *Tisoa* has not been recognized from core, but has been attributed to other trace fossils due to similarities. For instance, the material shown in Fig. 5.165 was originally described as large *Chondrites* (Knaust 2009b). *Diplocraterion* is another burrow with similar features as *Tisoa*, particularly when the limbs of long burrows are very close together and no spreite can be recognized between them. Such forms were described as *Diplocraterion* and *D. habichi* but some of them may belong to *T. siphonalis* for the reasons given above (e.g. Heinberg and Birkelund 1984; Bradley and Pemberton 1992; de Gibert and Martinell 1998; Bann and Fielding 2004; Bann et al. 2004; Hubbard and Schultz 2008; Riahi et al. 2014). Confusion may also arise with *Rhizocorallium jenense*, which like *Diplocraterion* bears a spreite and has a much lower length/width ratio compared to *Tisoa* (Knaust 2013). The U-shaped turn of *Tisoa* resembles that of some *Arenicolites*, but again their limbs are much closer to each other and the length/width ratio is greater. Similar to *Tisoa*, the vertically oriented trace fossils *Paratisoa* Gaillard, 1972 (Macsotay et al. 2003) and *Bathichnus* Bromley et al., 1975 (Nygaard 1983) are diagenetically enhanced (e.g. by carbonate and flint concretions) and can reach extraordinary length, but differ from *Tisoa* by having only one tube.

*Producers:* Based on modern analogs, tube-dwelling polychaetes, such as the giant pogonophoran worms, could have produced *Tisoa*-like traces. In the Pacific Ocean, 1.5 m long pogonophorans grow in heated, sulfur-rich water around warm-water vents (Ruppert et al. 2004).

*Ethology:* A preferred dwelling (domichnial) behavior can be inferred for the *Tisoa* producers.

*Depositional Environment:* A striking feature of *Tisoa* is its frequent occurrence in the central part of carbonate concretions related to hydrocarbon seep deposits (Breton 2006), preferably occurring in deep-marine basins and on continental slopes. The *Tisoa* producer seems to have an

**Fig. 5.164** *Tisoa siphonalis* in outcrop. *Scale bars* = 1 cm. **a** Large sandstone clast with cross sections of passively filled *T. siphonalis* with a dumbbell to figure-of-eight shape. Late Miocene Urenui Formation (deep marine, continental slope), Waiau coastal section north of New Plymouth, North Island, New Zealand. **b** Elongate *T. siphonalis* within a calcite-cemented sandstone. Same locality as in (**a**). **c** Bottom fragment of a pyritic specimen, with the U-turn to the left. Lower Cretaceous (Hauterivian), Ulyanovsk District, Volga River (coll. Hecker, Paleontological Museum Moscow). **d** Cross section. Same specimen as in (**c**)

affinity for sediment with a high organic content and thus can also be found in mass-transport deposits and even in coal seams. Callow et al. (2013) describe abundant, >1 m deep U-shaped trace fossils from cohesive mudstone below bypass surfaces within the channel axis of a deep-marine channel, which are similar to those forms reported by Knaust (2009b, Fig. 5.165).

*Ichnofacies: Tisoa* does not seem to have a preference to a particular ichnofacies because of its close association with deposits with a high content of organic matter. Such conditions, however, are typically met on deep-marine basin floors (*Nereites* Ichnofacies) and continental slopes (*Zoophycos* Ichnofacies), both preferred sites of seepage. In many cases, the sediment already became consolidated before its deep penetration by the *Tisoa* producer, and the resulting traces can be assigned the *Glossifungites* Ichnofacies.

*Age: Tisoa* is known from the Early Jurassic (van de Schootbrugge et al. 2010) to the Miocene (Frey and Cowles 1972).

*Reservoir Quality:* The passive fill (commonly consisting of sand), together with relatively large tubes and extraordinary penetration depth into mudstone (non-reservoir, source rock), make *Tisoa* very suitable for overcoming thin barriers and baffles within a given reservoir.

## 5.34 *Trichichnus* Frey, 1970b

*Morphology, Fill and Size: Trichichnus* refers to sparsely branched or unbranched, hairlike and exceptionally long burrows with a more or less vertical orientation (Fig. 5.166). The burrows can be lined and are typically filled with sulfide minerals (e.g. pyrite). The morphology of the cylindrical burrows is straight or gently winding to slightly sinuous. Burrow diameter is in the range of 0.1–1.0 mm, whereas burrow length of complete burrows commonly exceeds 20 cm but can reach several meters. Scholle (1971) reported specimens with a length of more than 6 m.

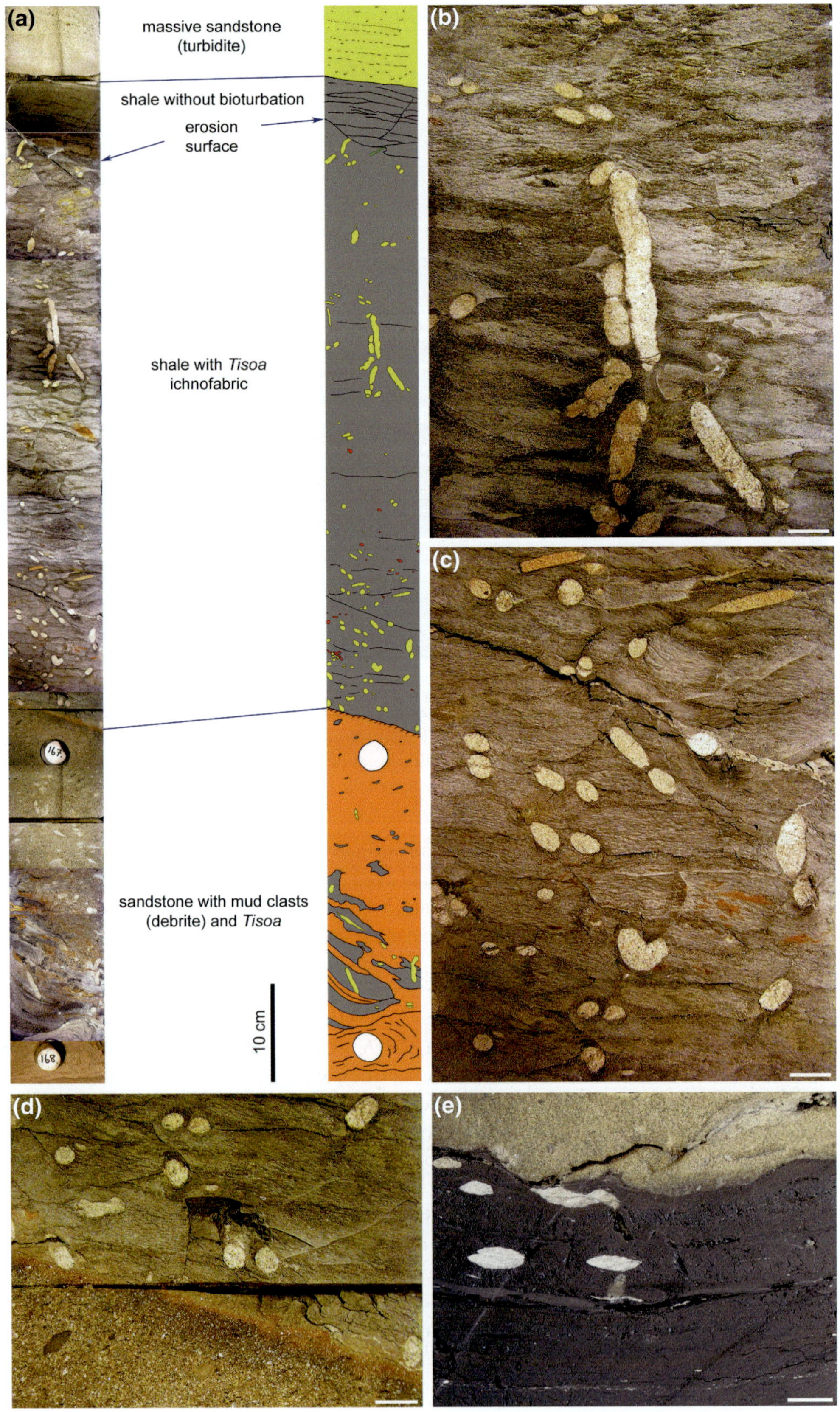


**Fig. 5.165** *Tisoa* in sectioned core. *Scale bars* = 1 cm. **a–d** *Tisoa* ichnofabric. Upper Cretaceous (Campanian) debris-flow deposit (deep marine, fan complex), Aasta Hansteen Field, Norwegian Sea (well 6707/10-1, 3058.7-3059.9 m). From Knaust (2009b), republished with permission of Elsevier; permission conveyed through Copyright Clearance Center, Inc. **a** Diagram of the *Tisoa* ichnofabric in photograph (*left*) and line drawing (*right*), indicating a maximum penetration depth on the order of 1 m. Note the occurrence of pyritized burrows (red) in addition to sand-filled burrows (yellow). The top of the ichnofabric is at a sharp erosion surface, below which the burrows appear either as vertical, oblique or horizontal, occasionally paired tunnels. Some burrows penetrate into the underlying sandy debris-flow deposit, where they are preferentially concentrated in large, reworked, sideritic mud clasts. **b** Close-up section from (**a**), showing subvertical, oblique and paired subhorizontal burrows. **c** Close-up section from (**a**), showing predominantly paired subhorizontal burrows. **d** Close-up section from (**a**), showing predominantly paired subhorizontal burrows within the shale. **e** Two pairs of subhorizontal and one oblique section of *Tisoa* on top of a compacted coal seam, all passively filled with sand from the layer above. Lower Jurassic (Pliensbachian) Åre Formation (top), Heidrun Field, Norwegian Sea (well 6507/7-A-27, ca. 3235.5 m)

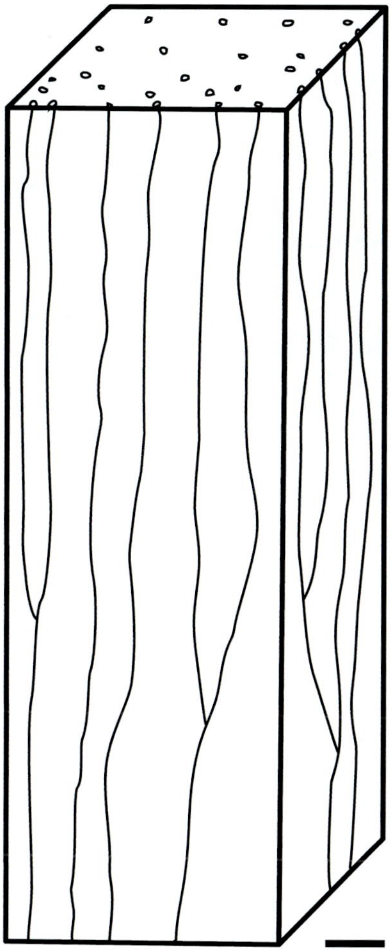

**Fig. 5.166** Sketch of modern *Trichichnus* from the Vøring Plateau, Norwegian Sea, formed by sipunculan worms. From Romero-Wetzel (1987), republished with permission of Springer

*Ichnotaxonomy: T. linearis* seems to be the only valid ichnospecies of *Trichichnus* (Fig. 5.167), with *T. simplex* being a preservational variant (Uchman 1999). *T. appendicus* Uchman, 1999 refers to horizontal and oblique burrows with short lateral appendages, and therefore is rather excluded from the ichnogenus *Trichichnus*.

*Substrate: Trichichnus* is common in fine-grained rocks such as mudstone, marlstone and limestone with an original soft to firm consistency, but may also occur in sandy substrate.

*Appearance in Core:* Based on its size, morphology and mineral staining, *Trichichnus* is easy to recognize in core (Fig. 5.168). Only fractions of the straight to slightly winding thin burrows are commonly preserved and appear as mud-filled or pyritic tubes.

*Similar Trace Fossils: Trichichnus* has close similarity to *Polykladichnus* and *Skolithos*, but however, differ from them by having a much smaller length/diameter ratio. In addition, *Polykladichnus* has Y- or U-shaped branching and *Skolithos* remains unbranched.

*Producers: Trichichnus* is a meiobenthic trace fossil (<1 mm in diameter) which has a modern analog produced by sipunculan worms (*Golfingia*, *Nephasoma*) on the Norwegian continental slope (Romero-Wetzel 1987; Shields and Kedra 2009; Fig. 5.166).

*Ethology: Trichichnus* has usually been interpreted as the dwelling (domichnion) of a sediment-feeding chemosymbiotic organism. Because of its exceptional length and common sulfide mineralization, *Trichichnus* can also be considered as chemichnion (Uchman 1995; Uchman and Wetzel 2012). These structures were produced by organisms feeding on chemosymbiotic microbes; the organisms maintain a

**Fig. 5.167** *Trichichnus linearis* in the Miocene Marnoso-arenacea Formation, Albignano roadcut (Santerno Valley), Italy. *Scale bars* = 1 cm. The hemipelagic background sediment (marlstone) is penetrated by *Trichichnus*, which is only partially preserved. **a** Overview picture with numerous fragmentary specimens (*arrow heads*). Note the brownish staining due to altering of sulfide minerals to limonite. **b** Close-up view of a burrow

**Fig. 5.168** *Trichichnus* in sectioned core. *Scale bars* = 1 cm. *Trichichnus* together with *Thalassinoides* and *Zoophycos* in the Lower Cretaceous (Berriasian) chalk of the Åsgard Formation (shelf), Johan Sverdrup Field, Norwegian North Sea. **a** Well 16/3-4 (ca. 1913.65 m). **b** Well 16/5-2S (ca. 1956.95 m)

connection to oxygenated water but penetrate into anoxic sediments rich in methan, the sulfides or ammonium that are required for microbial growth.

*Depositional Environment: Trichichnus* is common in deep-marine deposits (turbidites and hemipelagites; Wetzel 1981, 1991; McBride and Picard 1991) and often associated with chalk deposits (shelf and deep-water; Frey 1970b; Savrda 2012). It preferably occurs in oxygen-deficient sediment (e.g. Monaco et al. 2012).

*Ichnofacies: Trichichnus* is associated with the *Cruziana*, *Zoophycos* and *Nereites* ichnofacies.

*Age: Trichichnus* is known from Early Ordovician (Fillion and Pickerill 1990) to Holocene (Romero-Wetzel 1987).

*Reservoir Quality:* In their letter to the journal *Nature*, Weaver and Schultheiss (1983) demonstrated the profound effect of open burrows (such as *Trichichnus*) in deep-sea sediments "… on the overall permeability, and consequently, on the possible flow rates through them in response to any excess pore pressures". "The effect of the burrows changes the calculated permeability from the equivalent of a clay to that of a coarse sand".

## 5.35 *Virgaichnus* Knaust, 2010a

*Morphology, Fill and Size: Virgaichnus* is a meiobenthic trace fossil (burrow diameter less than 1 mm; Knaust, 2007a) with a complex, three-dimensional architecture (Fig. 5.169). The irregular burrow system consists of horizontal and inclined elements with Y- and/or T-shaped branching. Burrow sections pinch and swell, which leads to bulbous enlargements and alternating blade-like contractions (Knaust 2010a; Fig. 5.170). The burrows are smooth-walled and passively filled. The average burrow diameter is about 0.5 mm, whereas burrow length and penetration depth can reach several centimeters.

*Ichnotaxonomy: V. undulatus* so far is the only described ichnospecies of *Virgaichnus*.

*Substrate: V. undulatus* has been described to occur in micritic or chalky limestone with an originally firm substrate, although diffuse burrow segments may also indicate local softground conditions. It is common in sandy substrates too.

*Appearance in Core:* In core, *V. undulatus* commonly occurs in form of dense boxworks with interconnecting burrow elements of submillimetric diameter (Fig. 5.171). Individual burrow segments are characterized by a bulbous and undulating appearance, whereas others contrast with a strong contraction that leads to blade-like burrows with a size less than 0.1 mm. Over larger distances, this feature resembles a kind of boudinage. Consequently, the cross section of the burrows is highly variable and ranges from circular to flat elliptical and vertically extended.

*Similar Trace Fossils:* Morphologically, *Virgaichnus* resembles some ichnospecies of *Thalassinoides* but differs from them by having a much smaller and inconstant burrow diameter (pinch-and-swell-like features). *Virgaichnus* shows some similarities with *Chondrites* and *Pilichnus*, from which it differs by the lack of dichotomous branching but the inclusion of T-shaped bifurcations and crossing tunnels. *Bornichnus*, another meiobenthic trace fossil, includes small burrows composed of a crowded tangle of lined tubes that are closely and tortuously branched and have a constant diameter. Finally, the twig-like morphology of *Virgaichnus*

(a)
Bedding plane
Vertical section
1 cm
1 cm
1 cm
(b)
(c)

◀**Fig. 5.169** *Virgaichnus undulatus* in outcrop. **a** Slab with the holotype specimen showing multiple branching indicated by an *arrow head*. Upper Permian Saiq Formation (carbonate platform, open lagoon), Wadi Bani Awf near Rustaq, Oman. From Knaust (2010a), republished with permission of Elsevier; permission conveyed through Copyright Clearance Center, Inc. **b** *V. undulatus* isp. in bedding-plane view. Glauconitic, completely bioturbated sandstone. Paleocene Grumantbyen Formation (shoreface), near Longyearbyen, Svalbard. **c** Intensely branched burrow system in limestone. Carboniferous Honaker Trail Formation (deep- to shallow-marine), north of Moab, Utah, USA

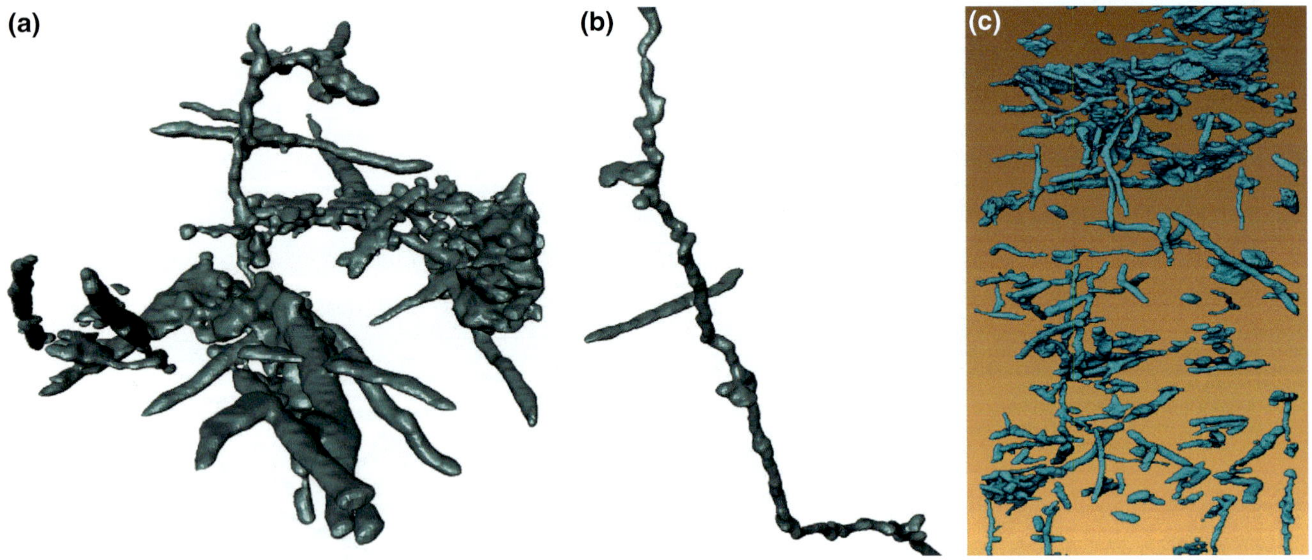


**Fig. 5.170** CT-scan of mud-filled *Virgaichnus undulatus* in glauconitic, completely bioturbated sandstone (full core). Paleocene Grumantbyen Formation (shoreface), near Longyearbyen, Svalbard (well BH 9-2006, ca. 389.7 m). Main burrow diameter varies between 0.3 and 0.6 mm. Images courtesy of Lars Rennan (Trondheim) and Ørjan Berge Øygard (Bergen). **a** Clusters of burrows trending in various directions. **b** Individual subvertical burrow element showing an undulating burrow section with bulbous enlargements. A short and flattened subhorizontal burrow element shows blade-like contractions. **c** Portion of an ichnofabric consisting of numerous burrows trending in various directions

may be confused with root traces (rootlets) which, however, preferably bifurcate downwards and may have carbonaceous material involved.

*Producers:* The pinching and swelling of the burrow segments is in agreement with a peristaltic movement of an unsegmented, highly deformable vermiform body as known in the nemerteans (Fig. 5.172).

*Ethology:* The passive fill of the burrow system indicates an open tunnel boxwork, suitable for a combined dwelling and feeding trace (domichnion and fodinichnion) of a deposit-feeding and subordinate suspension-feeding animal. Assuming nemerteans as tracemakers, a predatory lifestyle could also be a possibility for the producer of *V. undulatus*.

*Depositional Environment:* The sparse reports of *V. undulatus* document its occurrence in shelf deposits with relative quiet sedimentation regime, such as in a shallow-marine environment on a stable inner shelf (specimens from Oman and Norway), open lagoon (specimens from Iran), and deep- to shallow-marine (specimens from Utah).

*Ichnofacies: Virgaichnus* can be regarded as component of the *Cruziana* Ichnofacies.

*Age:* The few reports of marine *Virgaichnus* range from Carboniferous to Paleocene.

*Reservoir Quality: Virgaichnus* boxworks may act as connected pore systems if remaining open (e.g. not passively filled or cemented), and thus lead to a considerable increase of porosity and permeability in otherwise tight rocks (such as micritic carbonates and chalk; Knaust 2014a). Mud-filled burrows in sandstone, however, reduce the reservoir quality.

**Fig. 5.171** *Virgaichnus undulatus* in sectioned core (**a–f**) and full core (**g**). *Scale bars* = 1 cm. **a** Dolomitic limestone with a *Virgaichnus* (*V*) ichnofabric, accompanied by *Asterosoma* (*A*) and *Cylindrichnus* (*C*). Upper Permian Khuff Formation (open lagoon), South Pars Field, Persian Gulf, Iran (well SP9, ca. 3084.4 m). After Knaust (2014a), republished with permission from EAGE. **b–d** *Virgaichnus* ichnofabric in chalky limestone. Lower Cretaceous (Berriasian) Åsgard Formation (shelf), Norwegian North Sea (well 16/4-6S, 1943.9–1949.6 m). **e** Loosely arranged tubes and clustered burrows, partly displaying T-like branching and dichotomy. Upper Jurassic (Oxfordian) Heather Formation (shelf turbidites), Fram Field, Norwegian North Sea (well 35/11-11, ca. 2583.5 m). **f** Dense accumulation of *Virgaichnus* tubes. Upper Jurassic (Oxfordian) Heather Formation (shelf turbidites), Fram H-Nord Field, Norwegian North Sea (well 35/11-15ST2, ca. 2979 m). **g** *Virgaichnus* isp. in glauconitic sandstone. Paleocene Grumantbyen Formation (shoreface), near Longyearbyen, Svalbard (well BH 10-2008, ca. 820.5 m). Compare with Fig. 5.169b and 5.170

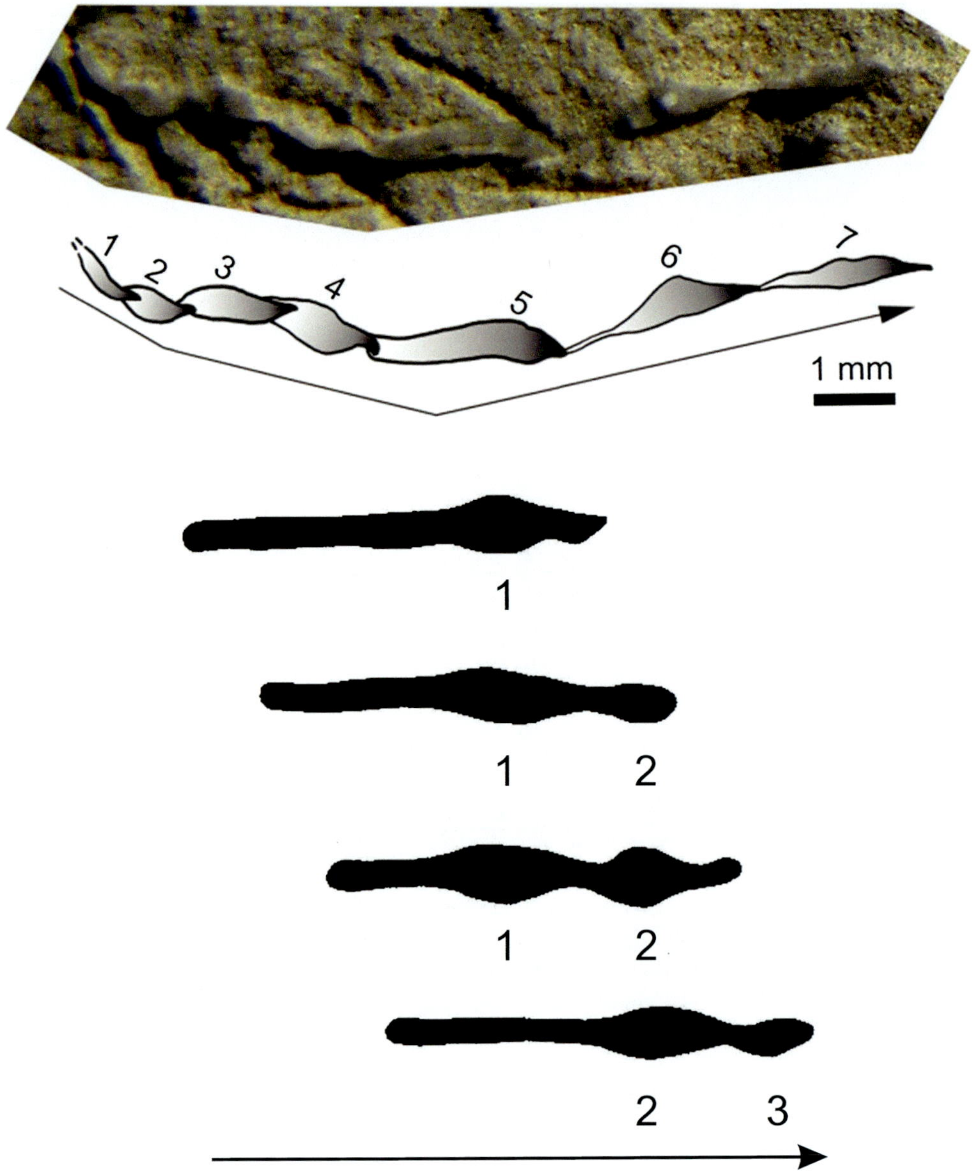


**Fig. 5.172** Selected branch of *Virgaichnus undulatus* from the holotype specimen (see Fig. 5.169a) with the diagnostic pinch-and-swell-like features (*photograph*), interpreted as the result of peristaltic movement of an unsegmented vermiform organism such as a nemertean (*line drawing*). The *arrow* indicates the moving direction and the numbers the peristaltic movements, made by narrowing and pushing out the proboscis and sending back a wave of alternate swellings and constrictions along the body. From Knaust (2010a), republished with permission of Elsevier; permission convoyed through Copyright Clearance Center, Inc. The sketch in the lower part illustrates peristaltic burrowing in *Carinoma tremaphoros*, where peristaltic waves (*numbered*) originate anteriorly and progress rearward as the animal burrows. Modified and redrawn from Turbeville and Ruppert (1983), with permission of Springer

## 5.36 *Zoophycos* Massalongo, 1855

*Morphology, Fill and Size: Zoophycos* is a complex trace fossil with wide distribution and a long history in research. It is a spreite burrow with a tremendous morphological variability (Bromley 1996), which in the past inspired the establishment of numerous ichnotaxa. The general characteristics of *Zoophycos* are as following (after Olivero and Gaillard 2007; Fig. 5.173):

- *Marginal tube*—A tubular structure bordering an area of bioturbated sediment and considered as a tunnel
- *Lamina*—Bioturbated sediment bordered by the marginal tube (also referred to as *spreite*)
- *Primary lamellae*—Arched grooves and ridges characterizing the lamina and interpreted as the subsequent positions of the marginal tube during its lateral displacement within the sediment.

Two main forms of *Zoophycos* can be classified: simple planar forms and complex spirally coiled forms (Olivero and Gaillard 2007; Fig. 5.174 and 5.175). Many *Zoophycos* appear with a lobate outline of their lamina. Another feature of some *Zoophycos* is the abundant occurrence of small ellipsoidal fecal pellets, ca. 1.5 by 0.5 mm in size. The complete *Zoophycos* burrow systems typically reach a size of few decimeters to more than 1 m in diameter and several decimeters in depth.

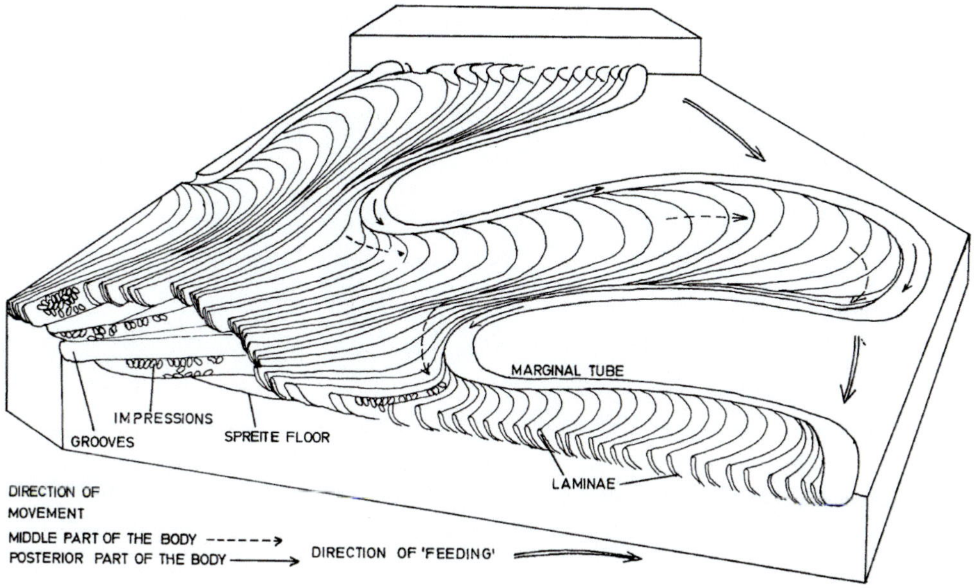


**Fig. 5.173** Schematic drawing of the *Zoophycos* spreite. The "laminae" correspond to the "primary lamellae" in the terminology of Olivero and Gaillard (2007), while the "impressions" refer to fecal pellets. From Bischoff (1968), © Paleontological Society, published by Cambridge University Press, reproduced with permission

*Ichnotaxonomy:* The ichnotaxonomic status of *Zoophycos* is far from being robust and much confusion exists nowadays when it comes to the delineation of individual ichnospecies. Work done by Olivero (2007) has confirmed that *Z. brianteus* can be regarded as the type ichnospecies. *Z. brianteus* comprises *Zoophycos* with a spirally coiled lamina and a slightly lobed outline. *Z. villae* is also regarded as valid and consists of a lamina furrowed by numerous sinuous and long lamellae radiating from a raised apex. Many other forms are described (Zhang and Gong 2012), some of which diverge considerably from the type ichnospecies and therefore would be best accomodated in other previously established ichnogenera such as *Taonurus* von Fischer-Ooster, 1858, *Cancellophycus* de Saporta, 1873 and *Echinospira* Girotti, 1970 (Bromley and Hanken 2003). Some workers refer to the '*Zoophycos* group' when dealing with those forms with a wide range of morphology (e.g. Uchman and Demircan 1999; Fig. 5.175). One might also call them the Alectoruridae.

The various forms of *Zoophycos* show a general evolutionary trend from relatively simple forms with semielliptical lobes and subcircular spreite fields in the Paleozoic to more spiral forms with continuous spreiten in the Mesozoic, and finally complex forms with discontinuous spreite fields and lobes in the Cenozoic (Seilacher 1977, 2007; Bottjer et al. 1988; Olivero 1996; Chamberlain 2000; Knaust 2004b; Zhang et al. 2015).

*Substrate: Zoophycos* is known from siliciclastic and carbonate rocks of originally soft to firm consistency. It preferably occurs in fine-grained sediments.

*Appearance in Core: Zoophycos* spreite burrows are relatively easy to recognize in core, although identification to the ichnospecies level requires knowledge about the three-dimensional burrow architecture. Assuming *Zoophycos* with a spirally coiled lamina as diagnostic for the type ichnospecies *Z. brianteus*, vertical core sections through such a burrow system can cut it either axially (relatively rare) or marginally (common). In axial expression, spreite burrows with a conical appearance alternate vertically (Fig. 5.176a, b). In the marginal expression, individual and more or less horizontally oriented spreiten become visible and may be stacked vertically (Fig. 5.176c–h; see also Figs. 5.39b, 5.132d and 5.168). The spreite burrows are typically a few millimeters to ca. 1 cm thick and display the characteristic internal lamellae composed of alternating sediment packages of different composition. The distance between individual lamellae can be short, leading to a height/width ratio of the lamellae of more than 1 (Fig. 5.176d, g), or may be more stretched horizontally, which results in a height/width ration of less than 1 (Fig. 5.176e, f). Some ichnospecies of *Zoophycos* contain ellipsoidal fecal pellets (about 1.5 mm in length and 0.5 mm in diameter), which belong to *Coprulus oblongus* and can be identified in core (Fig. 5.176c). Because of subsequent alteration (e.g. reburrowing, diagenetic processes), parts of the spreiten or entire burrows can lack their lamellae (Fig. 5.35b, d–f) and thus appear to be homogeneously filled with sediment (Fig. 5.176h).

*Similar Trace Fossils:* There are few other burrows with which *Zoophycos* could be confused. *Rhizocorallium* is another spreite burrow with a U-shaped morphology, and the horizontal to inclined *R. commune* constitutes a spreite commonly filled with fecal pellets *C. oblongus* (Knaust 2013). However, it does not develop a spirally coiled lamina and therefore is laterally restricted. Furthermore, the spreite of *R. commune* is generally thicker than in *Zoophycos* and is bounded by a proportionately large marginal tube. Other constituents of the broader '*Zoophycos* group' may resemble

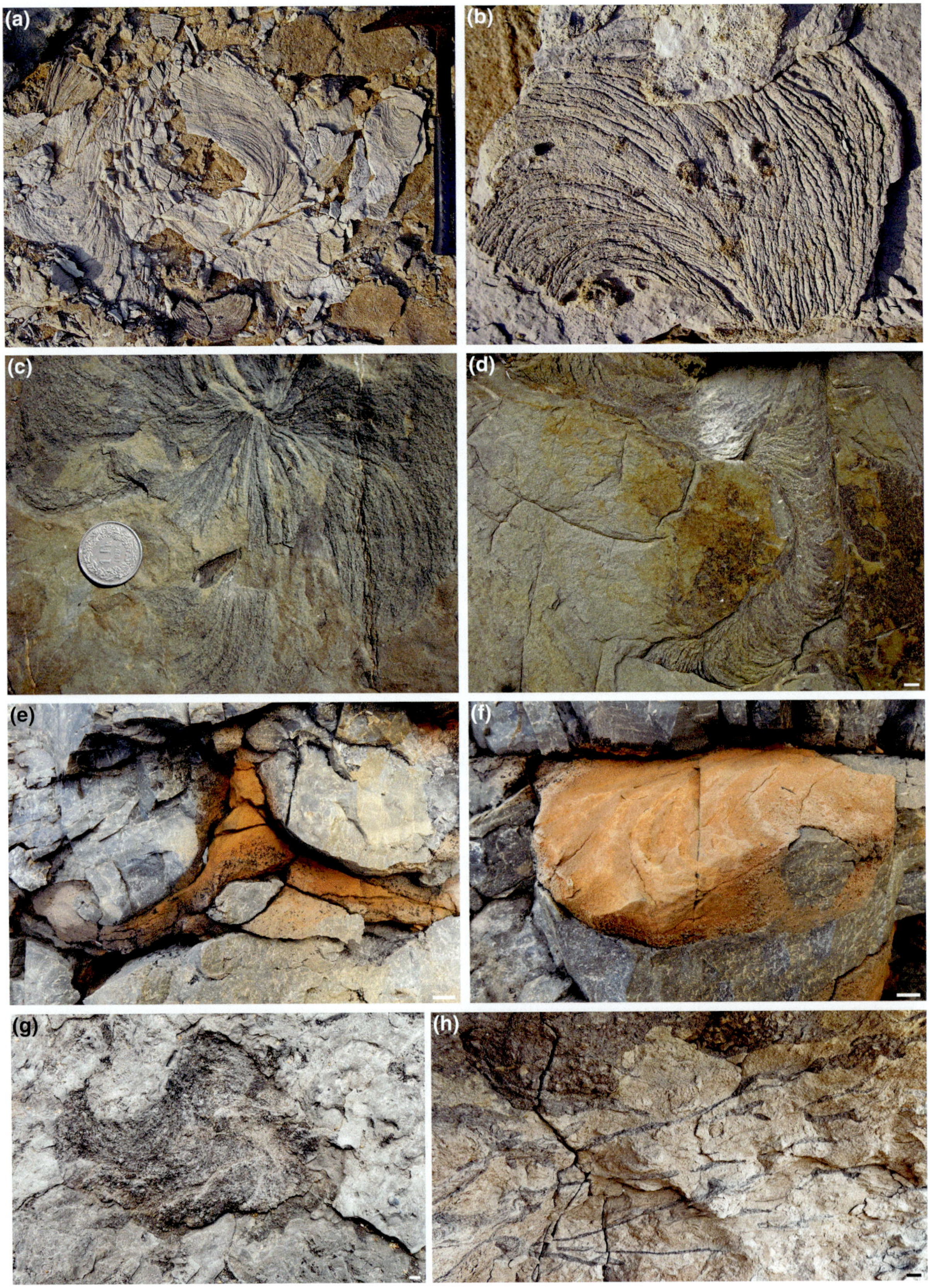
(a)
(b)
(c)
(d)
(e)
(f)
(g)
(h)

◀**Fig. 5.174** *Zoophycos* in outcrop. *Scale bars* = 1 cm. **a**, **b** Relatively simple, planar form of *Zoophycos* in bioclastic limestone. Middle Permian Khuff Formation (shallow marine, carbonate platform), Huqf-Haushi Uplift, Oman. From Knaust (2009c), republished with permission of Wiley; permission conveyed through Copyright Clearance Center, Inc. **c** Spiral form of *Zoophycos* in fine-grained carbonates. Middle Jurassic (slope deposits), southeastern France. See Olivero and Gaillard (1996), Olivero (2003). **d** Lobate part of a larger *Zoophycos* system, preserved on a sandstone bedding plane of the Eocene Grès d'Annot Formation (deep marine, turbiditic), southeastern France. From Knaust (2013) and Knaust et al. (2014), republished with permission of Elsevier and Wiley; permission conveyed through Copyright Clearance Center, Inc. **e** Vertical section of *Zoophycos* close to the whorl. Middle Permian Kapp Starostin Formation (mixed siliciclastic carbonate ramp), Akseløya, Svalbard. **f** Lobate spreite part in oblique section. Same as in (**e**). **g** Lobate spreite on carbonate bedding plane. Upper Jurassic (Kimmeridgian) Alcobaça Formation (restricted lagoon), coastal cliffs at Praia do Salgado, western Portugal. **h** Same as in (**g**), cross section of numerous stacked spreiten

*Zoophycos* sensu stricto, including *Spirophyton* and *Echinospira*. The corkscrew-like *Spirophyton* consists of tightly arranged whorls winding around an axial shaft, and has an unlobed edge without marginal tube with an upwards bent margin (Miller 1991; Gaillard et al. 1999; Seilacher 2007). *Echinospira* is characterized by a semicircular outline with numerous long, narrow U-shaped spreite burrows that are related to a central cylindrical structure. *Echinospira* is common in Upper Cretaceous to Miocene deep-marine deposits. Individual spreiten laminae of *Zoophycos* might be critical to distinguish from elongate sections of the cylindrical and meniscate backfilled burrow *Taenidium*, in which case several angles of observation are necessary to differentiate the two.

*Producers:* Originally interpreted as marine algae (Massalongo 1855; Olivero 2007) and later as remains of sedentary marine worms (Plička 1968), most workers now agree that the spreite burrow *Zoophycos* results from the feeding activity of a vermiform animal (Wetzel and Werner 1981). However, what kind of worm has produced *Zoophycos* remains debatable, and probably organisms of different phyla must be considered with regard to particular forms of this complex trace fossil. Among them, polychaetes (Bischoff 1968, Knaust 2009c), echiurans (Kotake 1992) and sipunculans (Wetzel and Werner 1981; Olivero and Gaillard 2007) are good candidates.

*Ethology:* The deposit-feeding behavior of a vermiform animal was the preferred interpretation of *Zoophycos* for a long time, until particular features of some forms suggested alternative interpretations, including the incorporation of sediment from the seafloor into the burrow by different processes (Löwemark 2012). It is likely that the broad morphological variability of *Zoophycos* is a reflection of contrasting ethologies.

*Depositional Environment: Zoophycos* can be regarded as a marine trace fossil. *Zoophycos* (sensu lato) has been proved to be a trace fossil with an evolutionary history through the Phanerozoic (Seilacher 1977, 2007; Bottjer et al. 1988; Olivero 1996; Neto de Carvalho and Rodrigues 2003). Whereas Paleozoic *Zoophycos* occur in nearshore deposits, Mesozoic forms are common on the shelf and later also move towards deeper marine environments, and Cenozoic *Zoophycos* seem to be restricted to the deep sea. Aside from its typical occurrence within the *Zoophycos* Ichnofacies on the slope, *Zoophycos* is frequently encountered in lagoonal environments (Fig. 5.176h).

*Ichnofacies: Zoophycos* is the namesake of the *Zoophycos* Ichnofacies, which as introduced by Seilacher (1967) was positioned on the continental slope between the shallow-marine *Cruziana* Ichnofacies and the deep-marine *Nereites* Ichnofacies. In many basins, *Zoophycos* is a constituent of both these adjacent ichnofacies as well.

*Age: Zoophycos* in the broad sense is a long-ranging trace fossil with worldwide distribution perhaps from the Cambrian (Alpert 1977; Jensen 1997) and certainly from the Ordovician to the Holocene (Wetzels 2008). The report of the oldest *Zoophycos* from the Early Cambrian (Sappenfield et al. 2012) must probably be rejected because of its likely inorganic origin.

*Reservoir Quality: Zoophycos* is a common constituent of mud-dominated facies such as lagoonal deposits (e.g. in the Khuff Formation, see Fig. 5.176h), which can be intensively perforated by complex spreite burrows. These burrows are actively filled with grainy (peloidal and oolitic) material introduced from outside the burrows. This phenomenon is related to the behavior of the *Zoophycos* maker, supposedly a worm-like animal, which excavates mud in the subsurface, transports it to the seafloor, and fills its burrow with grainy material from the surface. This model was discussed by Kotake (1989), Bromley (1991) and Löwemark et al. (2004), and the resulting introduction of grainy material in the host rock was recognized by Pemberton and Gingras (2005) and Knaust (2009a). In the given example, permeability is created by the activity of the *Zoophycos* producer, which turns the otherwise tight mudstone into a reservoir (Knaust 2009a, 2014a; Fig. 5.177). A similar situation does also occur in some chalk reservoirs of the North Sea.

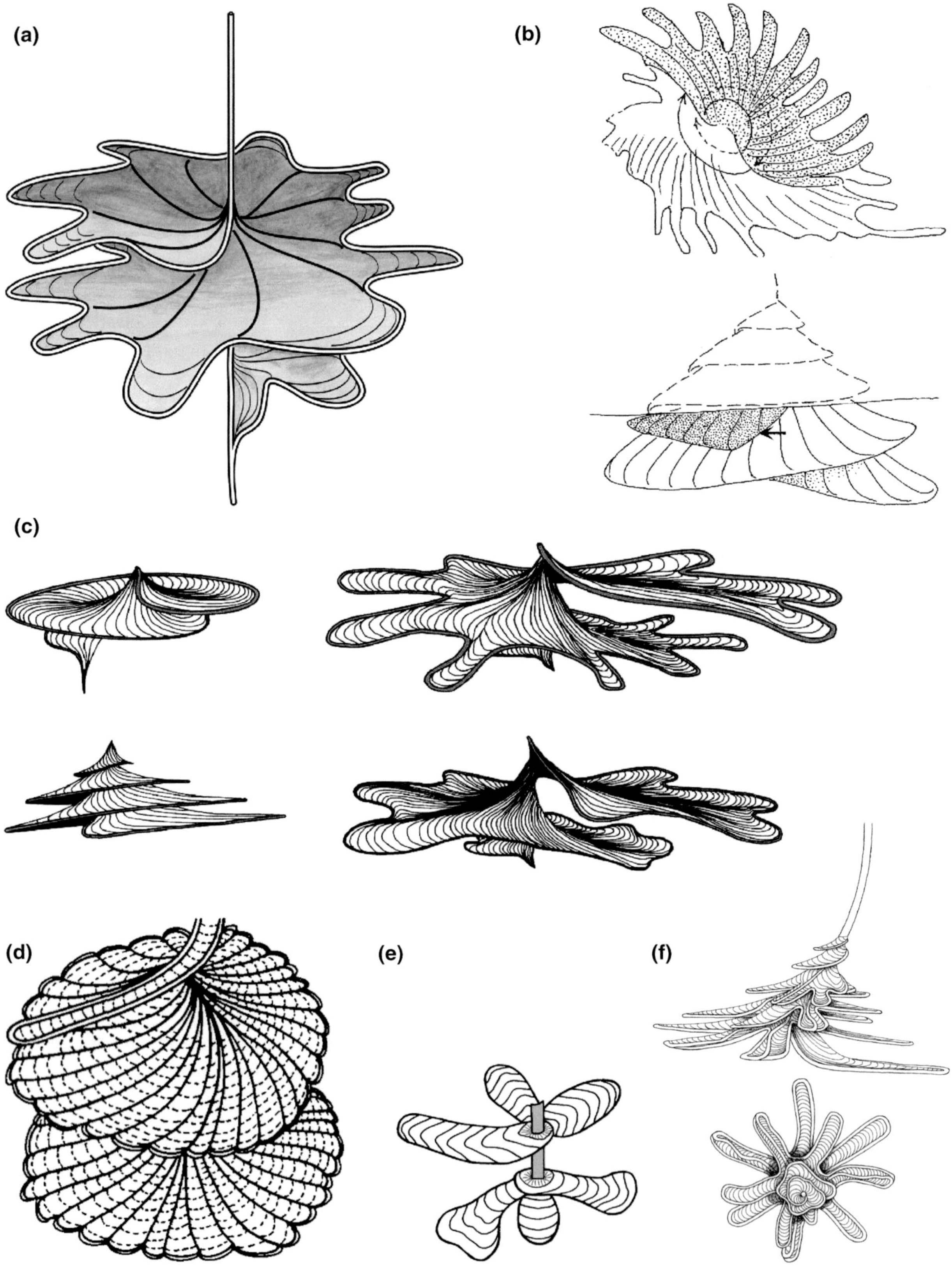
(a)
(b)
(c)
(d)
(e)
(f)

◀**Fig. 5.175** Sketches of representative examples of *Zoophycos* as described in the literature. Not to scale. **a** Helicoidally coiled form with semicircular to highly lobate spreiten, suggesting upwards construction. Late Quaternary (deep marine), Celebes Sea. After Löwemark et al. (2004). **b** *Z. rhodensis*, a large spiral form comprising a skirt-like zone of spreite surrounded by a zone of numerous marginal lobes. Plan view (*top*) and side view (*bottom*). Pliocene carbonates (deeper marine), Rhodes, Greece. After Bromley and Hanken (2003), republished with permission of Elsevier; permission conveyed through Copyright Clearance Center, Inc. **c** Four morphotypes of *Zoophycos* with increasing complexity. Lower Jurassic to Upper Cretaceous (shelf-to-basin limestones), southeastern France. After Olivero (2003), republished with permission of Elsevier; permission conveyed through Copyright Clearance Center, Inc. **d** Coiled retrusive spreite burrow built by successively overlapping U-shaped burrows. Similar forms were described under the ichnogenus name *Echinospira* Girotti, 1970. Upper Cretaceous to Miocene limestone (deep marine), New Zealand. After Ekdale and Lewis (1991), reprinted by permission of the publisher (Taylor & Francis Ltd., http://www.tandfonline.com). **e** Simple, lobate form with planar spreiten extending from a vertical shaft. Middle Triassic limestone (carbonate ramp), Thuringia, Germany. After Knaust (2004b), republished with permission of Wiley; permission conveyed through Copyright Clearance Center, Inc. **f** Coiled protrusive spreite burrow with convolute margin and long lobes. The spreite is cut by a tubular structure. Side view (*top*) and plan view (*bottom*). Upper Cretaceous White Chalk (shelf), Denmark and southernmost Sweden. After Bromley et al. (1999).

## 5.37 Diffuse Bioturbate Texture

*Morphology, Fill and Size:* The process of bioturbation not only results in more or less well defined discrete trace fossils, but often leads to an intensely or completely reworked sediment with a diffuse texture (e.g. mottled fabric), or even complete homogenization of the substrate after repeated bioturbation. In that case, the resultant rock contains a diffuse bioturbate texture (Richter 1952; Frey 1973; Frey and Pemberton 1990, 1991b). Although some burrows are still discernible, they hardly can be identified on an ichnotaxonomical basis (Fig. 5.178). If such remains of burrows are preserved, their size can correspond to the size of all major tracemakers and would fall into the two process-related categories of macrobioturbation (burrow diameter or width >1 mm) and meiobioturbation (burrow diameter or width between 0.06 and 1 mm). Diffuse bioturbate textures with less discernible burrows (typically of smaller size and as a result of the burrowing method) and well expressed physical sedimentary structures may also be referred to cryptic bioturbate textures, which results from cryptic bioturbation (Howard and Frey 1975, 1985; Pemberton et al. 2001, 2008).

*Ichnotaxonomy:* Even highly bioturbated sedimentary rocks sometimes reveal burrow parts that can be attributed to particular ichnotaxa. Completely homogenized substrate can experience repeated colonization by benthic organisms, which in turn leads to so-called elite trace fossils (Bromley 1996), which may overprint the pre-existing bioturbate texture. Bioturbate textures with a more organized style (e.g. tiering, discrete trace fossils) revealing insights into the colonization process of the substrate are commonly dealt with as ichnofabrics (Ekdale et al. 2012).

*Substrate:* Bioturbate texture is common in softground, looseground and occasionally firmground substrates of various lithologies, where sediment consistency is highly flexible and thus allows for multiple textural reorganizations.

*Appearance in Core:* The term bioturbate texture is typically applied to moderately to completely bioturbated rocks that do not allow for systematic description and interpretation of individual trace-fossil components. While in some cases discrete burrow features are still intact and visible (Figs. 4.5 and 5.121c), in other examples the sediment was subject of repeated bioturbation which resulted in a homogenized texture (Fig. 5.179). Yet another example contains cryptic bioturbate texture, in which the original bedding features are still partly intact but were disturbed by the burrowing animal (Figs. 5.179d, e and 5.180).

*Depositional Environment:* Bioturbate texture is commonly related to environments with favorable conditions for colonization, such as good oxygenation, food availability and sediment starvation (e.g. offshore environments, flooding events), but may also occur in conjunction with stressed environments where rapid and widespread colonization of a limited number of species takes place (e.g. lagoonal deposits).

*Ichnofacies:* Bioturbate texture may occur in a wide range of ichnofacies, especially the *Nereites, Cruziana* and *Skolithos* ichnofacies (e.g. piperock) within the marine realm. Paleosols within the continental *Scoyenia* Ichnofacies often also show bioturbate texture.

*Age:* Bioturbate texture occurs throughout the Phanerozoic.

*Reservoir Quality* Bioturbate texture may have a strong impact on reservoir quality (e.g. Knaust 2014a). In most cases, the reservoir quality is diminished because intense bioturbation leads to the incorporation of fine-grained material in form of mud particles and fecel material. In some case, however, the contrasting effect takes place, if coarser-grained particles are churn with a dense matrix and increase vertical connectivity (e.g. cryptic bioturbate texture, Fig. 5.180).

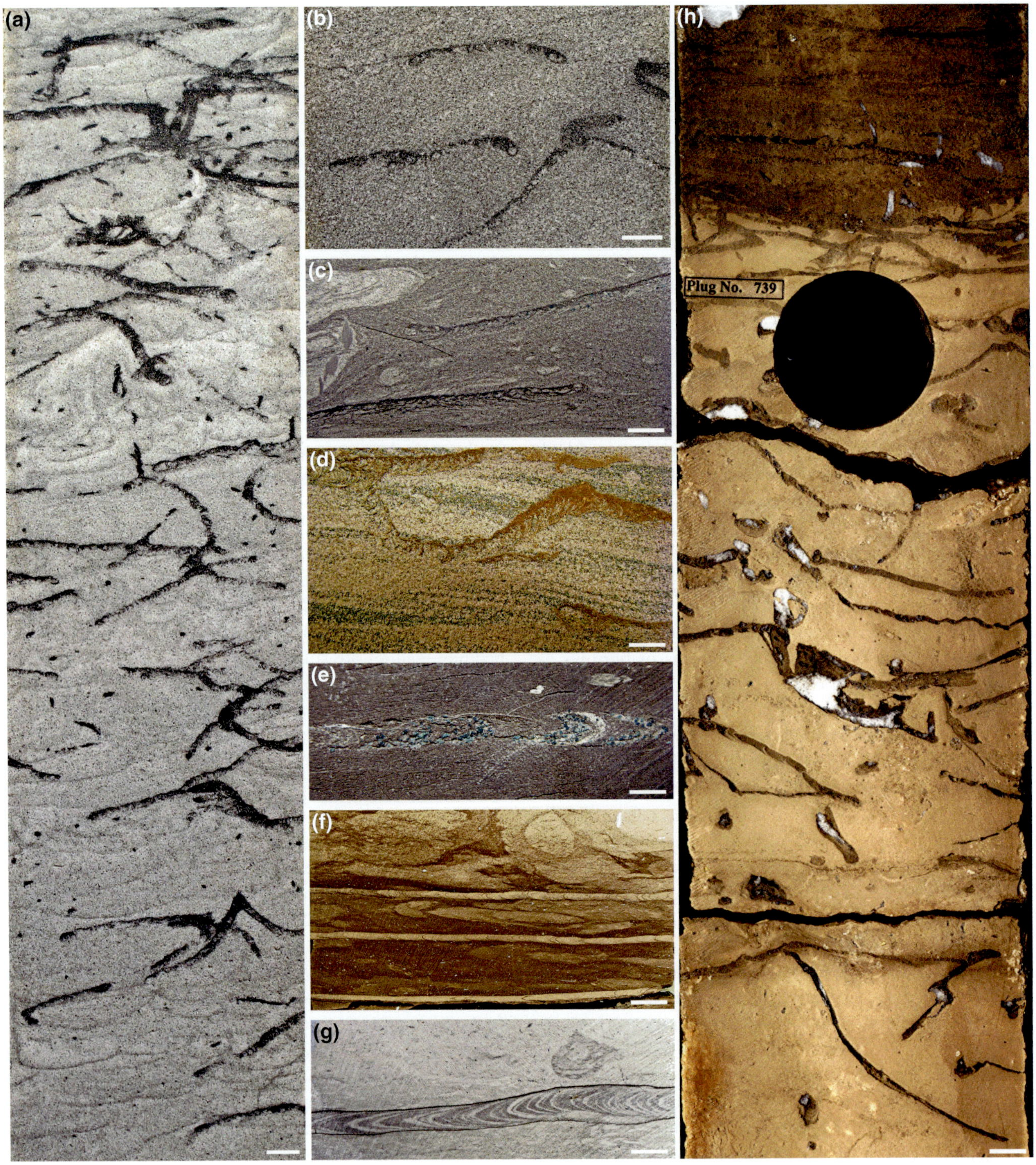
(a)
(b)
(c)
(d)
(e)
(f)
(g)
(h)
Plug No. 739

◄**Fig. 5.176** *Zoophycos* in sectioned core. *Scale bars* = 1 cm. **a** Homogeneous sandstone with dish-and-pillar structures due to rapid dewatering. Spreiten are readily visible due to their partial fill of black mud and the resulting color contrast with the white to gray sandstone. Some conical spreiten indicate close proximity to the axis of a *Zoophycos* burrow system. Lower Cretaceous (Albian, deep marine, channel system), off Tanzania. **b** Homogeneous sandstone with conical and planar spreiten, the latter displaying the marginal tube to the right. Lower Cretaceous (Albian, deep marine, channel system), off Tanzania. **c** Sandy debrite with two planar spreiten consisting of alternating mud and sand lamellae, incorporated glauconite grains and sandy fecal pellets. Upper Cretaceous (Maastrichtian) Springar Formation (deep marine, fan system), Gro Discovery, Norwegian Sea (well 6603/12-1, ca. 3724.5 m). **d** Ripple-laminated glauconitic sandstone with undulating spreiten, which are enhanced in visibility by diagenetic iron staining. Upper Cretaceous (Campanian) Nise Formation (deep marine, channel-levee system), Aasta Hansteen Field, Norwegian Sea (well 6707/10-1, ca. 3050.1 m). **e** Detail of a spreite showing discrete lamellae with different composition, mainly alternating between mud and glauconitic sand. Upper Cretaceous (Maastrichtian) Springar Formation (deep marine, fan system), Norwegian Sea (well 6604/10-1, ca. 3628.5 m). **f** Marly limestone with three successive, horizontal spreite burrows, overprinting extensively bioturbated background sediment. Lower Cretaceous (Berriasian) Åsgard Formation (carbonate shelf), Johan Sverdrup Field, Norwegian North Sea (well 16/5-2S, ca. 1951.5 m). **g** Arenitic limestone with a thick spreite section, consisting of alternating dark-gray and light-gray lamellae. Lower Cretaceous (Berriasian) Åsgard Formation (carbonate shelf), Johan Sverdrup Field, Norwegian North Sea (well 16/5-2S, ca. 1945.5 m). **h** Micritic limestone constituting a *Zoophycos* ichnofabric with spreite burrows partly diagenetically replaced by anhydrite (white color). The top is erosional truncated and overlain with grainstone, also containing *Zoophycos*. Upper Permian Khuff Formation (carbonate platform with open lagoon), South Pars Field, Persian Gulf, Iran (well SP-9, 3097.3-3097.6 m). After Knaust (2009a, 2014a), republished with permission of GulfPetroLink and EAGE

## 5.38 Plant Roots and Their Traces

*Morphology, Fill and Size:* Fossil plant roots and their traces can appear in a broad range of morphology and size, which reflects the variability in shape, size and behavior of their tracemakers (Retallack 1988; Fig. 5.181). The filamentous or tubular roots and root traces range in diameter from millimeter- to meter-size and taper downwards a few centimeters up to several meters (Fig. 5.182). The diagenetic history and maturity of rooted surfaces and paleosols results in different styles of preservation, ranging from body preservation or permineralization to deflection of bedding planes (Pfefferkorn and Fuchs 1991). Rhizoconcretions (or rhizoliths) are diagenetically encrusted root systems and have a high preservation potential (Owen et al. 2008).

*Ichnotaxonomy:* Fossil plant roots are covered by the *International Code of Nomenclature for Algae, Fungi, and Plants* and can be named independent from other parts of the plant (e.g. Uchman et al. 2012). There is a transition between fossil plant roots (body fossils) and root traces (trace fossils). Fossil root traces are fossilized work of organisms and thus are covered by the *International Code of Zoological Nomenclature*, but so far only have received few names, such as *Rhizoichnus* D'Alessandro and Iannone, 1982. Attempts to classify fossil root plants and their traces have been proposed by Klappa (1980), Pfefferkorn and Fuchs (1991), Bockelie (1994), Wright et al. (1995) and White and Curran (1997). The classification key proposed by Bockelie (1994) is mainly based on Mesozoic core material from the Norwegian North Sea. The hierarchical structure of that key involves sediment fill, presence and complexity of branching, size, orientation and morphology.

*Substrate:* Plants can grow on different substrates including soft and hard rock of different nature, and not only siliciclastics and carbonates. By their nature, plant roots are commonly associated with various kinds of paleosols, including caliche in carbonate rocks.

*Appearance in Core:* The diverse appearance of roots and root traces in rocks depends on the age of the rock (evolutionary aspect), paleoenvironment (kind of plants, substrate), and diagenesis (preservational aspect) (Fig. 5.181). In core, many roots appear in form of unbranched or downward-branched, irregularly distributed features (Fig. 5.183, see also Fig. 5.28b, d). Their fill can be coaly, carbonaceous, sandy or a combination of these. A typical preservation is sand-filled root traces with a thin carbonaceous lining. Other roots can be enhanced by diagenesis to build rhizoconcretions, some of which hardly reveal the remaining root structure at their center.

*Similar Trace Fossils:* Plant-root traces are often the subject of confusion with morphologically similar animal burrows, particularly in cases where they penetrate marine substrates (Curran 2015). The simple vertical burrow *Skolithos* is a good candidate for confusion with root traces, and some *Skolithos* actually originate from root penetrations (Gregory et al. 2006; Knaust 2014a), while tiny rootlets can be mistaken for small burrows (e.g. *Chondrites*, *Bornichnus* and *Virgaichnus*). More complex root traces, for instance those radiating from a tree stem, may resemble complex burrow systems such as *Phoebichnus* (e.g. Gregory et al. 2004).

*Producers:* Various groups of plants are potential producers of root traces (Bockelie 1994), although some compound trace fossils and paleosols can be formed by plant and animal (e.g. insect) interactions (e.g. Gregory et al. 2004; Strullu-Derrien et al. 2012).

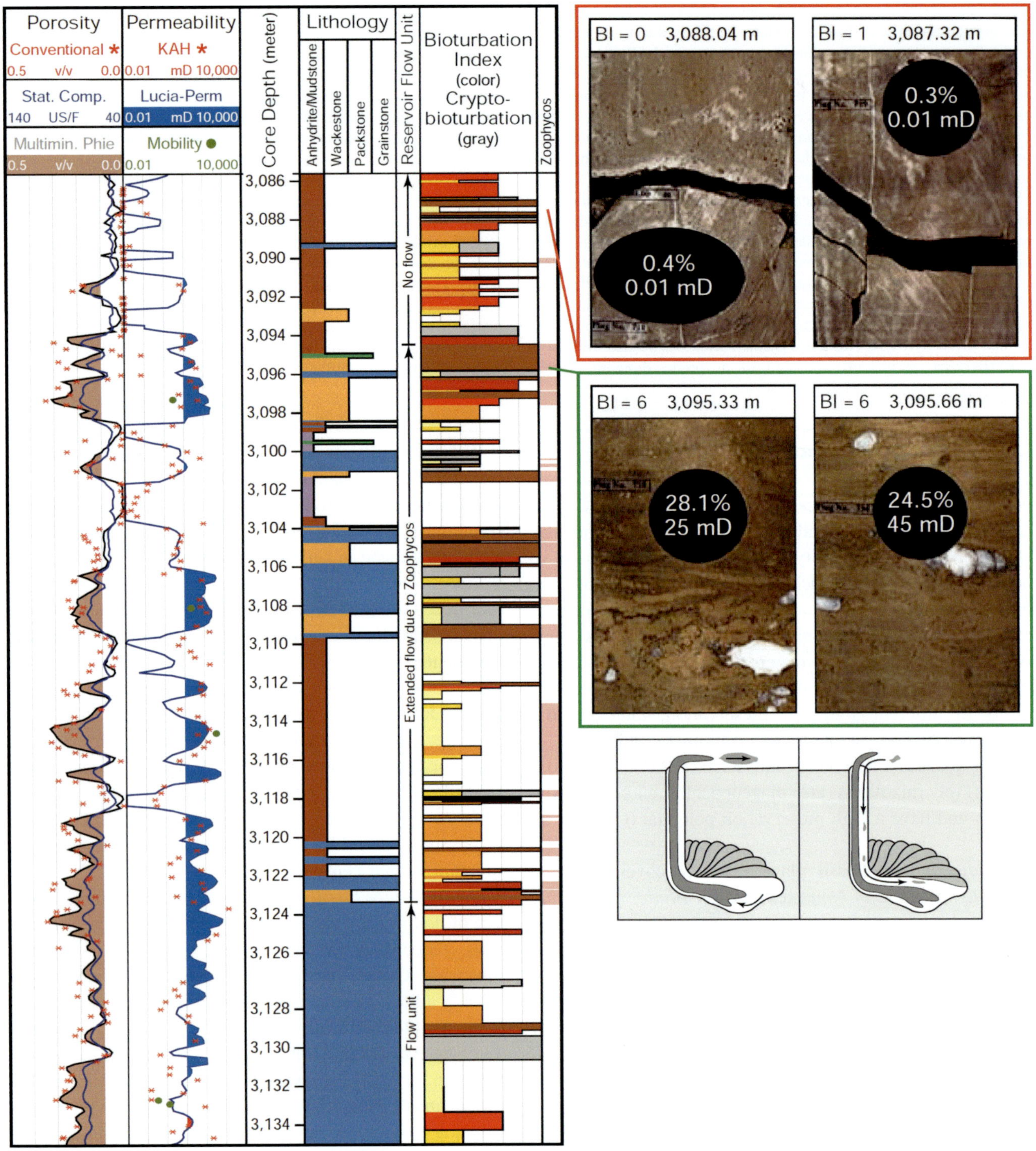


**Fig. 5.177** Impact of *Zoophycos* bioturbation on reservoir quality. A lagoonal mudstone unit is commonly tight if not bioturbated (*upper images*, core image width is 9.8 cm). However, complete bioturbation and discrete *Zoophycos* burrows in the deep tier transfer the mudstone barrier into a mudstone-wackestone flow unit (*images in the middle*, core image width is 9.8 cm). Reason for this is the behavior of the *Zoophycos* producer, a supposed worm-like animal, which excavates and feeds the subsurface mud, defecates on the sea floor and fills its burrow with grainy (oolitic) material (refuse dump model in the *lower image*; after Löwemark et al. 2004). In this particular case, the existing flow unit within the grainstone facies extends upwards with ca. 30 m. Upper Permian Khuff Formation (carbonate platform with open lagoon), South Pars Field, Persian Gulf, Iran. From Knaust (2009a), republished with permission of GulfPetroLink

**Fig. 5.178** Bioturbate texture in outcrop. *Scale bars* = 1 cm. **a** Oblique section with a few discernible burrows in heterolithic sandstone-siltstone (shallow marine), in which discrete *Scalichnus* and *Schaubcylindrichnus* are visible. Upper Jurassic (Early Kimmeridgian), Brora (Lothberg Point area), Scotland, UK. **b** Bedding-plane view of a micritic limestone (carbonate platform) with abundant pencil-like burrows resembling *Planolites*. Middle Triassic (Anisian) Muschelkalk Group, south of Roda, southern Spain. **c** Vertical section of totally bioturbated glauconitic sand (shelf) with some horizontal tubular burrows visible, probably belonging to *Macaronichnus segregatis*. Upper Cretaceous (Cenomanian) Arnager Greensand Formation, Arnager near Rønne, Bornholm, Denmark. **d** Vertical section of sand with high amount of bioturbation overprinting partly preserved cross bedding (lower part). Some discrete *Teichichnus rectus* and *Ophiomorpha nodosa* are visible. Lower Cretaceous (Berriasian) Robbedale Formation (shallow marine), Madsegrav near Rønne, Bornholm, Denmark. See Nielsen et al. (1996)

*Ethology:* Substrate penetration by roots happens for the purpose of plant stabilization, living and obtaining water and nutrition.

*Depositional Environment:* Plants commonly colonize terrestrial and aquatic environments in continental setting, such as alluvial, fluvial, lacustrine and eolian deposits (e.g. Glennie and Evamy 1968; Kraus and Hasiotis 2006; Knaust 2015b). Mangrove and other roots also occur in marginal-marine (paralic) environments including swamps, lagoons and tidal flats (e.g. Whybrow and McClure 1980; Knaust 2009a). Rooted horizons are important indicators of subaerial exposure and thus can be used to delineate sequence boundaries (Husinec and Read 2011) as well as unconformities (Fig. 2.6a).

*Ichnofacies:* A wide range of continental and marginal-marine ichnofacies typically contains plant roots, for instance *Scoyenia* and *Psilonichnus* ichnofacies.

(a)
(b)
(c)
(d)
(e)
(f)

**◄Fig. 5.179** Bioturbate texture in sectioned core. *Scale bars* = 1 cm. **a** Completely bioturbated argillaceous sandstone containing *Teichichnus* and *Phycosiphon*. Upper Jurassic (Oxfordian-Kimmeridgian) Spekk Formation (shelf), Norwegian Sea (well 6406/12-1S, ca. 3629.9 m). **b** Intensely bioturbated sandstone (with *Planolites*?), in which remains of the original bedding are poorly preserved. Upper Jurassic (Oxfordian-Kimmeridgian) Rogn Formation (offshore sand bar), Norwegian Sea (well 6406/12-1S, ca. 3622.9 m). **c** Moderately bioturbated cross-bedded sandstone. Middle Jurassic (Callovian) Fensfjord Formation (tidally influenced delta), Gjøa Field, Norwegian North Sea (well 36/7-1, ca. 2393.1 m). **d** Argillaceous sandstone with cryptic bioturbate texture, in which the mud laminae appear to be intact but are disturbed by numerous sand-filled burrows (*Planolites*). The sandstone includes vertical sand-filled *Skolithos* burrows (weakly visible). Upper Jurassic (Oxfordian) Heather Formation (shelf turbidites), Fram Field, Norwegian North Sea (well 35/11-11, ca. 2586.5 m). **e** Argillaceous sandstone with cryptic bioturbate texture as revealed by many tiny (meiobenthic) burrows. Upper Jurassic (Oxfordian-Kimmeridgian) Sognefjord Formation (marginal marine, deltaic), Vega Field, Norwegian North Sea (well 35/11-6, ca. 3184.2 m). **f** Wholly bioturbated, homogenized sandstone with diffuse burrow structures. Upper Jurassic (Oxfordian-Kimmeridgian) Heather Formation (submarine fan delta), Fram H-Nord Field, Norwegian North Sea (well 35/11-15ST2, ca. 2975.0 m)

*Age:* The oldest records of primitive plants are from the Ordovician, but diversification and complexity gradually increased through Silurian and Devonian time. From the Devonian until today, root traces have been common constituents of the rock record.

*Reservoir Quality:* Different preservation of plant roots may result in contrasting behavior with respect to reservoir quality and performance. For instance, large and sand-filled roots in a tight matrix considerably enhance vertical communication within the reservoir (Knaust 2014a), while carbonaceous roots of similar shape and size influence the reservoir in the opposite way.

## 5.39 Borings

*Morphology, Fill and Size:* Borings belong to a group of trace fossils that comprises excavations made by organisms in hard substrate in contrast to burrows, which originate in soft or firm substrate. They appear with different shapes and degrees of destruction, and include irregular morphologies, networks, pouches, grooves, as well as clavate, bulbous, tubular and clubbed forms (Figs. 5.184 and 5.185). The margin of a boring, which may be lined, is typically sharp and well-defined, creating a high contrast between the host rock and its fill. Borings are passively filled with sediment differing from the host rock, occur as open cavities, or are cemented. A wide size range of borings is known, commonly from micrometric to centimetric in size, with 1 mm as the arbitrary threshold size for the distinction of microborings and macroborings.

*Ichnotaxonomy:* Bioerosion trace fossils (including borings) comprise more than hundred ichnogenera, which correspond to about 17% of all invertebrate trace fossils (Knaust 2012a). Most common and widespread macroborings are *Entobia* (borings with networks of chambers and canals made by sponges), *Gastrochaenolites* (clavate bivalve borings in lithic substrates), *Teredolites* (clavate bivalve borings in lignic substrates), *Trypanites* and *Palaeosabella* (tubular borings made by worms), *Rogerella* (pouches made by cirripeds), and *Talpina* (networks of phoronid worms) (Taylor and Wilson 2003; Bromley 2004; Figs. 5.184 and 5.185).

*Substrate:* Borings occur in hard substrates that can be of lithic (e.g. rocks; skeletal material such as shells, corals, stromatolites, etc.), xylic (i.e. wood), or osteogen (i.e. bone) nature (Taylor and Wilson 2003). Lithified substrates can be formed by the exhumation of deeply buried sediment or by synsedimentary lithification of sediment (Savrda 2012).

*Appearance in Core:* Borings are relatively easy to recognize in core, although their assignment to particular ichnogenera often remains ambiguous as the whole morphology is not known. However, serrated surfaces, sharp boundaries, cross-cut particles and passive fills are unique features that prove a bioerosion nature (Fig. 5.186). Bioeroded surfaces of hardgrounds are often discontinuity surfaces (such as omission surfaces and unconformities) and appear as rather sharply defined boundaries with contrasting lithologies below and above. Because such surfaces often became subject of prolonged subsequent reworking by currents, they are commonly altered, stained, and disrupted or eroded. Consequently, reworked litho- and bioclasts may also show evidences of repeated bioerosion phases in form of one or more generations of borings.

*Similar Trace Fossils:* A continuous transition exists between cohesive (firm) and hard substrates, and firmgrounds may also contain incipient borings characterized by sharp boundaries and passive fill (Knaust 2008; Knaust and Dronov 2013). Therefore, a clear distinction between both trace-fossil categories (e.g. burrows in firm and borings in hard substrates) is not always possible. This is particularly true in cases where the trace fossils have similar shapes and/or were produced by similar kinds of tracemakers that are capable of both, burrowing and bioeroding (e.g. some polychaete and bivalve species).

*Producers:* Many groups of organisms comprise bioeroding species, of which cyanobacteria and algae, sponges, polychaetes, sipunculides, bivalves, echinoids, cirripedes and bryozoans are most relevant for the origin of borings (Warme 1970; Taylor and Wilson 2003; de Gibert et al. 2012; Tapanila and Hutchings 2012).

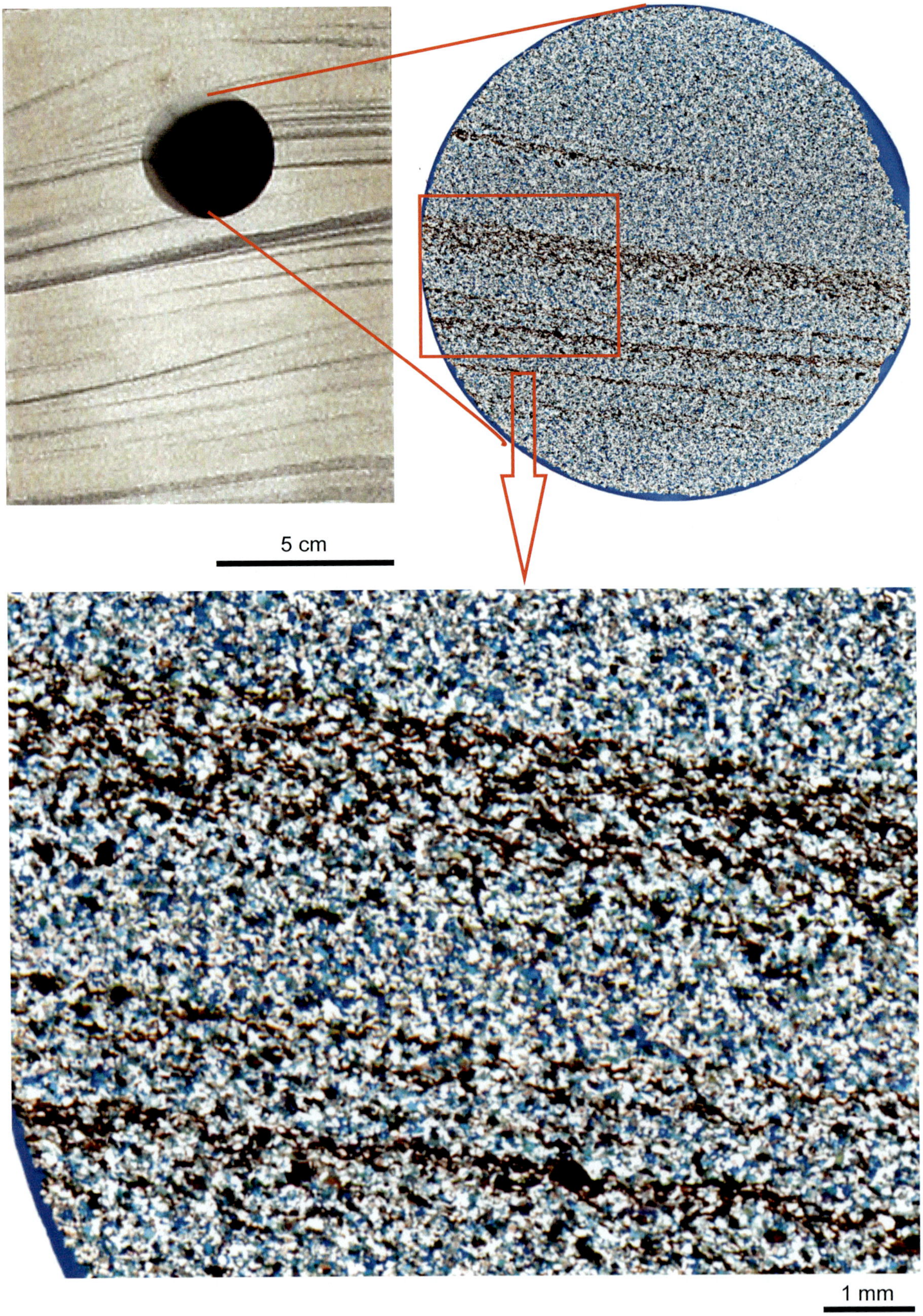


**Fig. 5.180** Cross-bedded sandstone, as seen in thin section, reveals disrupted mud-rich laminae due to the process of cryptobioturbation. This phenomenon results in increased reservoir quality. Upper Cretaceous (Maastrichtian) Springar Formation (deep marine, fan system), Norwegian Sea (well 6604/10-1)

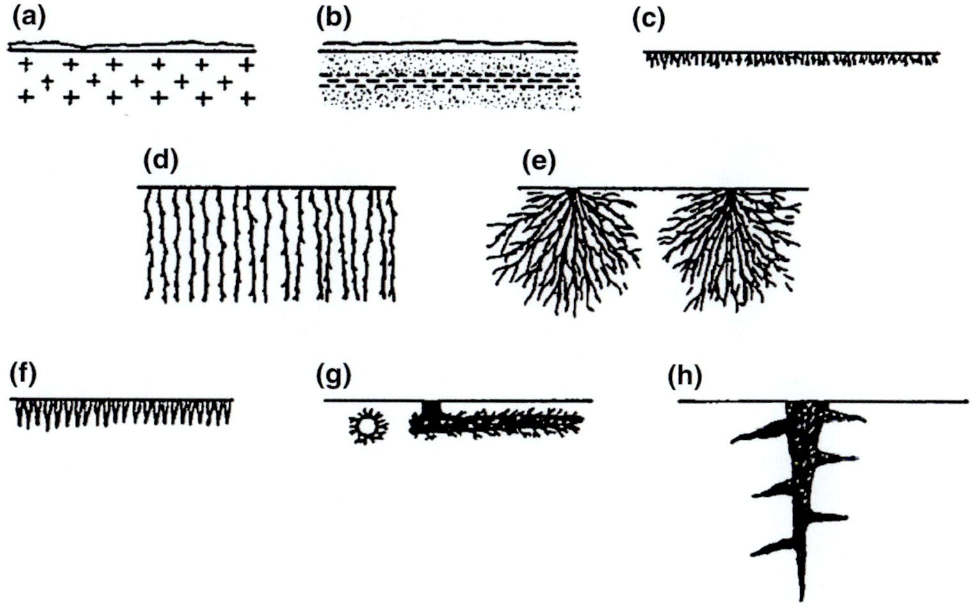


**Fig. 5.181** Types of rooting structures based on appearance. After Pfefferkorn and Fuchs (1991), republished with permission of Schweizerbart (www.schweizerbart.de/journals/njgpa). **a** Millimeter-thin crust on rocks. **b** Millimeter-thin crust on sediment. **c** Thin layer affected by rhizoids. **d** Fibrous roots system with more or less parallel or irregular roots. **e** Fibrous root system with radial root arrangement. **f** Vertical or oblique lateral roots coming from surface root, for instance buttress. **g** Shallow horizontal roots or rhizomes with lateral roots. **h** Tap root system

*Ethology:* Most bioeroders produce their borings for dwelling (domichnia) by mechanical or chemical bioerosion, or a combination of both.

*Depositional Environment:* The occurrence of borings and other bioerosion traces depends on the availability of suitable hardground to be colonized. Thus, preferred loci for enhanced bioerosion include reefs, bioherms and biostromes (Tapanila and Hutchings 2012), rocky shorelines (de Gibert et al. 2012; Fig. 5.187), and shallow-marine carbonate platforms (Knaust et al. 2012; Fig. 5.188), although local hardgrounds due to synsedimentary cementation (e.g. in delta-front or deep-marine environments) must be also considered. Bioeroded surfaces can be indicators of exhumation or synsedimentary lithification with low or no sedimentation over a longer period of time. Thus, borings are valuable indicators in the recognition of omission surfaces and unconformities (Fig. 2.6a). Borings in continental environments are known from living and dead wood, seeds and other plant material (e.g. Sutherland 2003; Feng et al. 2010; Genise et al. 2012).

*Ichnofacies:* Marine, bioeroded, rocky or skeletal hardgrounds are typically associated with the substrate-controlled *Trypanites* Ichnofacies (Frey and Seilacher 1980), although transitions to firmgrounds in the *Glossifungites* Ichnofacies (Seilacher 1967) may occur. Bromley and Asgaard (1993) proposed to replace or subdivide the *Trypanites* Ichnofacies with the *Entobia* Ichnofacies for deep-tier borings and the *Gnathichnus* Ichnofacies including superficial sculptures, a concept which still has been debated (MacEachern et al. 2012; de Gibert et al. 2012; Knaust et al. 2012). Borings in marine woodgrounds are characterized by the *Teredolites* Ichnofacies (Bromley et al. 1984). In continental settings, borings occur in a variety of unspecified ichnofacies.

*Age:* Borings are known from the Early Cambrian (James et al. 1977) to the Holocene (Tapanila and Hutchings 2012).

*Reservoir Quality:* Due to their tightness, hardgrounds commonly act as barriers and baffles to fluid flow and can contribute to layered reservoirs. Similar to structural elements such as fractures, borings have also the ability penetrating such tight horizons and to enhance the vertical connection for fluid flow. Passive fill with sediment grains or even open borings just contribute to the enhancement of this effect, which is comparable to the firmground burrows of the *Glossifungites* Ichnofacies (Gingras et al. 2012b).

**Fig. 5.182** Plant-root traces in outcrop. **a** Sandstone bedding plane with a large tabular root system with bifurcations. Eocene Aspelintoppen Formation (fluvial), Brongniartfjellet, Van Keulenfjorden, Svalbard. **b** Vertical section of coaly mudstone (marsh) overlying clean sandstone (shoreface) with deeply penetrating root traces. Middle Jurassic Scarborough Formation (marginal marine), coastal cliff near Scarborough, Yorkshire, UK. **c** Sandstone bedding plane (paleosol) with vertical root traces modified by a diagenetic halo. Same locality as in (**a**). *Scale bar* = 1 cm. **d** Dense occurrence of rhizomorphs in a Late Pleistocene regressive eolianite. Near Whale Point, North Eleuthera, Bahamas. From Knaust et al. (2012), republished with permission of Elsevier; permission conveyed through Copyright Clearance Center, Inc. **e** Large, cylindrical, vertical rhizolith in calichified (pedogenized) dolomitic preservation and subsequent cementation around the root structure with calcite (rhizoconcretion). Upper Triassic Kågeröd Formation (fluvial), Bornholm, Denmark. *Scale bar* = 1 cm. From Knaust (2015b). **f** Calichified root system within eolian sand. Pleistocene, coastal cliff at Cap Ghir, western Morocco. *Scale bar* = 5 cm

(a)
(b)
(c)
(d)
(e)
(f)

**◀Fig. 5.183** Plant-root traces in sectioned core. *Scale bars* = 1 cm. **a** Branched root system with coaly substrate originating from a thin coal seam and penetrating heterolithic sandstone with marine burrows. Lower Jurassic (Rhaetian to Pliensbachian) Åre Formation (tidal flat), Skuld Field, Norwegian Sea (well 6608/10-14S, ca. 2689.5 m). **b** Large fragments of root traces (partly sand-filled and partly coaly) penetrating ripple-laminated sandstone. Lower Jurassic (Hettangian to Sinemurian) Nansen Formation (marginal marine), Norwegian North Sea (well 25/10-11T2, ca. 4317.5 m). **c** Dense system of root traces of variable shape, size and fill. Upper Triassic (Rhaetian) Tubåen Formation (marginal marine), Skavl Discovery, Norwegian Barents Sea (well 7220/7-2S, ca. 1152 m). **d** Paleosol with large system of branched and sand-filled root system. Middle Jurassic (Bajocian-Bathonian) Hugin Formation (marginal marine), Norwegian North Sea (well 15/6-4, ca. 10631 ft). **e** Large vertical root trace penetrating inclined, ripple-laminated sandstone with marine bioturbation. Upper Triassic (Rhaetian) Tubåen Formation (marginal marine), Skavl Discovery, Norwegian Barents Sea (well 7220/7-2S, ca. 1152 m). **f** Individual vertical roots partly sand-filled (resembling *Skolithos*) and partly containing carbonaceous material. Upper Triassic (Rhaetian) Tubåen Formation (marginal marine), Skavl Discovery, Norwegian Barents Sea (well 7220/7-2S, ca. 1152 m)

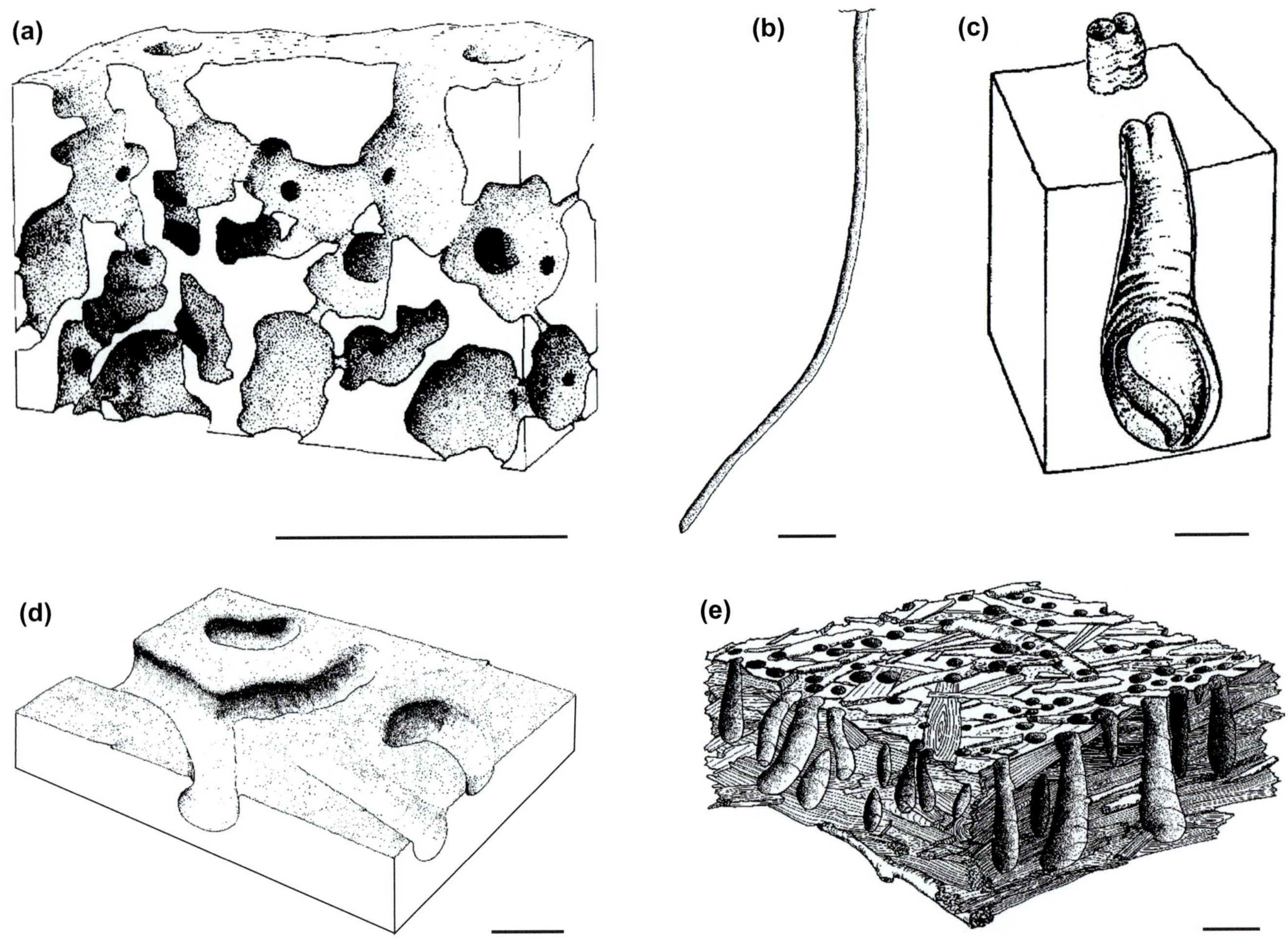


**Fig. 5.184** Line drawings of representative examples of borings, illustrating their morphology, size, substrate, producer, and ichnogenus. *Scale bars* = 1 cm except in (**d**) = 10 cm. Modified after Bromley (1978; **a**, **b** and **d**), republished with permission of Elsevier; permission conveyed through Copyright Clearance Center, Inc.; Kelly and Bromley (1984, **c**), republished with permission of Wiley; permission conveyed through Copyright Clearance Center, Inc.; and Bromley et al. (1984, **e**), © Paleontological Society, published by Cambridge University Press, reproduced with permission. **a** Boring with large rounded chambers and interconnections of the sponge *Cliona vermifera* in coral (ichnogenus *Entobia*). **b** Elongate, narrow boring of the polychaete *Hypsicomus elegans* in coral (ichnogenus *Trypanites*). **c** Clavate boring of the bivalve *Gastrochaena dubia* (with producer shown) in limestone (ichnogenus *Gastrochaenolites*). **d** Grooves bored by the echinoid *Echinometra lucunter* in limestone (ichnogenus *Ericichnus*). **e** Clavate borings of the bivalve *Martesia* sp. in woodground (ichnogenus *Teredolites*)

(a)
(b)
(c)
(d)
(e)
(f)

◀**Fig. 5.185** Borings and other bioerosion trace fossils in outcrop. *Scale bars* = 1 cm. **a** Thin hardground layer (Zhelty Bed, ochre) within an arenitic limestone (Dikari Limestone, pinkish-gray) in vertical section, with a complex *Balanoglossites triadicus* ichnofabric, including burrows and borings from different phases of colonization. Middle Ordovician (Dapingian) cool-water carbonates, St. Petersburg Region, NW Russia. From Knaust et al. (2012), republished with permission of Elsevier; permission conveyed through Copyright Clearance Center, Inc.; and Knaust and Dronov (2013), republished with permission of Springer. **b** Bedding plane of micritic limestone (dark blue) containing *Entobia* isp. (large rounded chambers) and *Trypanites weisei* (minute circular cross sections), both filled with dolomite (ochre). Cretaceous (Albian to Cenomanian) Natih Formation (carbonate platform), Misfah, Hajar Mountains, Oman. **c** Micritic hardground (lower part) in vertical section, with large *B. triadicus* and needle-like *T. weisei*. The top surface is eroded and overlain by arenitic limestone (upper part). Middle Triassic (Anisian) Jena Formation (carbonate ramp), Thuringia, Germany. **d** Same hardground as in (**c**) with deeply penetrating *T. weisei* and one specimen of *T. fosteryeomani* (next to the left margin). **e** Surface of a micritic limestone clast with openings of *Caulostrepsis* isp. (small) and *Gastrochaenolites* isp. (large). Eocene conglomerate with reworked Cretaceous limestone clasts, southeastern France. **f** Another clast with large *G. torpedo* of different orientation (longitudinal and cross sections) due to clast movement and repeated colonization. Same locality as in (**e**)

◀**Fig. 5.186** Borings in sectioned core. *Scale bars* = 1 cm. **a** and **b** republished with permission of Wiley and EAGE; permission conveyed through Copyright Clearance Center, Inc. **a** Reworked and bioeroded limestone clast within grainy substrate. Note the bioerosion on upper and lower surface of clast. Lower Triassic Khuff Formation (storm-reworked sand shoals and sand waves, inner to outer carbonate ramp), South Pars Field, Persian Gulf, Iran (well SP9, ca. 2918.3 m). From Knaust (2010b, 2014a). **b** Biolaminated dolomitic limestone (stromatolite) with bioeroded surface and *Gastrochaenolites* (*arrows*). Lower Triassic Khuff Formation (tidal flat, inner carbonate ramp), South Pars Field, Persian Gulf, Iran (well SP9, ca. 2909.8 m). From Knaust (2010b, 2014a). **c** Turbiditic sandstone with thin mudstone layers, which underwent synsedimentary lithification and bioerosion. Reworked clasts are concentrated in the turbidite bed above the mudstone layers, one clast (*arrow*) shows minute borings (*Entobia*) all around and a larger shaft. Lower Cretaceous (Albian, deep marine, channel-overbank), off Tanzania. **d** Omission surface (*arrow heads*) with reworked phosphatic (brown) pebbles and bioerosion (e.g. *Gastrochaenolites*, *arrow*). Upper Jurassic (Tithonian) Draupne Formation (shallow marine), Norwegian North Sea (well 25/10-12ST2, ca. 2123 m). **e** Reworked sideritic mudclast having an intensely bioeroded margin within sandstone matrix. Middle Jurassic (Callovian) Fensfjord Formation (delta front), Gjøa Field, Norwegian North Sea (well 36/7-1, ca. 2373.5 m)

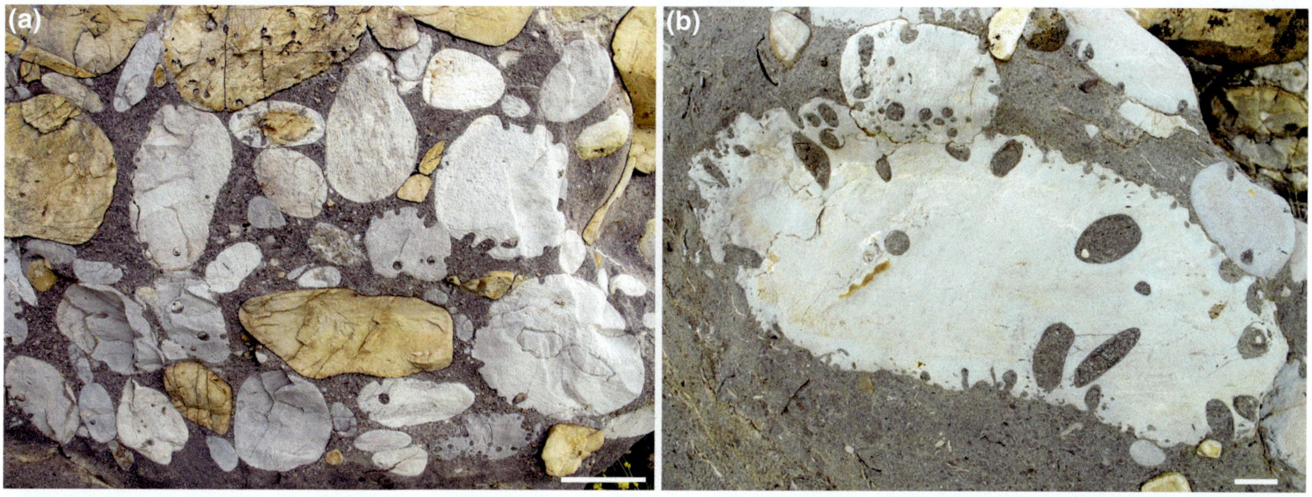


**Fig. 5.187** Rocky shoreline deposits. Eocene, southeastern France. *Scale bar* = 10 cm (**a**) and 1 cm (**b**). **a** The well rounded limestone clasts of Cretaceous age are fairly bioeroded and embedded in medium- to coarse-grained sandstone of shallow marine origin. **b** Close-up view of clasts with shallow borings (e.g. *Caulostrepsis*) and deeper penetrations (e.g. *Gastrochaenolites*), both distinct with their contrasting passive sand fill

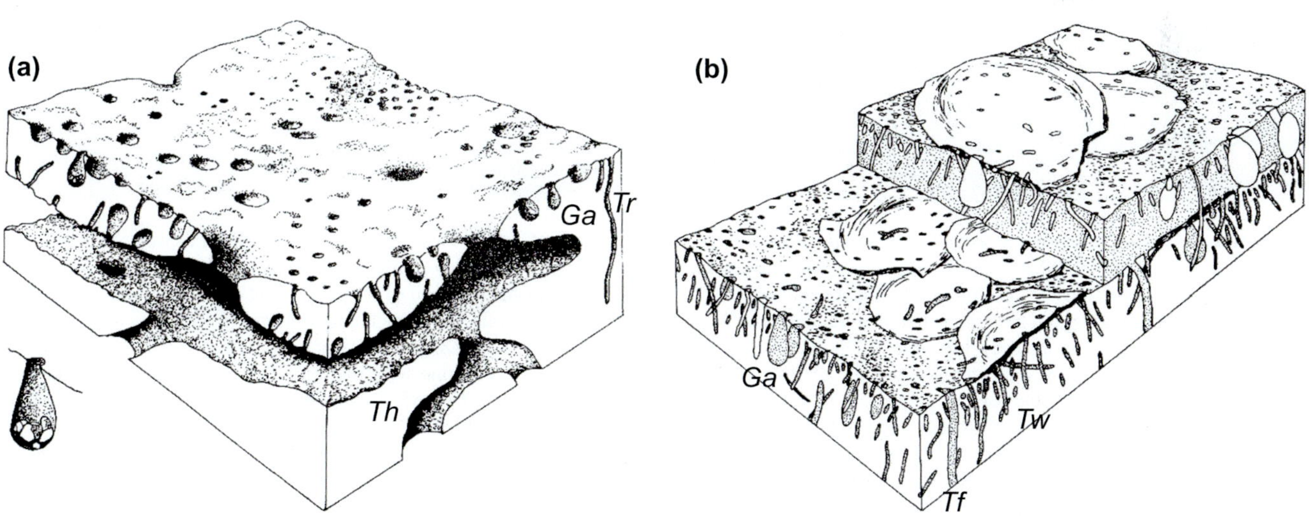


**Fig. 5.188** Omission suite assemblages from two hardgrounds. From Bromley (1975), republished with permission of Springer. **a** Cretaceous chalk in a relatively shallow-marine setting. Pre-lithification *Thalassinoides (Th)* is prevalent, accompanied with post-lithification *Gastrochaenolites (Ga)* and *Trypanites (Tr)*. Borings are partly trapped with intraclasts and debris, which are otherwise unpreserved (see enlarged *Gastrochaenolites*). Upper Cretaceous (Campanian-Maastrichtian transition), Belgium. Height of block ca. 8 cm. **b** Two hardground surfaces in limestone are plastered with oysters and contain three types of borings, *Gastrochaenolites* isp. (*Ga*), *Trypanites weisei (Tw)* and *Trypanites fosteryeomani (Tf)*. Carboniferous/Middle Jurassic unconformity, England, UK. Height of block ca. 4 cm

**Fig. 5.189** Examples of common pseudo-trace fossils as encountered in sectioned core. *Scale bars* = 1 cm. **a** Sandstone with dewatering structures (pipes) superficially resembles the ichnogenus *Skolithos*. Upper Jurassic (Kimmeridgian-Tithonian) Kimmeridgian Clay Formation (deep marine, fan system), Kingfisher Field, UK North Sea (well 16/8a-4, ca. 13515.6 ft). **b** Sandstone-mudstone alternation with synaeresis cracks, which could be mistaken as burrows. Middle Jurassic (Bajocian-Bathonian) Tarbert Formation (marginal marine, tidal flat), Valemon Field, Norwegian North Sea (well 34/11-B-13, ca. 4619.4 m). **c** Small-scale slump fold (recumbent fold) within a mudstone-sandstone alternation, accompanied by other deformation structures, somehow resembling a meniscate burrow such as *Taenidium*. Eocene (deep marine, channel-levee system), off Tanzania. **d** Small sandstone dyke (injectite) within a thin mudstone layer. Upper Jurassic (Tithonian) Draupne Formation (deep marine, channel-levee complex), Gudrun Field, Norwegian North Sea (well 15/3-9T2, ca. 4131.55 m). **e** Muddy debris-flow deposit (debrite) with floating and poorly consolidated sand clasts mimicking a partly bioturbate texture. Upper Cretaceous (Maastrichtian) Springar Formation (deep marine, fan system), Gro Discovery, Norwegian Sea (well 6603/12-1, ca. 3724.5 m). **f** Sandstone with microfaults resembling vertical burrows such as *Skolithos*. Upper Triassic (Rhaetian) Statfjord Group (fluvial), Johan Sverdrup Field, Norwegian North Sea (well 16/2-11, ca. 1940.5 m). **g** Branched tubular structure filled with coarse sand and tentatively interpreted to be originated from methane or hydrocarbon seepage (cold seep) in shallow subsurface position. Such structures, if regarded in isolation, are easily mistaken as large burrows such as shafts of *Thalassinoides* systems. Upper Jurassic (Tithonian) Rogn Formation (lower shoreface), Norwegian Sea, well 6406/12-1S, ca. 3631.9 m

## 5.40 Pseudo-trace Fossils

Distinction of trace fossils from physical, diagenetic and tectonic sedimentary structures is not always straightforward and in many cases requires careful analysis before judgment (Fig. 5.189). Such structures can originate synsedimentarily during sedimentation, in an early stage of the substrate's diagenesis, either during lithification or in a late stage after burial and under tectonic influence (Boyd 1975; Häntzschel 1975; Seilacher 2007).

## References


Abad M, Ruiz F, Pendón JG et al (2006) Escape and equilibrium trace fossils in association with *Conichnus conicus* as indicators of variable sedimentation rates in Tortonian littoral environments of SW Spain. Geobios 39:1–11 [In Spanish]

Aceñolaza GF, Alonso RN (2001) Ichno-associations of the Precambrian/Cambrian transition in the north-west Argentina. J Iberian Geol 27: 11–22 [In Spanish]

Ager DV, Wallace P (1970) The distribution and significance of trace fossils in the uppermost Jurassic rocks of the Boulonnais, northern France. In: Crimes TP, Harper JC (eds) Trace fossils. Geol J 3 (Special Issue):1–18

Ahn SY, Babcock LE (2012) Microorganism-mediated preservation of *Planolites*, a common trace fossil from the Harkless Formation, Cambrian of Nevada, USA. Sed Geol 263–264:30–35

Alpert SP (1973) *Bergaueria* Prantl (Cambrian and Ordovician), a probable actinian trace fossil. J Paleontol 47:919–924

Alpert SP (1974) Systematic review of the genus *Skolithos*. J Paleontol 48:661–669

Alpert SP (1975) *Planolites* and *Skolithos* from the Upper Precambrian-Lower Cambrian, White-Inyo Mountains, California. J Paleontol 49:508–521

Alpert SP (1977) Trace fossils and the basal Cambrian boundary. In: Crimes TP, Harper JC (eds) Trace fossils 2. Geol J 9 (Special Issue):1–8

Anderson BG, Droser ML (1998) Ichnofabrics and geometric configurations of *Ophiomorpha* within a sequence stratigraphic framework: an example from the Upper Cretaceous US Western Interior. Sedimentology 45:379–396

Angulo S, Buatois LA (2012) Ichnology of a Late Devonian-Early Carboniferous low-energy seaway: the Bakken Formation of subsurface Saskatchewan, Canada: assessing paleoenvironmental controls and biotic responses. Palaeogeogr Palaeoclimatol Palaeoecol 315–316:46–60

Ayranci K, Dashtgard SE (2013) Infaunal holothurian distributions and their traces in the Fraser River delta front and prodelta, British Columbia, Canada. Palaeogeogr Palaeoclimatol Palaeoecol 392:232–246

Ayranci K, Dashtgard SE, MacEachern JA (2014) A quantitative assessment of the neoichnology and biology of a delta front and prodelta, and implications for delta ichnology. Palaeogeogr Palaeoclimatol Palaeoecol 409:114–134

Bann KL, Fielding CR (2004) An integrated ichnological and sedimentological comparison of non-deltaic shoreface and subaqueous delta deposits in Permian reservoir units of Australia. In: McIlroy D (ed) The application of ichnology to palaeoenvironmental and stratigraphic analysis, vol 228. Geological Society of London (Special Publications), pp 273–310

Bann KL, Fielding CR, MacEachern JA et al (2004) Differentiation of estuarine and offshore marine deposits using integrated ichnology and sedimentology: Permian Pebbley Beach Formation, Sydney Basin, Australia. In: McIlroy D (ed) The application of ichnology to palaeoenvironmental and stratigraphic analysis, vol 228. Geological Society of London (Special Publications), pp 179–211

Basan PB, Scott RW (1979) Morphology of *Rhizocorallium* and associated traces from the lower Cretaceous Purgatoire Formation, Colorado. Palaeogeogr Palaeoclimatol Palaeoecol 28:5–23

Baucon A, Felletti F (2013) Neoichnology of a barrier-island system: the Mula di Muggia (Grado lagoon, Italy). Palaeogeogr Palaeoclimatol Palaeoecol 375:112–124

Baucon A, Ronchi A, Felletti F et al (2014) Evolution of crustaceans at the edge of the end-Permian crisis: ichnonetwork analysis of the fluvial succession of Nurra (Permian–Triassic, Sardinia, Italy). Palaeogeogr Palaeoclimatol Palaeoecol 410:74–103

Bedatou E, Melchor RN, Bellosi E et al (2008) Crayfish burrows from Late Jurassic-Late Cretaceous continental deposits of Patagonia: Argentina. Their palaeoecological, palaeoclimatic and palaeobiogeographical significance. Palaeogeogr Palaeoclimatol Palaeoecol 257:169–184

Bedatou E, Melchor RN, Genise JF (2009) Complex palaeosol ichnofabrics from Late Jurassic-Early Cretaceous volcaniclastic successions of central Patagonia, Argentina. Sed Geol 218:74–102

Bednarz M (2014) 3D ichnofabrics in shale gas reservoirs. PhD Thesis, University of St. John's, 230 pp. http://research.library.mun.ca/8190/1/thesis.pdf

Bednarz M, McIlroy D (2009) Three-dimensional reconstruction of "phycosiphoniform" burrows: implications for identification of trace fossils in core. Palaeontol Electr 12(13A):15

Bednarz M, McIlroy D (2012) Effect of phycosiphoniform burrows on shale hydrocarbon reservoir quality. AAPG Bull 96:1957–1980

Bednarz M, McIlroy D (2015) Organism–sediment interactions in shale-hydrocarbon reservoir facies—three-dimensional reconstruction of complex ichnofabric geometries and pore-networks. Int J Coal Geol 150–151:238–251

Belaústegui Z, de Gibert JM (2013) Bow-shaped, concentrically laminated polychaete burrows: a *Cylindrichnus concentricus* ichnofabric from the Miocene of Tarragona, NE Spain. Palaeogeogr Palaeoclimatol Palaeoecol 381–382:119–127

Belaústegui Z, de Gibert JM, Domènech R et al (2011) Taphonomy and palaeoenvironmental setting of cetacean remains from the Middle Miocene of Tarragona (NE Spain). Geobios 44:19–31 [In Spanish, with English summary]

Belaústegui Z, Ekdale AA, Domènech R et al (2016a) Paleobiology of firmground burrowers and cryptobionts at a Miocene omission surface, Alcoi, SE Spain. J Paleontol 90:721–733

Belaústegui Z, Muñiz F, Mángano MG et al (2016b) *Lepeichnus giberti* igen. nov. isp. nov. from the upper Miocene of Lepe (Huelva, SW Spain): evidence for its origin and development with proposal of a new concept, ichnogeny. Palaeogeogr Palaeoclimatol Palaeoecol 452:80–89

Benton MJ, Gray DI (1981) Lower Silurian distal shelf storm-induced turbidites in the Welsh borders: sediments, tool marks and trace fossils. J Geol Soc Lond 138:675–694

Bertling M, Braddy SJ, Bromley RG et al (2006) Names for trace fossils: a uniform approach. Lethaia 39:265–286

Bischoff B (1968) *Zoophycos*, a polychæte annelid, Eocene of Greece. J Paleontol 42:1439–1443

Bjerstedt TW (1988) Trace fossils from the early Mississippian Price Delta, southeast West Virginia. J Paleontol 62:506–519

Bockelie JF (1994) Plant roots in core. In: Donovan SK (ed) The palaeobiology of trace fossils. Wiley, Chichester, pp 177–199

Bottjer DJ, Droser ML, Jablonski D (1988) Palaeoenvironmental trends in the history of trace fossils. Nature 333:252–255

Bourgeois J (1980) A transgressive shelf sequence exhibiting hummocky stratification: the Cape Sebastian Sandstone (Upper Cretaceous), southwestern Oregon. J Sediment Petrol 50:681–702

Boyd DW (1975) False or misleading traces. In: Frey RW (ed) The study of trace fossils: a synthesis of principles, problems and procedures in ichnology. Springer, New York, pp 65–83

Bradley TL, Pemberton SG (1992) Examples of ichnofossil assemblages in the lower Cretaceous Wabiskaw Member and the Clearwater Formation of the Marten Hills gas field, north-central Alberta, Canada. In: Pemberton SG (ed) Applications of ichnology to petroleum exploration. A core workshop. SEPM Core Workshop, vol 17, pp 383–399

Bradshaw MA (2002) A new ichnogenus *Catenarichnus* from the Devonian of the Ohio Range, Antarctica. Antarct Sci 14:422–424

Bradshaw MA (2010) Devonian trace fossils of the Horlick Formation, Ohio Range, Antarctica: systematic description and palaeoenvironmental interpretation. Ichnos 17:58–114

Breton G (2006) Paramoudras … and other concretions around a burrow. Bulletin d'Information Géologues du Bassin Paris 43: 18–43

Bromley RG (1967) Some observations on burrows of thalassinidean Crustacea in chalk hardgrounds. Q J Geol Soc 123:157–177

Bromley RG (1978) Bioerosion of Bermuda reefs. Palaeogeogr Palaeoclimatol Palaeoecol 23:169–197

Bromley RG (1991) *Zoophycos*: strip mine, refuse dump, cache or sewage farm? Lethaia 24:460–462

Bromley RG (1996) Trace fossils: biology, taphonomy and applications. Chapman and Hall, London, 361 pp

Bromley RG (2004) A stratigraphy of marine bioerosion. In: McIlroy D (ed) The application of ichnology to palaeoenvironmental and stratigraphic analysis, vol 228. Geol Soc Lond (Special Publications), pp 455–479

Bromley RG, Asgaard U (1972) Notes on Greenland trace fossils, 1. Freshwater *Cruziana* from the Upper Triassic of Jameson Land, East Greenland. Grønlands Geologiske Undersøgelse, Rapport 49:7–13

Bromley RG, Asgaard U (1975) Sediment structures produced by a spatangoid echinoid: a problem of preservation. Bull Geol Soc Denm 24:261–281

Bromley RG, Asgaard U (1979) Triassic freshwater ichnocoenoses from Carlsberg Fjord, East Greenland. Palaeogeogr Palaeoclimatol Palaeoecol 28:39–80

Bromley RG, Asgaard U (1991) Ichnofacies: a mixture of taphofacies and biofacies. Lethaia 24:153–163

Bromley RG, Asgaard U (1993) Two bioerosion ichnofacies produced by early and late burial associated with sea-level change. Geol Rundsch 82:276–280

Bromley RG, Ekdale AA (1984) *Chondrites*: a trace fossil indicator of anoxia in sediments. Science 224:872–874

Bromley RG, Frey RW (1974) Redescription of the trace fossil *Gyrolithes* and taxonomic evaluation of *Thalassinoides*, *Ophiomorpha* and *Spongeliomorpha*. Bull Geol Soc Denm 23:311–335

Bromley RG, Hanken N-M (1991) The growth vector in trace fossils: examples from the Lower Cambrian of Norway. Ichnos 1:261–276

Bromley RG, Hanken N-M (2003) Structure and function of large, lobed *Zoophycos*, Pliocene of Rhodes, Greece. Palaeogeogr Palaeoclimatol Palaeoecol 192:79–100

Bromley RG, Mørk A (2000) The trace fossil *Phoebichnus trochoides* in the condensed Triassic-Jurassic-boundary strata of Svalbard. In: Bachmann GH, Lerche I (eds) Epicontinental Triassic, vol 2. Zentralblatt für Geologie und Paläontologie, Teil 1, 1998, pp 1431–1439

Bromley RG, Uchman A (2003) Trace fossils from the Lower and Middle Jurassic marginal marine deposits of the Sorthat Formation, Bornholm, Denmark. Bull Geol Soc Denm 50:185–208

Bromley RG, Schulz M-G, Peake NB (1975) Paramoudras: giant flints, long burrows and the early diagenesis of chalks. Det Kongelige Danske Videnskabers Selskab, Biologiske Skrifter 20, 31 pp

Bromley RG, Pemberton SG, Rahmani RA (1984) A Cretaceous woodground: the *Teredolites* ichnofacies. J Paleontol 58:488–498

Bromley RG, Ekdale AA, Richter B (1999) New *Taenidium* (trace fossil) in the Upper Cretaceous chalk of northwestern Europe. Bull Geol Soc Denm 46:47–51

Bromley RG, Uchman A, Gregory MR et al (2003) *Hillichnus lobosensis* igen. et isp. nov., a complex trace fossil produced by tellinacean bivalves, Paleocene, Monterey, California, USA. Palaeogeogr Palaeoclimatol Palaeoecol 192:157–186

Bromley RG, Uchman A, Milàn J et al (2009) Rheotactic *Macaronichnus*, and human and cattle trackways in Holocene beachrock, Greece: reconstruction of paleoshoreline orientation. Ichnos 16: 103–117

Buatois LA, Mángano MG (2004) Terminal Proterozoic-Early Cambrian ecosystems: ichnology of the Puncoviscana Formation, northwest Argentina. Fossils Strata 51:1–16

Buatois LA, Mángano MG (2011) Ichnology. Organism-substrate interactions in space and time. Cambridge University Press, Cambridge, pp 347

Buatois LA, Mángano MG, Alissa A et al. (2002) Sequence stratigraphic and sedimentologic significance of biogenic structures from a late Paleozoic marginal- to open-marine reservoir, Morrow Sandstone, subsurface of southwest Kansas, USA. Sedimentary Geology 152:99–132

Buatois LA, Gingras MK, MacEachern J et al (2005) Colonization of brackish-water systems through time: evidence from the trace-fossil record. Palaios 20:321–347

Buatois LA, Saccavino LL, Zavala C (2011) Ichnologic signatures of hyperpycnal flow deposits in Cretaceous river-dominated deltas, Austral Basin, southern Argentina. In: Slatt RM, Zavala C (eds) Sediment transfer from shelf to deep water—revisiting the delivery system. AAPG Stud Geol 61:153–170

Buck SG, Goldring R (2003) Conical sedimentary structures, trace fossils or not? Observations, experiments, and review. J Sediment Res 73:338–353

Buckman JO (1992) Palaeoenvironment of a Lower Carboniferous sandstone succession northwest Ireland: ichnological and sedimentological studies. In: Parnell J (ed) Basins on the Atlantic seaboard: petroleum sedimentology and basin evolution, vol 62. Geological Society of London (Special Publications), pp 217–241

Buckman JO (1996) An example of 'deep' tier level *Teichichnus* with vertical entrance shafts, from the Carboniferous of Ireland. Ichnos 4:241–248

Callow RHT, McIlroy D (2011) Ichnofabrics and ichnofabric-forming trace fossils in Phanerozoic turbidites. Bull Can Pet Geol 59: 103–111

Callow RHT, McIlroy D, Kneller B et al (2013) Integrated ichnological and sedimentological analysis of a Late Cretaceous submarine channel-levee system: the Rosario Formation, Baja California, Mexico. Mar Pet Geol 41:277–294

Campbell KA, Nesbitt EA, Bourgeois J (2006) Signatures of storms, oceanic floods and forearc tectonism in marine shelf strata of the Quinault Formation (Pliocene), Washington. Sedimentology 53: 945–969

Campbell SG, Botterill SE, Gingras MK et al (2016) Event sedimentation, deposition rate, and paleoenvironment using crowded assemblages of the Bluesky Formation, Alberta, Canada. J Sediment Res 86:380–393

Carmona NB, Buatois LA, Mángano MG (2004) The trace fossil record of burrowing decapod crustaceans: evaluating evolutionary radiations and behavioural convergence. Fossils Strata 51:141–153

Carmona NB, Buatois LA, Mángano MG et al (2008) Ichnology of the Lower Miocene Chenque Formation, Patagonia, Argentina: animal—substrate interactions and the modern evolutionary fauna. Ameghiniana 45:93–122

Carmona NB, Buatois LA, Ponce JJ et al (2009) Ichnology and sedimentology of a tide-influenced delta, lower Miocene Chenque Formation, Patagonia, Argentina: trace-fossil distribution and response to environmental stresses. Palaeogeogr Palaeoclimatol Palaeoecol 273:75–86

Chakraborty A, Bhattacharya HN (2013) Spreiten burrows: a model based study on *Diplocraterion parallelum*. In: Mukhopadhyay S, Ray D, Kundu A (eds) Geospectrum, pp 296–299

Chamberlain CK (1971) Morphology and ethology of trace fossils from the Ouachita Mountains, southeast Oklahoma. J Paleontol 45: 212–246

Chamberlain CK (2000) Prologue to the study of *Zoophycos*. Ichnol Newslett 22:13–23

Chamberlain CK, Baer JL (1973) *Ophiomorpha* and a new thalassinid burrow from the Permian of Utah. Geol Stud 20:79–94

Chen Y, Wang J, Bai P et al (2005) Trace fossils of the Middle Devonian Mazongling Formation at Wudang Guiyang Guizhou Province. Guizhou Geol 22:273–279

Chen Z-Q, Tong J, Fraiser ML (2011) Trace fossil evidence for restoration of marine ecosystems following the end-Permian mass extinction in the Lower Yangtze region, South China. Palaeogeogr Palaeoclimatol Palaeoecol 299:449–474

Cherns L, Wheeley JR, Karis L (2006) Tunneling trilobites: Habitual infaunalism in an Ordovician carbonate seafloor. Geology 34: 657–660

Chlupáč I (1987) Ordovician ichnofossils in the metamorphic mantle of the Central Bohemian Pluton. Časopis pro Meineralogii a Geologii 32:249–260

Clausen CK, Vilhjálmsson M (1986) Substrate control of Lower Cambrian trace fossils from Bornholm, Denmark. Palaeogeogr Palaeoclimatol Palaeoecol 56:51–68

Clemmensen LB, Bromley RG, Holm PM (2011) Glauconitic deposits at Julegård on the south coast of Bornholm, Denmark dated to the Cambrian. Bull Geol Soc Denm 59:1–12

Clifton HE, Thompson JK (1978) *Macaronichnus segregatis*: a feeding structure of shallow marine polychaetes. J Sediment Petrol 48: 1293–1302

Corner GD, Fjalstad A (1993) Spreite trace fossils (*Teichichnus*) in a raised Holocene fjord-delta, Breidvikeidet, Norway. Ichnos 2: 155–164

Cornish FG (1986) The trace-fossil *Diplocraterion*: evidence of animal-sediment interactions in Cambrian tidal deposits. Palaios 1: 478–491

Cotillon P (2010) Sea bottom current activity recorded on the southern margin of the Vocontian Basin (southeastern France) during the Lower Aptian. Evidence for a climate signal. Bulletin de la Société Géologique de France 181:3–18

Crimes TP, Legg I, Marcos A et al (1977)? Late Precambrian-Lower Cambrian trace fossils from Spain. In: Crimes TP, Harper JC (eds) Trace fossils 2. Geol J 9 (Special Issue):91–138

Crimes TP, Goldring R, Homewood P et al (1981) Trace fossil assemblages of deep-sea fan deposits, Gurnigel and Schlieren flysch (Cretaceous-Eocene), Switzerland. Eclogae Geol Helv 74:953–995

Cummings JP, Hodgson DM (2011) Assessing controls on the distribution of ichnotaxa in submarine fan environments, the Basque Basin, northern Spain. Sed Geol 239:162–187

Cunningham KJ, Sukop MC (2012) Megaporosity and permeability of *Thalassionoides*-dominated ichnofabrics in the Cretaceous karst-carbonate Edwards-Trinity Aquifer System, Texas. U.S. Geological survey, Open-file report 2012–1021, 4 pp

Cunningham KJ, Sukop MC, Huang H et al (2009) Prominence of ichnologically influenced macroporosity in the karst Biscayne aquifer: Stratiform "super-K" zones. GSA Bull 121:164–186

Cunningham KJ, Sukop MC, Curran HA (2012) Carbonate aquifers. In: Knaust D, Bromley RG (eds), Trace fossils as indicators of sedimentary environments. Developments in Sedimentology, vol 164, pp 869–896

Curran HA (1976) A trace fossil brood structure of probable callianassid origin. J Paleontol 50:249–259

Curran HA (2015) Sinuous rhizoliths mimic invertebrate trace fossils on Upper Pleistocene caliche surfaces, San Salvador Island, Bahamas. In: McIlroy D (ed) Ichnology: Papers from ICHNIA III, vol 9. Geological Association of Canada, Miscellaneous Publication, pp 63–72

Curran HA, Frey RW (1977) Pleistocene trace fossils from North Carolina (U.S.A.), and their Holocene analogues. In: Crimes TP, Harper JC (eds) Trace fossils 2. Geol J 9 (Special Issue):139–162

D'Alessandro A, Iannone A (1982) Pleistocene carbonate deposits in the area of Monopoli (Bari Province): sedimentology and palaeoecology. Geol Romana 21:603–653

da Silva ID, Jensen S, González-Clavijo E (2014) Trace fossils from the Desejosa Formation (Schist and Greywacke Complex, Douro Group, NE Portugal): new Cambrian age constraints. Geol Acta 12:109–120

Dafoe LT, Gingras MK, Pemberton SG (2008a) Determinating *Euzonus mucronata* burrowing rates with application to ancient *Macaronichnus segregatis* tracemakers. Ichnos 15:78–90

Dafoe LT, Gingras MK, Pemberton SG (2008b) Analysis of mineral segregation in *Euzonus mucronata* burrow structures: one possible method in the construction of ancient *Macaronichnus segregates*. Ichnos 15:91–102

Dafoe LT, Gingras MK, Pemberton SG (2010) Wave-influenced deltaic bodies and offshore deposits in the Viking Formation, Hamilton Lake area, south-central Alberta, Canada. Bull Can Pet Geol 58: 173–201

Dahmer G (1937) Lebensspuren aus dem Taunusquarzit und aus den Siegener Schichten (Unterdevon). Jahrbuch der Preußisch-Geologischen Landesanstalt 57:523–539

D'Alessandro A, Bromley RG (1986) Trace fossils in Pleistocene sandy deposits from Gravina area, southern Italy. Rivista Italiana di Paleontologia e Stratigrafia 92:67–102

D'Alessandro A, Bromley RG (1987) Meniscate trace fossils and the *Muensteria-Taenidium* problem. Palaeontology 30:743–763

D'Alessandro A, Fürsich FT (2005) *Tursia*—a new ichnogenus from Pleistocene shallow water settings in southern Italy. Ichnos 12: 65–73

D'Alessandro A, Bromley RG, Stemmerik L (1987) *Rutichnus*: a new ichnogenus for branched, walled, meniscate trace fossils. J Paleontol 61:1112–1119

Dam G (1990) Taxonomy of trace fossils from the shallow marine Lower Jurassic Neill Klinter Formation, East Greenland. Bull Geol Soc Denm 38:119–144

Dando PR, Southward AJ (1986) Chemoautotrophy in bivalve molluscs of the genus *Thyasira*. J Mar Biol Assoc UK 66:915–929

Dashtgard SE, Gingras MK (2012) Marine invertebrate neoichnology. In: Knaust D, Bromley RG (eds) Trace fossils as indicators of Sedimentary environments. Developments in Sedimentology, vol 64, pp 273–295

Dashtgard SE, Gingras MK, Pemberton SG (2008) Grain-size controls on the occurrence of bioturbation. Palaeogeogr Palaeoclimatol Palaeoecol 257:224–243

Dawson WC (1981) Secondary burrow porosity in quartzose biocalcarenites, Upper Cretaceous, Texas: U.S.A. VIII Congreso

Geológico Argentino, San Luis (20–26 Setiembre, 1981), Actas II, pp 637–649

de Gibert JM, Martinell J (1998) Ichnofabric analysis of the Pliocene marine sediments of the Var Basin (Nice, SE France). Geobios 31: 271–281

de Gibert JM, Netto RG, Tognoli FMW et al (2006) Commensal worm traces and possible juvenile thalassinidean burrows associated with *Ophiomorpha nodosa*, Pleistocene, southern Brazil. Palaeogeogr Palaeoclimatol Palaeoecol 230:70–84

de Gibert JM, Domènech R, Martinell J (2012) Rocky shorelines. In: Knaust D, Bromley RG (eds) Trace fossils as indicators of sedimentary environments. Developments in Sedimentology, vol 64, pp 441–462

de Quatrefages MA (1849) Note sur la *Scolicia prisca* (A. de Q.) annélide fossile de la Craie. Annales des Sciences Naturalles, 3 série. Zoologie 12:265–266

de Saporta G (1873) Plantes Jurassiques, tome I, Algues, Equisétaceés, Characées, Fougères. Paléontologie Française ou Description des Fossiles de la France, Serie 2, Végétaux. Masson, Paris

de Serres M (1840) Description de quelques mollusques fossiles nouveaux des terrains infra-liassiques et de la craie compacte inférieure du Midi de la France, Annales des sciences naturelles, Zoologie 14:5–26 (pl 1)

Demírcan H, Uchman A (2012) The miniature echinoid trace fossil *Bichordites kuzunensis* isp. nov. from early Oligocene prodelta sediments of the Mezardere Formation, Gökçeada Island, NW Turkey. Acta Geol Polonica 62:205–215

Desai BG, Shukla R, Saklani RD (2010) Ichnology of the Early Cambrian Tal Group, Nigalidhar Syncline, Lesser Himalaya, India. Ichnos 17:233–245

Desjardins PR, Mángano MG, Buatois LA et al (2010) *Skolithos* pipe rock and associated ichnofabrics from the southern Rocky Mountains, Canada: colonization trends and environmental controls in an early Cambrian sand-sheet complex. Lethaia 43:507–528

Desjardins PR, Buatois LA, Mángano MG (2012) Tidal flats and subtidal sand bodies. In: Knaust D, Bromley RG (eds) Trace fossils as indicators of sedimentary environments. Developments in Sedimentology, vol 64, pp 529–561

Dronov A, Mikuláš R (2010) Paleozoic ichnology of St. Petersburg region. In: Excursion guidebook. 4th Workshop on Ichnotaxonomy, Moscow, St. Petersburg, vol 596. Transactions of the Geological Institute, pp 1–70

Dronov A, Tolmacheva T, Raevskaya E et al (2005) Cambrian and Ordovician of St. Petersburg region. In: 6th Baltic Stratigraphical Conference, IGCP 503 Meeting, Guidebook of the pre-conference field trip, 64 pp

Droser ML (1991) Ichnofabric of the Paleozoic *Skolithos* Ichnofacies and the nature and distribution of *Skolithos* piperock. Palaios 6: 316–325

Dufour SC, Feldbeck H (2003) Sulphide mining by the superextensile foot of symbiotic thyasirid bivalves. Nature 426:65–67

Ehrenberg K (1944) Ergänzende Bemerkungen zu den seinerzeit aus dem Miozän von Burgschleinitz beschriebenen Gangkernen und Bauten dekapoder Krebse. Paläontologische Zeitschrift 23:354–359

Ekdale AA, Bromley RG (1991) Analysis of composite ichnofabrics: an example in the uppermost Cretaceous chalk of Denmark. Palaios 6:232–249

Ekdale AA, Bromley RG (2003) Paleoethologic interpretation of complex *Thalassinoides* in shallow-marine limestones, Lower Ordovician, southern Sweden. Palaeogeogr Palaeoclimatol Palaeoecol 192:221–227

Ekdale AA, Harding SC (2015) *Cylindrichnus concentricus* Toots in Howard, 1966 (trace fossil) in its type locality, Upper Cretaceous, Wyoming. Ann Soc Geol Pol 85:427–432

Ekdale AA, Lewis DW (1991) The New Zealand *Zoophycos* revisited: morphology, ethology and paleoecology. Ichnos 1:183–194

Ekdale AA, Bromley RG, Knaust D (2012) The ichnofabric concept. In: Knaust D, Bromley RG (eds) Trace fossils as indicators of sedimentary environments. Developments in Sedimentology, vol 64. Elsevier, Amsterdam, pp 139–155

Ekdale AA, Bromley RG, Pemberton SG (1984) Ichnology: the use of trace fossils in sedimentology and stratigraphy. SEPM Short Course Notes 15:1–317

Emig CC (1982) The biology of Phoronida. Adv Mar Biol 19:1–89

Emig CC, Gall J-C, Pajaud D et al (1978) Réfexions critiques sur l'écologie et la systématique des lingules actuelles et fossils. Geobios 11:573–609

Evans JN, McIlroy D (2015) Ichnology and palaeobiology of *Phoebichnus trochoides* from the Middle Jurassic of north-east England. Papers Palaeontol 2015:1–16

Evans JN, McIlroy D (2016) Palaeobiology of *Schaubcylindrichnus heberti* comb. nov. from the Lower Jurassic of Northeast England. Palaeogeogr Palaeoclimatol Palaeoecol 449:246–254

Farrow GE (1966) Bathymetric zonation of Jurassic trace fossils from the coast of Yorkshire, England. Palaeogeogr Palaeoclimatol Palaeoecol 2:103–151

Fedonkin MA (1981) White Sea biota of the Vendian (Precambrian non-skeletal fauna of the Russian Platform north). Transactions of the Geological Institute, vol 342. Nauka, Moscow, pp 1–100 [in Russian]

Feng Z, Wang J, Liu L-J (2010) First report of oribatid mite (arthropod) borings and coprolites in Permian woods from the Helan Mountains of northern China. Palaeogeogr Palaeoclimatol Palaeoecol 288:54–61

Fenton MA, Fenton CL (1934) *Scolithus* as a fossil phoronid. Pan-American Geologist 61:341–348, 1 pl

Fillion D, Pickerill RK (1990) Ichnology of the Upper Cambrian? to Lower Ordovician Bell Island and Wabana groups of eastern Newfoundland, Canada. Palaeontogr Canadiana 7:1–119

Fiorillo AR, McCarthy PL, Hasiotis ST (2016) Crayfish burrows from the latest Cretaceous lower Cantwell Formation (Denali National Park, Alaska): their morphology and paleoclimatic significance. Palaeogeogr Palaeoclimatol Palaeoecol 441:352–359

Fischer-Ooster C (1858) Die fossilen Fucoiden der Schweizer Alpen, nebst Erörterungen über deren geologisches Alter. Huber, Bern

Forbes AT (1973) An unusual abbreviated larval life in the estuarine burrowing prawn *Callianassa kraussi* (Crustacea: Decapoda: Thalassinidea). Mar Biol 22:361–365

Frébourg G, Davaud E, Gaillot J et al (2010) An aeolianite in the Upper Dalan Member (Khuff Formation), South Pars Field, Iran. J Petroleum Geol 33:41–154

Frey RW (1970a) The lebensspuren of some common marine invertebrates near Beaufort, North Carolina. II. Anemone burrows. J Paleontol 44:308–311

Frey RW (1970b) Trace fossils of Fort Hays Limestone Member of Niobrara Chalk (Upper Cretaceous), west-central Kansas. University of Kansas Paleontological Contributions 53:1–41, 10 pl

Frey RW (1973) Concepts in the study of biogenic sedimentary structures. J Sediment Petrol 43:6–19

Frey RW (1990) Trace fossils and hummocky cross-stratification, Upper Cretaceous of Utah. Palaios 5:203–218

Frey RW, Bromley RG (1985) Ichnology of American chalks: the Selma Group (Upper Cretaceous), western Alabama. Can J Earth Sci 22:801–828

Frey RW, Cowles JG (1972) The trace fossil *Tisoa* in Washington and Oregon. Oregon Department of Geology and Mineral Industries, The Ore Bin, 34:113–119

Frey RW, Howard JD (1981) *Conichnus* and *Schaubcylindrichnus*: redefined trace fossils from the Upper Cretaceous of the Western Interior. J Paleontol 55:800–804

Frey RW, Howard JD (1985) Trace fossils from the Panther Member, Star Point Formation (Upper Cretaceous), Coal Creek Canyon, Utah. J Paleontol 59:370–404

Frey RW, Howard JD (1990) Trace fossils and depositional sequences in a clastic shelf setting, Upper Cretaceous of Utah. J Paleontol 64:803–820

Frey RW, Pemberton SG (1990) Bioturbate texture or ichnofabric? Palaios 5:385–386

Frey RW, Pemberton SG (1991a) The ichnogenus *Schaubcylindrichnus*: morphological, temporal, and environmental significance. Geol Mag 128:595–602

Frey RW, Pemberton SG (1991b) Or, is it 'bioturbate texture'? Ichnos 1:327–329

Frey RW, Howard JD, Pryor WA (1978) *Ophiomorpha*: its morphologic, taxonomic, and environmental significance. Palaeogeogr Palaeoclimatol Palaeoecol 23:199–229

Frey RW, Seilacher A (1980) Uniformity in marine invertebrate ichnology. Lethaia 13:183–207

Frey RW, Pemberton SG, Fagerstrom JA (1984) Morphological, ethological, and environmental significance of the ichnogenera *Scoyenia* and *Ancorichnus*. J Paleontol 58:511–528

Frieling D (2007) *Rosselia socialis* in the Upper Marine Molasse of southwestern Germany. Facies 53:479–492

Fu S (1991) Funktion, Verhalten und Einteilung fucoider und lophocteniider Lebensspuren. Courier Forschungs-Institut Senckenberg 135:1–79

Fu S, Werner F (2000) Distribution, ecology and taphonomy of the organism trace, *Scolicia*, in northeast Atlantic deep-sea sediments. Palaeogeogr Palaeoclimatol Palaeoecol 156:289–300

Fürsich FT (1973) *Thalassinoides* and the origin of nodular limestone in the Corallian Beds (Upper Jurassic) of Southern England. Neues Jahrbuch für Geologie und Paläontologie, Abhandlungen 140:33–48

Fürsich FT (1974a) Corallian (Upper Jurassic) trace fossils from England and Normandy. Stuttgarter Beiträge zur Naturkunde Serie B (Geologie und Paläontologie) 13:1–52

Fürsich FT (1974b) On *Diplocraterion* Torell 1870 and the significance of morphological features in vertical, spreiten-bearing, U-shaped trace fossils. J Paleontol 48:952–962

Fürsich FT (1974c) Ichnogenus *Rhizocorallium*. Paläontologische Zeitschrift 48:16–28

Fürsich FT, Mayr H (1981) Non-marine *Rhizocorallium* (trace fossil) from the Upper Freshwater Molasses (Upper Miocene) of southern Germany. Neues Jahrbuch für Geologie und Paläontologie, Monatshefte 6:321–333

Fürsich FT, Wilmsen M, Seyed-Emami K (2006) Ichnology of Lower Jurassic beach deposits in the Shemshak Formation at Shahmirzad, southeastern Alborz Mountains, Iran. Facies 52:599–610

Gaillard C (1972) *Paratisoa contorta* n. gen., n. sp., trace fossil nouvelle de l'Oxfordian du Jura: Archives des Sciences de Gènéve 25:149–160

Gaillard C, Hennebert M, Olivero D (1999) Lower Carboniferous *Zoophycos* from the Tournai area (Belgium): environmental and ethologic significance. Geobios 32:513–524

Gaillard C, Racheboeuf PR (2006) Trace fossils from nearshore to offshore environments: Lower Devonian of Bolivia. J Paleontol 80:1205–1226

Gámez Vintaned JA, Liñán E, Mayoral E et al (2006) Trace and soft body fossils from the Pedroche Formation Ovetian, Lower Cambrian of the Sierra de Córdoba, S Spain) and their relation to the Pedroche event. Geobios 39:443–468

Gani MR, Bhattacharya JP, MacEachern JA (2007) Using ichnology to determine relative influence of waves, storms, tides, and rivers in deltaic deposits: examples from Cretaceous Western Interior Seaway, U.S.A. In: MacEachern JA, Bann KL, Gingras MK et al (eds) Applied ichnology. SEPM Short Course Notes, vol 52, pp 209–225

Genise JF, Garrouste R, Nel P et al (2012) *Asthenopodichnium* in fossil wood: different trace makers as indicators of different terrestrial palaeoenvironments. Palaeogeogr Palaeoclimatol Palaeoecol 365–366: 184–191

Gerard JRF, Bromley RG (2008) Ichnofabrics in clastic sediments—application to sedimentological core studies: a practical guide. Jean R.F. Gerard, Madrid, pp 97

Gingras MK, MacEachern JA (2012) Tidal ichnology of shallow-water clastic settings. In: Davis RA Jr, Dalrymple RW (eds) Principles of tidal sedimentology. Springer Science+Business Media, Berlin, pp 57–77

Gingras MK, Pemberton SG, Saunders T (2001) Bathymetry, sediment texture, and substrate cohesiveness; their impact on modern *Glossifungites* trace assemblages at Willapa Bay, Washington. Palaeogeogr Palaeoclimatol Palaeoecol 169:1–21

Gingras MK, MacEachern JA, Dashtgard SE et al (2012a) Estuaries. In: Knaust D, Bromley RG (eds) Trace fossils as indicators of sedimentary environments. Developments in Sedimentology, vol 64, pp 463–505

Gingras MK, Baniak G, Gordon J et al (2012b) Porosity and permeability in bioturbated sediments. In: Knaust D, Bromley RG (eds) Trace fossils as indicators of sedimentary environments. Developments in Sedimentology, vol 64, pp 837–868

Gingras MK, MacEachern JA, Dashtgard SE (2012c) The potential of trace fossils as tidal indicators in bays and estuaries. Sedimentary Geology 279:97–106

Gingras MK, McMillan B, Balcom BJ et al (2002) Using magnetic resonance imaging and petrographic techniques to understand the textural attributes and porosity distribution in *Macaronichnus*-burrowed sandstone. J Sediment Res 72:552–558

Gingras MK, Dashtgard SE, MacEachern JA et al (2008) Biology of shallow-marine ichnology: a modern perspective. Aquatic Biol 2: 255–268

Girotti O (1970) *Echinospira pauciradiata* g. n., sp. n., ichnofossil from the Serravallian-Tortonian of Ascoli Piceno (central Italy). Geol Romana 9:59–62

Glennie KW, Evamy BD (1968) Dikaka: plants and plant-root structures associated with aeolian sand. Palaeogeogr Palaeoclimatol Palaeoecol 4:77–87

Głuszek A (1998) Trace fossils from Late Carboniferous storm deposits, Upper Silesia Coal Basin, Poland. Acta Palaeontol Pol 43:517–546

Goldring R (1962) The trace fossils of the Baggy Beds (Upper Devonian) of North Devon, England. Paläontologische Zeitschrift 36:232–251

Goldring R (1964) Trace fossils and the sedimentary surface in shallow water marine sediments. Dev Sedimentol 1:136–143

Goldring R (1996) The sedimentological significance of concentrically laminated burrows from Lower Cretaceous Ca-bentonites, Oxfordshire. J Geol Soc London 53:255–263

Goldring R, Pollard JE (1995) A re-evaluation of *Ophiomorpha* burrows in the Wealden Group (Lower Cretaceous) of southern England. Cretac Res 16:665–680

Goldring R, Pollard JE, Taylor AM (1991) *Anconichnus horizontalis*: a pervasive ichnofabric-forming trace fossil in post-Paleozoic offshore siliciclastic facies. Palaios 6:250–263

Goldring R, Gruszczynski M, Gatt PA (2002) A bow-form burrow and its sedimentological and paleoecological significance. Palaios 17: 622–630

Goldring R, Taylor AM, Hughes GW (2005) The application of ichnofabrics towards bridging the dichotomy between siliciclastic and carbonate shelf facies: examples from the Upper Jurassic Fulmar Formation (UK) and Jubaila Formation (Saudi Arabia). Proc Geol Assoc 116:235–249

Goldring R, Layer MG, Magyari A et al (1998) Facies variation in the Corallian Group (U. Jurassic) of the Faringdon-Shellingford area (Oxfordshire) and the rockground base to the Faringdon Sponge Gravels (L. Cretaceous). Proc Geol Assoc 109:115–125

Gordon JB, Pemberton SG, Gingras MK et al (2010) Biogenically enhanced permeability: a petrographic analysis of *Macaronichnus segregatus* in the Lower Cretaceous Bluesky Formation, Alberta, Canada. AAPG Bull 94:1779–1795

Gottis C (1954) Sur un *Tisoa* très abondants dans le Numidien de Tunisie, Bull Soc Sci Nat Tunisie 7:184–195

Gowland S (1996) Facies characteristics and depositional models of highly bioturbated shallow marine siliciclastic strata: an example from the Fulmar Formation (Late Jurassic), UK Central Graben. In: Hurst A, Johnson HD, Urley DB et al (eds) Geology of the Humber Group: Central Graben and Moray Firth, UKCS, vol 114. Geological Society of London (Special Publications), pp 185–214

Greb SF, Chesnut DR (1994) Paleoecology of an estuarine sequence in the Breathitt Formation (Pennsylvanian), central Appalachian Basin. Palaios 9:388–402

Gregory MR (1985) Taniwha footprints or fossilised starfish impressions? A reinterpretation: the fodinichnial trace fossil *Asterosoma*. Newslett Geol Soc N Z 70:61–64

Gregory MR, Campbell KA (2003) A '*Phoebichnus* look-alike': a fossilised root system from Quaternary coastal dune sediments, New Zealand. Palaeogeogr Palaeoclimatol Palaeoecol 192:247–258

Gregory MR, Martin AJ, Campbell KA (2004) Compound trace fossils formed by plant and animal interactions: Quaternary of northern New Zealand and Sapelo Island, Georgia (USA). Fossils Strata 51:88–105

Gregory MR, Campbell KA, Zuraida R et al (2006) Plant traces resembling *Skolithos*. Ichnos 13:205–216

Griffis RB, Suchanek TH (1991) A model of burrow architecture and trophic modes in thalassinidean shrimp (Decapoda: Thalassinidea). Mar Ecol Prog Ser 79:171–183

Hakes WG (1976) Trace fossils and depositional environment of four clastic units, Upper Pennsylvanian megacyclothems, northeast Kansas, vol 63. University of Kansas Paleontological Contributions, pp 1–60

Haldeman SS (1840) Supplement to number one of "A monograph of the Limniades, and other freshwater univalve shells of North America," containing descriptions of apparently new animals in different classes, and the names and characters of the subgenera in Paludina and Anculosa. J. Dobson, Philadelphia, 3 pp

Hall J (1847) Palaeontology of New-York, vol 1. C. Van Benthuysen, Albany

Häntzschel W (1960) Spreitenbauten (*Zoophycos* Massal.) im Septarienton Nordwest-Deutschlands. Mitteilungen aus dem Geologischen Staatsinstitut in Hamburg 29:95–100

Häntzschel W (1975) Trace fossils and problematica. In: Teichert C (ed) Treatise on invertebrate paleontology (Part W, Miscellanea Supplement 1). Geological Society of America/University of Kansas Press, Boulder/Lawrence, pp W1–W269

Hasiotis ST (2008) Reply to the comments by Bromley et al. of the paper "Reconnaissance of the Upper Jurassic Morrison Formation ichnofossils, Rocky Mountain Region, USA: paleoenvironmental, stratigraphic, and paleoclimatic significance of terrestrial and freshwater ichnocoenoses" by Stephen T. Hasiotis. Sediment Geol 208:61–68

Hasiotis ST (2010) Continental trace fossils. SEPM Short Course Notes 51:1–132

Hasiotis ST, Honey JG (2000) Paleohydrologic and stratigraphic significance of crayfish burrows in continental deposits: examples from several Paleocene Laramide basins in the Rocky Mountains. J Sediment Res 70:127–139

Hasiotis ST, Mitchell CE (1993) A comparison of crayfish burrow morphologies: Triassic and Holocene fossil, paleo- and neo-ichnological evidence, and the identification of their burrowing signatures. Ichnos 2:291–314

Heard TG, Pickering KT (2008) Trace fossils as diagnostic indicators of deep-marine environments, Middle Eocene Ainsa-Jaca Basin, Spanish Pyrenees. Sedimentology 55:809–844

Heer O (1877) Flora fossilis Helvetiae. Die vorweltliche Flora der Schweiz. J. Würster & Co., Zurich, 182 pp, LXX pl

Heinberg C, Birkelund T (1984) Trace-fossil assemblages and basin evolution of the Vardekløft Formation (Middle Jurassic, central East Greenland). J Paleontol 58:362–397

Hembree DI, Hasiotis ST (2008) Miocene vertebrate and invertebrate burrows defining compound paleosols in the Pawnee Creek Formation, Colorado, U.S.A. Palaeogeogr Palaeoclimatol Palaeoecol 270:349–365

Hertweck G (1972) Georgia coastal region, Sapelo Island, U.S.A.: sedimentology and biology. Senckenb Marit 4:125–167

Hertweck G, Wehrmann A, Liebezeit G (2007) Bioturbation structures of polychaetes in modern shallow marine environments and their analogues to *Chondrites* group traces. Palaeogeogr Palaeoclimatol Palaeoecol 245:382–389

Higgs KT, Higgs BM (2015) New discoveries of *Diplocraterion* and tidal rhythmites in the Upper Devonian rocks of Grab-all Bay, Cork Harbour: palaeoenvironmental implications. Irish J Earth Sci 33:35–54

Hobbs HH (1981) The crayfishes of Georgia. Smithson Contributions Zool 318:1–549

Howard JD (1966) Characteristic trace fossils in Upper Cretaceous sandstones of the Book Cliffs and Wasatch Plateau. Bull Utah Geol Mineral Surv 80:35–53

Howard JD (1972) Trace fossils as criteria for recognizing shorelines in stratigraphic record. In: Rigby JK, Hamblin WK (eds) Recognition of ancient sedimentary environments. SEPM Special Publications 16:215–225

Howard JD, Frey RW (1975) Estuaries of the Georgia coast, U.S.A.: sedimentology and biology. II. Regional animal-sediment characteristics of Georgia estuaries. Senckenb Marit 7:33–103

Howard JD, Frey RW (1984) Characteristic trace fossils in nearshore to offshore sequences, Upper Cretaceous of east-central Utah. Can J Earth Sci 21:200–219

Howard JD, Frey RW (1985) Physical and biogenic aspects of backbarrier sedimentary sequences, Georgia coast, U.S.A. Marine Geology 63:77–127

Hu B, Wang G, Goldring R (1998) *Nereites* (or *Neonereites*) from Lower Jurassic lacustrine turbidites of Henan, central China. Ichnos 6:203–209

Hubbard SM, Gingras MK, Pemberton SG (2004) Palaeoenvironmental implications of trace fossils in estuarine deposits of the Cretaceous Bluesky Formation, Cadotte region, Alberta, Canada. Fossils Strata 51:68–87

Hubbard SM, Shultz MR (2008) Deep burrows in submarine fan-channel deposits of the Cerro Toro Formation (Cretaceous), Chilean Patagonia: implications for firmground development and colonization in the deep sea. Palaios 23:223–232

Hubbard SM, MacEachern JA, Bann KL (2012) Slopes. In: Knaust D, Bromley RG (eds) Trace fossils as indicators of sedimentary environments. Developments in Sedimentology, vol 64, pp 607–642

Hubert JF, Dutcher JA (2010) *Scoyenia* escape burrows in fluvial pebbly sand: Upper Triassic Sugarloaf Arkose, Deerfield Rift Basin, Massachusetts, USA. Ichnos 17:20–24

Husinec A, Read JF (2011) Microbial laminite versus rooted and burrowed caps on peritidal cycles: Salinity control on parasequence development, Early Cretaceous isolated carbonate platform, Croatia. GSA Bull 123:1896–1907

Hyman LH (1959) The invertebrates: smaller Coelomate groups Chaetognatha, Hemichordata, Pogonophora, Phoronida, Ectoprocta, Brachiopoda, Sipunculida. In: The Coelomate Bilateria, vol 5. McGraw-Hill Book Company, New York, pp 1–51

Izumi K (2012) Formation process of the trace fossil *Phymatoderma granulata* in the Lower Jurassic black shale (Posidonia Shale, southern Germany) and its paleoecological implications. Palaeogeogr Palaeoclimatol Palaeoecol 353–355:116–122

Izumi K (2014) Utility of geochemical analysis of trace fossils: case studies using *Phycosiphon incertum* from the Lower Jurassic shallow-marine (Higashinagano Formation, southwest Japan) and Pliocene deep-marine deposits (Shiramazu Formation, central Japan). Ichnos 21:62–72

James NP, Kobluk DR, Pemberton SG (1977) The oldest macroborers: Lower Cambrian of Labrador. Science 197:980–983

Jensen S (1997) Trace fossils from the Lower Cambrian Mickwitzia Sandstone, south-central Sweden. Fossils Strata 42:1–110

Jessen W (1950) "Augenschiefer"-Grabgänge, ein Merkmal für Faunenschiefer-Nähe im westfälischen Oberkarbon. Zeitschrift der Deutschen Geologischen Gesellschaft 101:23–43

Joeckel RM, Korus JT (2012) Bayhead delta interpretation of an Upper Pennsylvanian sheetlike sandbody and the broader understanding of transgressive deposits in cyclothems. Sed Geol 275–276:22–37

Jordan DW (1985) Trace fossils and depositional environments of Upper Devonian black shales, east-central Kentucky, U.S.A. In: Curran HA (ed) Biogenic structures: their use in interpreting depositional environments, vol 35 (SEPM Special Publication), pp 279–298

Joseph JK, Patel SJ, Bhatt NY (2012) Trace fossil assemblages in mixed siliciclastic-carbonate sediments of the Kaladongar Formation (Middle Jurassic), Patcham Island, Kachchh, Western India. J Geol Soc India 80:189–214

Keighley DG, Pickerill RK (1994) The ichnogenus *Beaconites* and its distinction from *Ancorichnus* and *Taenidium*. Palaeontology 37: 305–337

Keighley DG, Pickerill RK (1995) The ichnotaxa *Palaeophycus* and *Planolites*: historical perspectives and recommendations. Ichnos 3:301–309

Kelly SRA, Bromley RG (1984) Ichnological nomenclature of clavate borings. Palaeontology 27:793–807

Kennedy WJ (1967) Burrows and surface traces from the Lower Chalk of Southern England. Bull Br Mus (Nat Hist) Geol 15:125–167

Kikuchi K, Kotake N, Furukawa N (2016) Mechanism and process of construction of tubes of the trace fossil *Schaubcylindrichnus coronus* Frey and Howard, 1981. Palaeogeogr Palaeoclimatol Palaeoecol 443:1–9

Kim J-Y, Paik IS (1997) Nonmarine *Diplocraterion luniforme* (Blanckenhorn 1916) from the Hasandong Formation (Cretaceous) of the Jinju area, Korea. Ichnos 5:131–138

Klappa CF (1980) Rhizoliths in terrestrial carbonates: classification, recognition, genesis and significance. Sedimentology 27:613–629

Knaust D (1998) Trace fossils and ichnofabrics on the Lower Muschelkalk carbonate ramp (Triassic) of Germany: tool for high-resolution sequence stratigraphy. Geol Rundsch 87:21–31

Knaust D (2004a) Cambro-Ordovician trace fossils from the SW-Norwegian Caledonides. Geol J 39:1–24

Knaust D (2004b) The oldest Mesozoic nearshore *Zoophycos*: evidence from the German Triassic. Lethaia 37:297–306

Knaust D (2007a) Meiobenthic trace fossils as keys to the taphonomic history of shallow-marine epicontinental carbonates. In: Miller III W (ed) Trace fossils: concepts, problems, prospects. Elsevier, Amsterdam, pp 502–517

Knaust D (2007b) Invertebrate trace fossils and ichnodiversity in shallow-marine carbonates of the German Middle Triassic (Muschelkalk). In: Bromley RG, Buatois LA, Mángano G et al (eds) Sediment-organism interactions: a multifaceted ichnology, vol 88. SEPM Special Publication, pp 221–238

Knaust D (2008) *Balanoglossites* Mägdefrau, 1932 from the Middle Triassic of Germany: part of a complex trace fossil probably produced by boring and burrowing polychaetes. Paläontologische Zeitschrift 82:347–372

Knaust D (2009a) Ichnology as a tool in carbonate reservoir characterization: a case study from the Permian—Triassic Khuff Formation in the Middle East. GeoArabia 14:17–38

Knaust D (2009b) Characterisation of a Campanian deep-sea fan system in the Norwegian Sea by means of ichnofabrics. Mar Pet Geol 26:1199–1211

Knaust D (2009c) Complex behavioural pattern as an aid to identify the producer of *Zoophycos* from the Middle Permian of Oman. Lethaia 42:146–154

Knaust D (2010a) Meiobenthic trace fossils comprising a miniature ichnofabric from Late Permian carbonates of the Oman Mountains. Palaeogeogr Palaeoclimatol Palaeoecol 286:81–87

Knaust D (2010b) The end-Permian mass extinction and its aftermath on an equatorial carbonate platform: insights from ichnology. Terra Nova 22:195–202

Knaust D (2012a) Trace-fossil systematics. In: Knaust D, Bromley RG (eds) Trace fossils as indicators of sedimentary environments. Developments in Sedimentology, vol 64, pp 79–101

Knaust D (2012b) Methodology and techniques. In: Knaust D, Bromley RG (eds) Trace fossils as indicators of sedimentary environments. Developments in Sedimentology, vol 64, pp 245–271

Knaust D (2013) The ichnogenus *Rhizocorallium*: classification, trace makers, palaeoenvironments and evolution. Earth Sci Rev 126:1–47

Knaust D (2014a) Classification of bioturbation-related reservoir quality in the Khuff Formation (Middle East): towards a genetic approach. In: Pöppelreiter MC (ed) Permo-Triassic Sequence of the Arabian Plate. EAGE, pp 247–267

Knaust D (2014b) Case 3662: *Siphonichnus* Stanistreet, le Blanc Smith and Cadle, 1980 (trace fossil): proposed conservation by granting precedence over the senior subjective synonym *Opthalmichnium* Pfeiffer, 1968. Bull Zool Nomenclature 71:147–152

Knaust D (2015a) Siphonichnidae (new ichnofamily) attributed to the burrowing activity of bivalves: ichnotaxonomy, behaviour and palaeoenvironmental implications. Earth Sci Rev 150:497–519

Knaust D (2015b) Trace fossils from the continental Upper Triassic Kågeröd Formation of Bornholm, Denmark. Ann Soc Geol Pol 85:481–492

Knaust D, Dronov A (2013) *Balanoglossites* ichnofabrics from the Middle Ordovician Volkhov Formation (St. Petersburg Region, Russia). Stratigr Geol Correl 21:265–279

Knaust D, Curran HA, Dronov A (2012) Shallow-marine carbonates. In: Knaust D, Bromley RG (eds) Trace fossils as indicators of sedimentary environments. Developments in Sedimentology, vol 64, pp 705–750

Knaust D, Uchman A, Hagdorn H (2016) The probable isopod burrow *Sinusichnus seilacheri* isp. n. from the Middle Triassic of Germany: an example of behavioral convergence. Ichnos 23:138–146

Knaust D, Warchoł M, Kane IA (2014) Ichnodiversity and ichnoabundance: revealing depositional trends in a confined turbidite system. Sedimentology 62:2218–2267

Kotake N (1989) Paleoecology of the *Zoophycos* producers. Lethaia 22:327–341
Kotake N (1991) Packing process for the filling material in *Chondrites*. Ichnos 1:277–285
Kotake N (1992) Deep-sea echiurans: possible producers of *Zoophycos*. Lethaia 25:311–316
Kotake N (2003) Ethologic and ecologic interpretation of complex stellate structures in Pleistocene deep-sea sediments (Otadai Formation), Boso Peninsula, central Japan. Palaeogeogr Palaeoclimatol Palaeoecol 192:143–155
Kotlarczyk J, Uchman A (2012) Integrated ichnology and ichthyology of the Oligocene Menilite Formation, Skole and Subsilesian nappes, Polish Carpathians: a proxy to oxygenation history. Palaeogeogr Palaeoclimatol Palaeoecol 331–332:104–118
Kowalewski M, Demko TM (1997) Trace fossils and population paleoecology: comparative analysis of size-frequency distributions derived from burrows. Lethaia 29:113–124
Kowalewski M, Demko TM, Hasiotis ST et al (1998) Quantitative ichnology of Triassic crayfish burrows (*Camborygma eumekenomos*): ichnofossils as linkages to population paleoecology. Ichnos 6:5–21
Krapovickas V, Ciccioli PL, Mángano MG et al (2009) Paleobiology and paleoecology of an arid–semiarid Miocene South American ichnofauna in anastomosed fluvial deposits. Palaeogeogr Palaeoclimatol Palaeoecol 284:129–152
Kraus MJ, Hasiotis ST (2006) Significance of different modes of rhizolith preservation to interpreting paleoenvironmental and paleohydrologic settings: examples from Paleogene paleosols, Bighorn Basin, Wyoming, U.S.A. J Sediment Res 76:633–646
Książkiewicz M (1977) Trace fossils in the flysch of the Polish Carpathians. Palaeontologia Polonica 36:1–208
La Croix AD, Gingras MK, Pemberton SG et al (2013) Biogenically enhanced reservoir properties in the Medicine Hat gas field, Alberta, Canada. Mar Pet Geol 43:464–477
Leaman M, McIlroy D (2016) Three-dimensional morphological permeability modelling of *Diplocraterion*. Ichnos, doi:10.1080/10420940.2016.1232650
Leaman M, McIlroy D, Herringshaw LG et al (2015) What does *Ophiomorpha irregulaire* really look like? Palaeogeogr Palaeoclimatol Palaeoecol 439:38–49
Leszczyński S (2010) Coniacian–? Santonian paralic sedimentation in the Rakowice Małe area of the North Sudetic Basin, SW Poland: sedimentary facies, ichnological record and palaeogeographical reconstruction of an evolving marine embayment. Ann Soc Geol Pol 80:1–24
Leszczynski S, Uchman A, Bromley RG (1996) Trace fossils indicating bottom aeration changes: Folusz Limestone, Oligocene, Outer Carpathians, Poland. Palaeogeogr Palaeoclimatol Palaeoecol 121: 79–87
Li Y, Yuan J-L, Lin T-R (1999) Lower Cambrian trace fossils from the Mantou Formation of Huainan, Anhui. Acta Palaeontol Sin 38: 114–124, (pl 1–3) [In Chinese, with English summary]
Linck O (1949) Lebens-Spuren aus dem Schilfsandstein (Mittl. Keuper km 2) NW-Württembergs und ihre Bedeutung für die Bildungs-Geschichte der Stufe. Jahreshefte des Vereins für vaterländische Naturkunde in Württemberg 97–101:1–100
Loughlin NJD, Hillier RD (2010) Early Cambrian *Teichichnus*-dominated ichnofabrics and palaeoenvironmental analysis of the Caerfai Group, Southwest Wales, UK. Palaeogeogr Palaeoclimatol Palaeoecol 297:239–251
Löwemark L (2012) Ethological analysis of the trace fossil *Zoophycos*: hints from the Arctic Ocean. Lethaia 45:290–298
Löwemark L, Hong E (2006) *Schaubcylindrichnus formosus* isp. nov. in Miocene sandstones from northeastern Taiwan. Ichnos 13: 267–276
Löwemark L, Nara M (2010) Morphology, ethology and taxonomy of the ichnogenus *Schaubcylindrichnus*: notes for clarification. Palaeogeogr Palaeoclimatol Palaeoecol 297:184–187
Löwemark L, Nara M (2013) Morphological variability of the trace fossil *Schaubcylindrichnus coronus* as a response to environmental forcing. Palaeontol Electron 16:14
Löwemark L, Lin I-T, Wang C-H et al. (2004) Ethology of the *Zoophycos*-producer: arguments against the gardening model from $\delta_{13}$ $C_{org}$ evidences of the spreiten material. TAO 15:713–725
Lundgren B (1891) Studier öfver fossilförande lösa block. Geol Fören Stockh Förh 13:111–121
MacEachern JA, Bann KL (2008) The role of ichnology in refining shallow marine facies models. In: Hampson GJ (ed) Recent advances in models of siliciclastic shallow-marine stratigraphy, vol 90 (SEPM Special Publication), pp 73–116
MacEachern JA, Gingras MK (2007) Recognition of brackish-water trace-fossil suites in the Cretaceous Western Interior Seaway of Alberta, Canada. In: Bromley RG, Buatois LA, Mángano G et al (eds) Sediment-organism interactions: a multifaceted ichnology, vol 88 (SEPM Special Publication), pp 149–193
MacEachern JA, Bann KL, Bhattacharya JP et al (2005) Ichnology of deltas: organisms responses to the dynamic interplay of rivers, waves, storms and tides. In: Giosan L, Bhattacharya JP (eds) River Deltas—Concepts, Models, and Examples, vol 83 (SEPM Special Publication), pp 49–85
MacEachern JA, Bann KL, Gingras MK et al (2012) The ichnofacies paradigm. In: Knaust D, Bromley RG (eds), Trace fossils as indicators of sedimentary environments. Developments in Sedimentology, vol 64, pp 103–138
MacSotay O, Erlich RN, Peraza T (2003) Sedimentary Structures of the La Luna, Navay and Querecual formations, Upper Cretaceous of Venezuela. Palaios 18:334–348
Mángano MG, Buatois LA (2004) Ichnology of Carboniferous tide-influenced environments and tidal flat variability in the North American Midcontinent. In: McIlroy D (ed) The application of ichnology to palaeoenvironmental and stratigraphic analysis. Geological Society of London, vol 228 (Special Publications), pp 157–178
Mangano MG, Buatois LA, Maples CG et al (2000) A new ichnospecies of *Nereites* from Carboniferous tidal-flat facies of eastern Kansas, USA: implications for the *Nereites-Neonereites* debate. J Paleontol 74:149–157
Männil RM (1966) O Vertikalnykh norkakh zaryvaniya v Ordovikskikh izvestinyakakh Pribaltiki [A small vertically excavated cavity in Baltic Ordovician limestone]. Organizm i Sreda v Geologischeskom Proshlom: Moscow, Akademiya Nauk SSSR, Paleontologischeskii Institut, pp 200–207 [In Russian]
Marenco KN, Bottjer DJ (2008) The importance of *Planolites* in the Cambrian substrate revolution. Palaeogeogr Palaeoclimatol Palaeoecol 258:189–199
Martin KD (2004) A re-evaluation of the relationship between trace fossils and dysoxia. In: McIlroy D (ed) The application of ichnology to palaeoenvironmental and stratigraphic analysis. Geological Society of London, vol 228 (Special Publications), pp 141–156
Martin MA, Pollard JE (1996) The role of trace fossil (ichnofabric) analysis in the development of depositional models for the Upper Jurassic Fulmar Formation of the Kittiwake Field (Quadrant 21 UKCS). In: Hurst A, Johnson HD, Urley DB et al (eds) Geology of the Humber Group: Central Graben and Moray Firth, UKCS, vol 114. Geological Society of London (Special Publications), pp 163–183
Martin AJ, Rindsberg AK (2007) Arthropod trace makers of *Nereites*? Neoichnological observations of juvenile limulids and their paleoichnological applications. In: Miller W III (ed) Trace fossils: concepts, problems, prospects. Elsevier, Amsterdam, pp 478–491

Martin AJ, Rich TH, Poore GCB et al (2008) Fossil evidence in Australia for oldest known freshwater crayfish of Gondwana. Gondwana Res 14:287–296

Martin AJ, Blair M, Dattilo BF et al (2016) The ups and downs of *Diplocraterion* in the Glen Rose Formation (Lower Cretaceous), Dinosaur Valley State Park, Texas (USA). Geodin Acta 28:101–119

Martinsson A (1965) Aspects of a Middle Cambrian thanatotope on Öland. Geologiska Föreningen i Stockholm, Förhandlingar 87:181–230

Mason TR, Christie ADM (1986) Palaeoenvironmental significance of ichnogenus *Diplocraterion* Torell from the Permian Vryheid Formation of the Karoo Supergroup, South Africa. Palaeogeogr Palaeoclimatol Palaeoecol 52:249–265

Massalongo A (1855) *Zoophycos*, novum genus plantarum fossilium. Monographia, Typis Antonellianis, Veronae, pp 45–52

Mata SA, Corsetti CL, Corsetti FA et al (2012) Lower Cambrian anemone burrows from the Upper Member of the Wood Canyon Formation, Death Valley Region, United States: paleoecological and paleoenvironmental significance. Palaios 27:594–606

Mayoral E (1986) *Ophiomorpha isabeli*; nov. icnosp. (Plioceno Marino) en el sector suroccidental del Valle del Guadalquivir (Palos de la Frontera, Huelva, España). Estud Geol 42:461–470

McBride EF, Picard MD (1991) Facies implications of *Trichichnus* and *Chondrites* in turbidites and hemipelagites, Marnoso-arenacea Formation (Miocene), northern Apennines, Italy. Palaios 6:281–290

McCall GJH (2006) The Vendian (Ediacaran) in the geological record: Enigmas in geology's prelude to the Cambrian explosion. Earth Sci Rev 77:1–229

McCarthy B (1979) Trace fossils from a Permian shoreface-foreshore environment, eastern Australia. J Paleontol 53:345–366

McIlroy D (2004) Ichnofabrics and sedimentary facies of a tide-dominated delta: Jurassic Ile Formation of Kristin Field, Haltenbanken, Offshore Mid-Norway. In: McIlroy D (ed) The application of ichnology to palaeoenvironmental and stratigraphic analysis, vol 228. Geological Society of London (Special Publications), pp 237–272

McIlroy D (2007) Ichnology of a macrotidal tide-dominated deltaic depositional system: Lajas Formation, Neuquén Province, Argentina. In: Bromley RG, Buatois LA, Mángano G et al (eds) Sediment-organism interactions: a multifaceted ichnology, vol 88. SEPM Special Publication, pp 195–211

Melchor RN, Genise JF, Farina JL et al (2010) Large striated burrows from fluvial deposits of the Neogene Vinchina Formation, La Rioja, Argentina: a crab origin suggested by neoichnology and sedimentology. Palaeogeogr Palaeoclimatol Palaeoecol 291:400–418

Melchor RN, Genise JF, Buatois LA et al (2012) Fluvial environments. In: Knaust D, Bromley RG (eds), Trace fossils as indicators of sedimentary environments. Developments in Sedimentology, vol 64, pp 329–378

Michalík J, Šimo V (2010) A new spreite trace fossil from Lower Cretaceous limestone (Western Carpathians, Slovakia). Trans R Soc Edinb, Earth Sci 100:417–427

Mikuláš R (1997) Ethological interpretation of the ichnogenus *Pragichnus* Chlupáč, 1987 (Ordovician, Czech Republic). Neues Jahrbuch für Geologie und Paläontologie, Monatshefte 1997:93–108

Mikuláš R (2006) Trace fossils in the collections of the Czech Republic (with emphasis on type material). A special publication for the Workshop on Ichnotaxonomy—III, Prague and Moravia, Sept 2006, 137 pp

Miller MF (1991) Morphology and paleoenvironmental distribution of Paleozoic *Spirophyton* and *Zoophycos*: implications for the *Zoophycos* ichnofacies. Palaios 6:410–425

Miller MF, Knox LW (1985) Biogenic structures and depositional environments of a Lower Pennsylvanian coal-bearing sequence, northern Cumberland Plateau, Tennessee, U.S.A. In: Curran HA (ed) Biogenic structures: their use in interpreting depositional environments, vol 35 (SEPM Special Publication), pp 67–97

Miller W (1995) Examples of Mesozoic and Cenozoic *Bathysiphon* (Foraminiferida) from the Pacific rim and the taxonomic status of *Terebellina* Ulrich, 1904. J Paleontol 69:624–634

Miller W (2011) A stroll in the forest of the fucoids: Status of *Melatercichnus burkei* Miller, 1991, the doctrine of ichnotaxonomic conservatism and the behavioral ecology of trace fossil variation. Palaeogeogr Palaeoclimatol Palaeoecol 307:109–116

Monaco P (2008) Taphonomic features of *Paleodictyon* and other graphoglyptid trace fossils in Oligo-Miocene thin-bedded turbidites, northern Apennines, Italy. Palaios 23:668–683

Monaco P (2014) Taphonomic aspects of the radial backfill of asterosomids in Oligo-Miocene turbidites of central Italia (northern Appenines). Riv Ital Paleontol Stratigr 120:215–224

Monaco P, Caracuel JE, Giannetti A et al (2009) *Thalassinoides* and *Ophiomorpha* as cross-facies trace fossils of crustaceans from shallow to-deep-water environments: Mesozoic and Tertiary examples from Italy and Spain. In: Garassino A, Feldmann RM, Teruzzi G (eds) 3rd Symposium on Mesozoic and Cenozoic Decapod Crustaceans—Museo di Storia Naturale di Milano, May 23–25, 2007. Memorie della Società Italiana di Scienze Naturali e del Museo Civico di Storia Naturale di Milano, vol 35, pp 79–82

Monaco P, Rodríguez-Tovar FJ, Uchman A (2012) Ichnological analysis of lateral environmental heterogeneity within the Bonarelli level (uppermost Cenomanian) in the classical localities near Gubbio, central Appenines, Italy. Palaios 27:48–54

Morris JE, Hampson GJ, Johnson HD (2006) A sequence stratigraphic model for an intensely bioturbated shallow-marine sandstone: the Bridport Sand Formation, Wessex Basin, UK. Sedimentology 53:1229–1263

Müller AH (1971) Zur Kenntnis von *Asterosoma* (Vestigia invertebratorum). Freiberger Forschungshefte C 267:7–17

Murchison RI (1839) The Silurian system. John Murray, London, pp 768

Myrow PM (1995) *Thalassinoides* and the enigma of early Paleozoic open-framework burrow systems. Palaios 10:58–74

Nara M (1995) *Rosselia socialis*: a dwelling structure of a probable terebellid polychaete. Lethaia 28:171–178

Nara M (2002) Crowded *Rosselia socialis* in Pleistocene inner shelf deposits: benthic paleoecology during rapid sea-level rise. Palaios 17:268–276

Nara M (2006) Reappraisal of *Schaubcylindrichnus*: a probable dwelling/feeding structure of a solitary funnel feeder. Palaeogeogr Palaeoclimatol Palaeoecol 240:439–452

Nara M, Haga M (2007) The youngest record of trace fossil *Rosselia socialis*: occurrence in the Holocene shallow marine deposits of Japan. Paleontol Res 11:21–27

Nara M, Seike K (2004) *Macaronichnus segregatis*-like traces found in the modern foreshore sediments of the Kujukurihama Coast, Japan. J Geol Soc Jpn 110:545–551 [In Japanese, with English abstract]

Narbonne GM, Hofmann HJ (1987) Ediacaran biota of the Wernecke Mountains, Yukon, Canada. Palaeontology 30:647–676

Naruse H, Nifuku K (2008) Three-dimensional morphology of the ichnofossil *Phycosiphon incertum* and its implication for paleoslope inclination. Palaios 23:270–279

Neto de Carvalho C, Baucon A (2010) *Nereites* trails and other sandflat trace fossils from Portas de Almourão geomonument (Lower Ordovician, Naturtejo Geopark). e-Terra 17:1–4

Neto de Carvalho C, Baucon A, Bayet-Goll A (2016) The ichnological importance and interest of the Geological Museum of Lisbon collections: *Cladichnus lusitanicum* in continental facies from the Lower Cretaceous of the Lusitanian Basin (Portugal). Comunicações Geológicas 103, Especial I:7–12

Neto de Carvalho C, Rodrigues NPC (2003) Los *Zoophycos* del Bajociense-Bathoniense de la Praia da Mareta (Algarve, Portugal): Arquitectura y finalidades en régimen de dominancia ecológica (The *Zoophycos* from the Bajocian-Bathonian of Praia da Mareta (Algarve, Portugal): Architecture and purposes in ecological dominance regime). Revista Española de Paleontologia 18:229–241 [In Portuguese]

Neto de Carvalho C, Rodrigues NPC (2007) Compound *Asterosoma ludwigae* Schlirf, 2000 from the Jurassic of the Lusitanian Basin (Portugal): conditional strategies in the behaviour of Crustacea. J Iber Geol 33:295–310

Neto de Carvalho C, Rodrigues NPC, Viegas PA et al (2010) Patterns of occurrence and distribution of crustacean ichnofossils in the Lower Jurassic-Upper Cretaceous of Atlantic occidental margin basins, Portugal. Acta Geol Pol 60:19–28

Netto RG, Benner JS, Buatois LA et al (2012) Glacial environments. In: Knaust D, Bromley RG (eds), Trace fossils as indicators of sedimentary environments. Developments in Sedimentology, vol 64, pp 299–327

Netto RG, Tognoli FMW, Assine ML et al (2014) Crowded *Rosselia* ichnofabric in the early Devonian of Brazil: an example of strategic behaviour. Palaeogeogr Palaeocl Palaeoecol 395:107–113

Nicholson HA (1873) Contributions to the study of the errant annelides of the older Palaeozoic rocks. R Soc Lond Proc 21:288–290 (also Geological Magazine 10:309–310)

Nickel LA, Atkinson RJA (1995) Functional morphology of burrows and trophic modes of three thalassinidean shrimp species, and a new approach to the classification of thalassinidean burrow morphology. Mar Ecol Prog Ser 128:181–197

Nielsen JK, Hansen KS, Simonsen L (1996) Sedimentology and ichnology of the Robbedale Formation (Lower Cretaceous), Bornholm, Denmark. Bull Geol Soc Denm 43:115–131

Nilsen TH, Kerr DR (1978) Turbidites, redbeds, sedimentary structures, and trace fossils observed in DSDP Leg 38 cores and the sedimentary history of the Norwegian-Greenland Sea. Initial report of the deep sea drilling project, vol 38 (part 1), pp 259–288

Nygaard E (1983) *Bathichnus* and its significance in the trace fossil association of Upper Cretaceous chalk, Mors, Denmark. Danmarks Geologiske Undersøgelser, Årbog 1982:107–137

Olariu C, Steel RJ, Dalrymple RW et al (2012) Tidal dunes versus tidal bars: the sedimentological and architectural characteristics of compound dunes in a tidal seaway, the lower Baronia Sandstone (Lower Eocene), Ager Basin, Spain. Sed Geol 279:134–155

Olivero D (1996) *Zoophycos* distribution and sequence stratigraphy. Examples from the Jurassic and Cretaceous deposits of southeastern France. Palaeogeogr Palaeoclimatol Palaeoecol 123:273–287

Olivero D (2003) Early Jurassic to Late Cretaceous evolution of *Zoophycos* in the French Subalpine Basin (southeastern France). Palaeogeogr Palaeoclimatol Palaeoecol 192:59–78

Olivero D (2007) *Zoophycos* and the role of type specimens in ichnotaxonomy. In: Miller W III (ed) Trace fossils: concepts, problems, prospects. Elsevier, Amsterdam, pp 219–231

Olivero D, Gaillard C (1996) Paleoecology of Jurassic *Zoophycos* from south-eastern France. Ichnos 4:249–260

Olivero D, Gaillard C (2007) A constructional model for *Zoophycos*. In: Miller W III (ed) Trace fossils: concepts, problems, prospects. Elsevier, Amsterdam, pp 466–477

Olivero EB, López Cabrera MI (2013) *Euflabella* n. igen.: Complex horizontal spreite burrows in Upper Cretaceous-Paleogene shallow-marine sandstones of Antarctica and Tierra del Fuego. J Paleontol 87:413–426

Olivero EB, Buatois LA, Scasso RA (2004) *Paradictyodora antarctica*: a new complex vertical spreite trace fossil from the Upper Cretaceous-Paleogene of Antarctica and Tierra del Fuego, Argentina. J Paleontol 78:783–789

Olóriz F, Rodríguez-Tovar FJ (2000) *Diplocraterion*: a useful marker for sequence stratigraphy and correlation in the Kimmeridgian, Jurassic (Prebetic Zone, Betic Cordillera, southern Spain). Palaios 15:546–552

Orłowski S (1989) Trace fossils in the Lower Cambrian sequence in the Świętokrzyskie Mountains, Central Poland. Acta Palaeontol Pol 34:211–231

Orłowski S, Radwański A (1986) Middle Devonian sea-anemone burrows, *Alpertia sanctacrucensis* ichnogen. et ichnosp. n., from the Holy Cross Mountains. Acta Geol Pol 36:233–249

Osgood RG (1970) Trace fossils of the Cincinnati area. Palaeontogr Am 6:281–444

Owen RA, Owen RB, Renaut RW et al (2008) Mineralogy and origin of rhizoliths on the margins of saline, alkaline Lake Bogoria, Kenya Rift Valley. Sed Geol 203:143–163

Pacześna J (2010) Ichnological record of the activity of Anthozoa in the early Cambrian succession of the Upper Silesian Block (southern Poland). Acta Geol Pol 60:93–103

Pazos PJ, Fernández DE (2010) Three-dimensionally integrated trace fossils from shallow-marine deposits in the Lower Cretaceous of the Neuquén Basin: *Hillichnus agrioensis* isp. nov. Acta Geol Pol 60:105–118

Pearson NJ, Mángano GM, Buatois LA et al (2013) Environmental variability of *Macaronichnus* ichnofabrics in Eocene tidal-embayment deposits of southern Patagonia, Argentina. Lethaia 46:341–354

Pemberton SG, Frey RW (1982) Trace fossil nomenclature and the *Planolites-Palaeophycus* dilemma. J Paleontol 56:843–881

Pemberton SG, Frey RW (1984) Ichnology of storm-influenced shallow marine sequence: Cardium Formation (Upper Cretaceous) at Seebe, Alberta. In: Stott DF, Glass DJ (eds) The Mesozoic of Middle North America. Canadian Society of Petroleum Geologists, Memoir 9, pp 281–304

Pemberton SG, Gingras MK (2005) Classification and characterizations of biogenically enhanced permeability. AAPG Bull 89:1493–1517

Pemberton SG, Wightman DM (1992) Ichnological characteristics of brackish water deposits. In: Pemberton SG (ed) Applications of ichnology to petroleum exploration. A core workshop. SEPM Core Workshop, vol 17, pp 141–167

Pemberton SG, Frey RW, Bromley RG (1988) The ichnotaxonomy of *Conostichus* and other plug-shaped ichnofossils. Can J Earth Sci 25:866–892

Pemberton SG, MacEachern JA, Ranger MJ (1992) Ichnology and event stratigraphy: the use of trace fossils in recognizing tempestites. In: Pemberton SG (ed) Applications of ichnology to petroleum exploration. A core workshop. SEPM Core Workshop, vol 17, pp 85–117

Pemberton SG, MacEachern JA, Dashtgard SE et al (2012) Shorefaces. In: Knaust D, Bromley RG (eds) Trace fossils as indicators of sedimentary environments. Developments in Sedimentology, vol 64, pp 563–604

Pemberton SG, MacEachern JA, Gingras MK et al (2008) Biogenic chaos: cryptobioturbation and the work of sedimentologically friendly organisms. Palaeogeogr Palaeoclimatol Palaeoecol 270: 273–279

Pemberton SG, Spila MV, Pulham AJ, et al (2001) Ichnology and sedimentology of shallow to marginal marine systems. Ben Nevis and Avalon Reservoirs, Jeanne d'Arc Basin. Geological Association of Canada, Short Course Notes, vol 15, 343 pp

Percival CJ (1981) Carboniferous quartz arenites and ganisters of the Northern Pennines. Durham Theses, Durham University, 353 pp http://etheses.dur.ac.uk/1103/

Pfefferkorn HW, Fuchs K (1991) A field classification of fossil plant substrate interactions. Neues Jahrbuch für Geologie und Paläontologie, Abhandlungen 183:17–36

Pickerill RK (1980) Phanerozoic flysch trace fossil diversity—observations based on Ordovician flysch ichnofauna from the Aroostook-Matapedia Carbonate Belt of northern New Brunswick. Can J Earth Sci 17:1259–1270

Plička M (1968) *Zoophycos*, and a proposed classification of sabellid worms. J Paleontol 42:836–849

Pollard JE, Goldring R, Buck SG (1993) Ichnofabrics containing *Ophiomorpha*: significance in shallow-water facies interpretation. J Geol Soc Lond 150:149–164

Powichrowski LK (1989) Trace fossils from the Helminthoid Flysch (Upper Cretaceous-Paleocene) of the Ligurian Alps (Italy): development of deep marine ichnoassociations in fan and basin plain environments. Eclogae Geol Helv 82:385–411

Prantl F (1946) Two new problematic trails from the Ordovician of Bohemia: Académie Tchèque des Sciences. Bull Int Classe des Sciences Mathématiques, Naturelles et de la Médecine 46:49–59

Price S, McCann T (1990) Environmental significance of *Arenicolites* ichnosp. In Pliocene lake deposits of southwest Turkey. Neues Jahrbuch für Geologie und Paläontologie, Monatshefte 1990:687–694

Quiroz LI, Buatois LA, Mángano MG et al (2010) Is the trace fossil *Macaronichnus segregatis* an indicator of temperate to cold waters? Exploring the paradox of its occurrence in tropical coasts. Geology 38:651–654

Reineck H-E (1958) Wühlbau-Gefüge in Abhängigkeit von Sediment-Umlagerungen. Senckenb Lethaea 39:1–23, 54–56

Retallack GJ (1988) Field recognition of paleosols. Geol Soc Am Spec Pap 216:1–20

Retallack GJ (2001) *Scoyenia* burrows from Ordovician palaeosols of the Juniata Formation in Pennsylvania. Palaeontology 44:209–235

Riahi S, Uchman A, Stow D et al (2014) Deep-sea trace fossils of the Oligocene-Miocene Numidian Formation, northern Tunisia. Palaeogeogr Palaeoclimatol Palaeoecol 414:155–177

Richter R (1928) Psychische Reaktionen fossiler Tiere. Palaeobiol 1:225–244 (1 pl)

Richter R (1931) Tierwelt und Umwelt im Hunsrückschiefer. Zur Entstehung eines schwarzen Schlammsteins. Senckenbergiana 13: 299–342

Richter R (1937) Marken und Spuren aus allen Zeiten. I-II. Senckenbergiana 19:150–169

Richter R (1952) Fluidal-Textur in Sediment-Gesteinen und über Sedifluktion überhaupt. Notizblatt des Hessischen Landesamtes für Bodenforschung zu Wiesbaden 6:67–81

Rindsberg AK (1994) Ichnology of the Upper Mississippian Hartselle Sandstone of Alabama, with notes on other Carboniferous formations. Geol Surv Alabama Bull 158:1–107

Rindsberg AK, Kopaska-Merkel DC (2005) *Treptichnus* and *Arenicolites* from the Steven C. Minkin Paleozoic footprint site (Langsettian, Alabama, USA). In: Buta RJ, Rindsberg AK, Kopaska-Merkel DC (eds) Pennsylvanian footprints in the Black Warrior Basin of Alabama, vol 1. Alabama Paleontological Society Monograph, pp 121–141

Rindsberg AK, Martin AJ (2003) *Arthrophycus* in the Silurian of Alabama (USA) and the problem of compound trace fossils. Palaeogeogr Palaeoclimatol Palaeoecol 192:187–219

Rodríguez-Tovar FJ, Aguirre J (2014) Is *Macaronichnus* an exclusively small, horizontal and unbranched structure? *Macaronichnus segregatis degiberti* isubsp. nov. Span J Palaeontol 29:131–142

Rodríguez-Tovar FJ, Pérez-Valera F (2008) Trace fossil *Rhizocorallium* from the Middle Triassic of the Betic Cordillera, Southern Spain: characterization and environmental implications. Palaios 23:78–86

Rodríguez-Tovar FJ, Pérez-Valera F (2013) Variations in population structure of *Diplocraterion parallelum*: hydrodynamic influence, food availability, or nursery settlement? Palaeogeogr Palaeoclimatol Palaeoecol 369:501–509

Rodríguez-Tovar FJ, Pérez-Valera F, Pérez-López A (2007) Ichnological analysis in high-resolution sequence stratigraphy: the *Glossifungites* Ichnofacies in Triassic successions from the Betic Cordillera (southern Spain). Sed Geol 198:293–307

Rodríguez-Tovar FJ, Alcalá L, Cobos A (2016) *Taenidium* at the lower Barremian El Hoyo dinosaur tracksite (Teruel, Spain): assessing palaeoenvironmental conditions for the invertebrate community. Cretac Res 65:48–58

Romero-Wetzel MB (1987) Sipunculans as inhabitants of very deep, narrow burrows in deep-sea-sediments. Mar Biol 96:87–91

Ruppert EE, Fox RS (1988) Seashore Animals of the Southeast: A Guide to Common Shallow-Water Invertebrates of the Southeastern Atlantic Coast Columbia, SC: University of South Carolina Press, 429 pp

Ruppert EE, Fox RS, Barnes RD (2004) Invertebrate zoology. A functional evolutionary approach. Brooks Cole, Belmont, pp xvii+989

Salter JW (1857) On annelide-burrows and surface-markings from the Cambrian rocks of the Longmynd, No. 2. Q J Geol Soc Lond 13:199–206, Pl V

Sappenfield A, Droser M, Kennedy M et al (2012) The oldest *Zoophycos* and implications for early Cambrian deposit feeding. Geol Mag 149:1118–1123

Savrda CE (2002) Equilibrium responses reflected in a large *Conichnus* (Upper Cretaceous Eutaw Formation, Alabama, USA). Ichnos 9: 33–40

Savrda CE (2012) Chalk and related deep-marine carbonates. In: Knaust D, Bromley RG (eds) Trace fossils as indicators of Sedimentary environments. Developments in Sedimentology, vol 64, pp 245–271

Savrda CE, Bottjer DJ (1991) Oxygen-related biofacies in marine strata: an overview and update. In: Tyson RV, Pearson TH (eds) Modern and ancient continental shelf anoxia, vol 58. Geological Society of London (Special Publications), pp 201–219

Savrda CE, Uddin A (2005) Large *Macaronichnus* and their behavioral implications (Cretaceous Eutaw Formation, Alabama, USA). Ichnos 12:1–9

Savrda CE, Blanton-Hooks AD, Collier JW et al (2000) *Taenidium* and associated ichnofossils in fluvial deposits, Cretaceous Tuscaloosa Formation, eastern Alabama, southeastern USA. Ichnos 7:777–806

Savrda CE, Krawinkel H, McCarthy FMG et al (2001) Ichnofabrics of a Pleistocene slope succession, New Jersey margin: relations to climate and sea-level dynamics. Palaeogeogr Palaeoclimatol Palaeoecol 171:41–61

Schäfer W (1962) Aktuo-Paläontologie nach Studien in der Nordsee. Kramer, Frankfurt am Main, pp VIII+666

Schieber J (1999) Distribution and deposition of mudstone facies in the Upper Devonian Sonyea Group of New York. J Sediment Res 69:909–925

Schieber J (2003) Simple gifts and buried treasures—implications of finding bioturbation and erosion surfaces in black shales. Sediment Rec 1:4–8

Schlirf M (2000) Upper Jurassic trace fossils from the Boulonnais (northern France). Geol Paleontol 34:145–213

Schlirf M (2003) Palaeoecologic significance of Late Jurassic trace fossils from the Boulonnais, N France. Acta Geol Pol 53:123–142

Schlirf M (2011) A new classification concept for U-shaped spreite trace fossils. Neues Jahrbuch für Geologie und Paläontologie, Abhandlungen 260:33–54

Schlirf M, Uchman A (2005) Revision of the ichnogenus *Sabellarifex* Richter, 1921 and its relationship to *Skolithos* Haldeman, 1840 and *Polykladichnus* Fürsich, 1981. J Syst Paleontol 3:115–131

Scholle PA (1971) Sedimentology of fine-grained deep-water carbonate turbidites, Monte Antola Flysch (Upper Cretaceous), northern Apennines Italy. Geol Soc Am Bull 82:629–658

Schweigert G (1998) Die Spurenfauna des Nusplinger Plattenkalks (Oberjura, Schwäbische Alb). Stuttgarter Beiträge zur Naturkunde Serie B (Geologie und Paläontologie) 262:1–47

Seike K (2007) Palaeoenvironmental and palaeogeographical implications of modern *Macaronichnus segregatis*-like traces in foreshore sediments on the Pacific coast of central Japan. Palaeogeogr Palaeoclimatol Palaeoecol 252:497–502

Seike K (2008) Burrowing behaviour inferred from feeding traces of the opheliid polychaete *Euzonus* sp. as response to beach morphodynamics. Mar Biol 153:1199–1206

Seike K, Yanagishima S, Nara M et al (2011) Large *Macaronichnus* in modern shoreface sediments: Identification of the producer, the mode of formation, and paleoenvironmental implications. Palaeogeogr Palaeoclimatol Palaeoecol 311:224–229

Seilacher A (1955) Spuren und Fazies im Unterkambrium. In: Schindewolf OH, Seilacher A (eds), Beiträge zur Kenntnis des Kambriums in der Salt Range (Pakistan). Akademie der Wissenschaften und der Literatur zu Mainz, Abhandlung Mathematisch-Naturwissenschaftliche Klasse 1955, pp 373–399

Seilacher A (1957) An-aktualistisches Wattenmeer? Paläontologische Zeitschrift 31:198–206

Seilacher A (1967) Bathymetry of trace fossils. Mar Geol 5:413–428

Seilacher A (1977) Evolution of trace fossil communities. In: Hallam A (ed) Patterns of evolution. Elsevier, Amsterdam, pp 359–376

Seilacher A (1986) Evolution of behavior as expressed in marine trace fossils. In: Nitecki MH, Kitchell JA (eds) Evolution of animal behavior. Paleontological and field approaches. Oxford University Press, Oxford, pp 62–87

Seilacher A (1990) Aberration in bivalve evolution related to photo- and chemosymbiosis. Hist Biol 3:289–311

Seilacher A (2007) Trace fossil analysis. Springer, Berlin, pp 226

Serpagli E, Serventi P, Monegatti P (2008) The ichnofossil genus *Paradictyodora* Olivero, Buatois and Scasso (2004) from the Pleistocene of the northern Apennines, Italy. Rivista Italiana Paleontologia e Stratigrafia 114:161–167

Shields MA, Kedra M (2009) A deep burrowing sipunculan of ecological and geochemical importance. Deep Sea Res I 56:2057–2064

Shinn EA (1968) Burrowing in recent lime sediments of Florida and the Bahamas. J Paleontol 42:879–894

Shuto T, Shiraishi S (1979) A Lower Miocene ichnofauna of the middle Ashiya Group, North Kyushu—ichnological study of the Ashiya Group-I. Trans Proc Palaeontol Soc Jpn New Ser 115:109–134

Simpson S (1956) On the trace fossil *Chondrites*. Q J Geol Soc 112:475–499

Smilek KR, Hembree DI (2012) Neoichnology of *Thyonella gemmata*: a case study for understanding holothurian ichnofossils. Open Paleontol J 4:1–10

Smith JJ (2007) Ichnofossils of the Paleogene Willwood Formation and the Paleocene-Eocene thermal maximum (PETM): response of an ancient soil ecosystem to transient global warming. PhD Thesis, University of Kansas, 184 pp. http://search.proquest.com/docview/304858899?accountid=142725

Smith JJ, Hasiotis ST, Kraus MJ et al (2008) *Naktodemasis bowni*: new ichnogenus and ichnospecies for adhesive meniscate burrows (AMB), and paleoenvironmental implications, Paleogene Willwood Formation, Bighorn Basin, Wyoming. J Paleontol 82:267–278

Stanistreet IG, Le Blanc Smith G, Cadle AB (1980) Trace fossils as sedimentological and palaeoenvironmental indices in the Ecca Group (Lower Permian) of the Transvaal. Trans Geol Soc S Afr 83:333–344

Stanley DCA, Pickerill RK (1994) *Planolites constriannulatus* isp. nov. from the Late Ordovician Georgian Bay Formation of southern Ontario, eastern Canada. Ichnos 3:119–123

Stanley DCA, Pickerill RK (1998) Systematic ichnology of the Late Ordovician Georgian Bay Formation of southern Ontario, eastern Canada, vol 162. Royal Ontario Museum Life Sciences Contributions, pp 1–55

Staub M (1899) Über die *Chondrites* benannten fossilen Algen. Földt Közl 29:110–121 [Hungarian 16–32]

Steinmann G (1907) Einführung in die Paläontologie. Wilhelm Engelmann, Leipzig, pp XII+542

Strullu-Derrien C, McLoughlin S, Philippe M et al (2012) Arthropod interactions with bennettitalean roots in a Triassic permineralized peat from Hopen, Svalbard Archipelago (Arctic). Palaeogeogr Palaeoclimatol Palaeoecol 348–349:45–58

Sundberg FA (1983) *Skolithos linearis* Haldeman from the Carrara Formation (Cambrian) of California. J Paleontol 57:145–149

Sutherland JI (2003) Miocene petrified wood and associated borings and termite faecal pellets from Hukatere Peninsula, Kaipara Harbour, North Auckland, New Zealand. J R Soc N Z 33:395–414

Swinbanks DD, Luternauer JL (1987) Burrow distribution of thalassinidean shrimp on a Fraser Delta tidal flat, British Columbia. J Paleontol 61:315–332

Tapanila L, Hutchings P (2012) Reefs and mounds. In: Knaust D, Bromley RG (eds) Trace fossils as indicators of sedimentary environments. Developments in Sedimentology, vol 64, pp 751–775

Tauber AF (1949) Paläobiologische Analyse von *Chondrites furcatus* Sternberg. Jahrbuch der Geologischen Bundesanstalt 93:141–154

Taylor AM, Gawthorpe RL (1993) Application of sequence stratigraphy and trace fossil analysis to reservoir description: examples from the Jurassic of the North Sea. In: Parker JR (ed) Petroleum geology of Northwest Europe, Proceedings of the 4th Conference. Geological Society of London, pp 317–335

Taylor PD, Wilson MA (2003) Palaeoecology and evolution of marine hard substrate communities. Earth Sci Rev 62:1–103

Taylor A, Goldring R, Gowland S (2003) Analysis and application of ichnofabrics. Earth Sci Rev 60:227–259

Tchoumatchenco P, Uchman A (2001) The oldest deep-sea *Ophiomorpha* and *Scolicia* and associated trace fossils from the Upper Jurassic-Lower Cretaceous deep-water turbidite deposits of SW Bulgaria. Palaeogeogr Palaeoclimatol Palaeoecol 169:85–99

Thayer CW, Steele-Petrović HM (1975) Burrowing of the lingulid brachiopod *Glottidia pyramidata*: its ecologic and paleoecologic significance. Lethaia 8:209–221

Tonkin NS (2012) Deltas. In: Knaust D, Bromley RG (eds) Trace fossils as indicators of sedimentary environments. Developments in Sedimentology, vol 64, pp 507–528

Tonkin NS, McIlroy D, Meyer R et al (2010) Bioturbation influence on reservoir quality: a case study from the Cretaceous Ben Nevis Formation, Jeanne d'Arc Basin, offshore Newfoundland, Canada. AAPG Bull 94:1059–1078

Torell O (1870) Petrificata suecana formationis Cambricæ. Lunds Universitets Årsskrift 6 (Afdelningen 2) 8:1–14

Tunis G, Uchman A (1996) Ichnology of Eocene flysch deposits of the Istria Peninsula, Croatia and Slovenia. Ichnos 5:1–22

Turbeville JM, Ruppert EE (1983) Epidermal muscles and peristaltic burrowing in *Carinoma tremaphoros* (Nemertini): correlates of effective burrowing without segmentation. Zoomorphology 103:103–120

Uchman A (1995) Taxonomy and palaeoecology of flysch trace fossils: the Marnoso-arenacea Formation and associated facies (Miocene, Northern Apennines, Italy). Beringeria 15:3–115

Uchman A (1998) Taxonomy and ethology of flysch trace fossils: a revision of the Marian Książkiewicz collection and studies of complementary material. Ann Soc Geol Pol 68:105–218

Uchman A (1999) Ichnology of the Rhenodanubian flysch (Lower Cretaceous-Eocene) in Austria and Germany. Beringeria 25:65–171

Uchman A (2009) The *Ophiomorpha rudis* ichnosubfacies of the *Nereites* ichnofacies: characteristics and constraints. Palaeogeogr Palaeoclimatol Palaeoecol 276:107–119

Uchman A (2010) A new ichnogenus for *Chondrites hoernesii* Ettingshausen, 1863, a deep-sea radial trace fossil from the Upper Cretaceous of the Polish Flysch Carpathians: its taxonomy and palaeoecological interpretation as a deep-tier chemichnion. Cretac Res 31:515–523

Uchman A, Demírcan H (1999) A *Zoophycos* group trace fossil from Miocene flysch in southern Turkey: evidence for a U-shaped causative burrow. Ichnos 6:251–259

Uchman A, Krenmayr HG (1995) Trace fossils from lower Miocene (Ottnangian) Molasse deposits of Upper Austria. Paläontologische Zeitschrift 69:503–524

Uchman A, Rattazzi B (2016) *Rhizocorallium hamatum* (Fischer-Ooster 1858), a *Zoophycos*-like trace fossil from deep-sea Cretaceous-Neogene sediments. Hist Biol. doi:10.1080/08912963.2016.1167481

Uchman A, Wetzel A (2011) Deep-sea ichnology: the relationships between depositional environment and endobenthic organisms. In: Hüneke H, Mulder T (eds) Deep-sea sediments. Developments in Sedimentology, vol 63, pp 517–556

Uchman A, Wetzel A (2012) Deep-sea fans. In: Knaust D, Bromley RG (eds) Trace fossils as indicators of sedimentary environments. Developments in Sedimentology, vol 64, pp 643–671

Uchman A, Ślączka A, Renda P (2012) Probable root structures and associated trace fossils from the lower Pleistocene calcarenites of Favignana Island, southern Italy: dilemmas of interpretation. Geol Q 56:745–756

Uchman A, Johnson ME, Rebelo AC et al (2016) Vertically-oriented trace fossil *Macaronichnus segregatis* from Neogene of Santa Maria Island (Azores; NE Atlantic) records vertical fluctuations of the coastal groundwater mixing zone on a small oceanic island. Geobios 49:229–241

van de Schootbrugge B, Harazim D, Sorichter K et al (2010) The enigmatic ichnofossil *Tisoa siphonalis* and widespread authigenic seep carbonate formation during the Late Pliensbachian in southern France. Biogeosciences 7:3123–3138

Verde M, Martínez S (2004) A new ichnogenus for crustacean trace fossils from the Upper Miocene Camacho Formation of Uruguay. Palaeontology 47:39–49

Sternberg KM Graf von (1833–1838) Versuch einer geognostisch–botanischen Darstellung der Flora der Vorwelt. Fr. Fleischer, Leipzig, Prague, vol 5–8

von Otto E (1854) Additamente zur Flora des Quadergebirges in Sachsen. II. Heft, Gustav Mayer, Leipzig

Ward JE, Shumway SE (2004) Separating the grain from the chaff: particle selection in suspension- and deposit-feeding bivalves. J Exp Mar Biol Ecol 300:83–130

Warme JE (1970) Traces and significance of marine rock borers. In: Crimes TP, Harper JC (eds) Trace fossils. Seel House Press, Liverpool, pp 515–526

Weaver PPE, Schultheiss PJ (1983) Vertical open burrows in deep-sea sediments 2 m in length. Nature 301:329–331

Webby BD (1984) Precambrian-Cambrian trace fossils from western New South Wales. Aust J Earth Sci 31:427–437

Werner F (2002) Bioturbation structures in marine Holocene sediments of Kiel Bay (western Baltic). Meyniana 54:41–72

Wetzel A (1981) Ökologische und stratigraphische Bedeutung biogener Gefüge in quartären Sedimenten am NW-afrikanischen Kontinentalrand. "Meteor" Forschungs-Ergebnisse C34:1–47

Wetzel A (1991) Ecologic interpretation of deep-sea trace fossil communities. Palaeogeogr Palaeoclimatol Palaeoecol 85:47–69

Wetzel A (2002) Modern *Nereites* in the South China Sea—ecological association with redox conditions in the sediment. Palaios 17:507–515

Wetzel A (2008) Recent bioturbation in the deep South China Sea: a uniformitarian ichnologic approach. Palaios 23:601–615

Wetzel A (2010) Deep-sea ichnology: observations in modern sediments to interpret fossil counterparts. Acta Geol Pol 60:125–138

Wetzel A, Bromley RG (1994) *Phycosiphon incertum* revisited: *Anconichnus horizontalis* is its junior subjective synonym. J Paleontol 68:1396–1402

Wetzel A, Uchman A (2001) Sequential colonization of muddy turbidites in the Eocene Beloveža Formation, Carpathians, Poland. Palaeogeogr Palaeoclimatol Palaeoecol 168:171–186

Wetzel A, Uchman A (2012) Hemipelagic and pelagic basin plains. In: Knaust D, Bromley RG (eds) Trace fossils as indicators of sedimentary environments. Developments in Sedimentology, vol 64, pp 673–701

Wetzel A, Werner F (1981) Morphology and ecological significance of *Zoophycos* in deep-sea sediments off NW Africa. Palaeogeogr Palaeoclimatol Palaeoecol 32:185–212

Wetzel A, Werner F, Stow DAV (2008) Bioturbation and biogenic sedimentary structures in contourites. In: Rebesco M, Camerlenghi A (eds) Contourites. Developments in sedimentolgy, vol 60, pp 183–202

White CD (1929) Flora of the Hermit Shale, Grand Canyon, Arizona, vol 405. Publications of the Carnegie Institution of Washington, 221 pp

White B, Curran HA (1997) Are the plant-related features in Bahamian quaternary limestones trace fossils? Discussion, answers, and a new classification system. In: Curran HA (ed) Guide to Bahamian ichnology: Pleistocene, Holocene and modern environments. Bahamian Field Station, San Salvador, pp 47–54

Whybrow PJ, McClure HA (1980) Fossil mangrove roots and palaeoenvironments of the Miocene of the eastern Arabian Peninsula. Palaeogeogr Palaeoclimatol Palaeoecol 32:213–225

Wignall PB (1991) Dysaerobic trace fossils and ichnofabrics in the Upper Jurassic Kimmeridge Clay of southern England. Palaios 6:264–270

Wikander PB (1980) Biometry and behaviour in *Abra nitida* (Müller) and *A. longicallus* (Scacchi) (Bivalvia, Tellinacea). Sarsia 65: 255–268

Winn K (2006) Bioturbation structures in marine Holocene sediments of the Great Belt (western Baltic). Meyniana 58:157–178

Worsley D, Mørk A (2001) The environmental significance of the trace fossil *Rhizocorallium jenense* in the Lower Triassic of western Spitsbergen. Polar Res 20:37–48

Wright VP, Platt NH, Marriott SB et al (1995) A classification of rhizogenic (root-formed) calcretes, with examples from the Upper Jurassic-Lower Cretaceous of Spain and Upper Cretaceous of southern France. Sed Geol 100:143–158

Xing L, Marty D, You H et al (2016) Complex in-substrate dinosaur (Sauropoda, Ornithopoda) foot pathways revealed by deep natural track casts from the Lower Cretaceous Xiagou and Zhonggou formations, Gansu Province, China. Ichnos, doi:10.1080/10420940.2016.1244054

Yang S, Zhang J, Yang M (2004) Trace fossils of China. Science Press, Beijing, 353 pp

Zenker JC (1836) Historisch-topographisches Taschenbuch von Jena und seiner Umgebung. Friedrich Frommann, Jena, pp 338

Zhang L, Gong Y (2012) Systematic revision and ichnotaxonomy of *Zoophycos*. Earth Sc J China Univ Geosci 37:60–79

Zhang G, Uchman A, Chodyn R et al (2008) Trace fossil *Artichnus pholeoides* igen. nov. isp. nov. in Eocene turbidites, Polish Carpathians: possible ascription to holothurians. Acta Geol Pol 58:75–86

Zhang L, Fan R, Gong Y (2015) *Zoophycos* macroevolution since 541 Ma. Scientific Reports, vol 5, 14954

Zhang L, Knaust D, Zhao Z (2016) Palaeoenvironmental and ecological interpretation of the trace fossil *Rhizocorallium* based on contained iron framoboids (Upper Devonian, South China). Palaeogeogr Palaeoclimatol Palaeoecol 446:144–151

Zonneveld J-P, Gingras MK (2013) The ichnotaxonomy of vertically oriented bivalve-generated equilibrichnia. J Paleontol 87:243–253

Zonneveld J-P, Pemberton SG (2003) Ichnotaxonomy and behavioral implications of lingulide-derived trace fossils from the Lower and Middle Triassic of western Canada. Ichnos 10:25–39

Zonneveld J-P, Beatty TW, Pemberton SG (2007) Lingulide brachiopods and the trace fossil *Lingulichnus* from the Triassic of western Canada: implications for faunal recovery after the end-Permian mass extinction. Palaios 22:74–97

Zorn ME, Muehlenbachs K, Gingras MK et al (2007) Stable isotopic analysis reveals evidence for groundwater-sediment-animal interactions in a marginal-marine setting. Palaios 22:546–553

# Index




D. Knaust, *Atlas of Trace Fossils in Well Core*,
DOI 10.1007/978-3-319-49837-9

MIX
Papier aus verantwortungsvollen Quellen
Paper from responsible sources
FSC® C105338

If you have any concerns about our products,
you can contact us on
**ProductSafety@springernature.com**

In case Publisher is established outside the EU,
the EU authorized representative is:
**Springer Nature Customer Service Center GmbH**
**Europaplatz 3, 69115 Heidelberg, Germany**

Printed by Libri Plureos GmbH
in Hamburg, Germany